BIOCOMPOSITES

Environmental and Biomedical Applications

BIOCOMPOSITES

Environmental and Biomedical Applications

Edited by

Omar Mukbaniani, DSc

Tamara Tatrishvili, PhD

Neha Kanwar Rawat, PhD

A. K. Haghi, PhD

AAP APPLE ACADEMIC PRESS

First edition published 2024

Apple Academic Press Inc.
1265 Goldenrod Circle, NE,
Palm Bay, FL 32905 USA

760 Laurentian Drive, Unit 19,
Burlington, ON L7N 0A4, CANADA

CRC Press
2385 NW Executive Center Drive,
Suite 320, Boca Raton FL 33431

4 Park Square, Milton Park,
Abingdon, Oxon, OX14 4RN UK

Library and Archives Canada Cataloguing in Publication

Title: Biocomposites : environmental and biomedical applications / edited by Omar Mukbaniani, DSc, Tamara Tatrishvili, PhD, Neha Kanwar Rawat, PhD, A.K. Haghi, PhD.
Other titles: Biocomposites (Apple Academic Press)
Names: Mukbaniani, O. V. (Omar V.), editor. | Tatrishvili, Tamara, editor. | Rawat, Neha Kanwar, editor. | Haghi, A. K., editor.
Description: First edition. | Includes bibliographical references and index.
Identifiers: Canadiana (print) 20230515320 | Canadiana (ebook) 20230515355 | ISBN 9781774913697 (hardcover) | ISBN 9781774913703 (softcover) | ISBN 9781003408512 (ebook)
Subjects: LCSH: Composite materials. | LCSH: Composite materials—Environmental aspects. | LCSH: Polymers in medicine.
Classification: LCC TA418.9.C6 B56 2023 | DDC 620.1/18—dc23

Library of Congress Cataloging-in-Publication Data

CIP data on file with US Library of Congress

ISBN: 978-1-77491-369-7 (hbk)
ISBN: 978-1-77491-370-3 (pbk)
ISBN: 978-1-00340-851-2 (ebk)

About the Editors

Omar Mukbaniani, DSc

Professor at Ivane Javakhishvili Tbilisi State University (TSU), Faculty of Exact and Natural Sciences, Department of Chemistry; Chair of Macromolecular Chemistry, Tbilisi, Georgia

Omar Mukbaniani, DSc, was a Professor at Ivane Javakhishvili Tbilisi State University (TSU), Faculty of Exact and Natural Sciences, Department of Chemistry; Chair of Macromolecular Chemistry, Tbilisi, Georgia. He is a member of the Academy of Natural Sciences of Georgia. For several years, he was a member of the advisory board of the journals *Proceedings of Ivane Javakhishvili Tbilisi State University* (Chemical Series), Contributing Editor of the journals *Polymer News*, *Polymers Research Journal,* and *Chemistry and Chemical Technology*. He was the author of more than 480 publications, 25 books, monographs, and 10 inventions. In 2007, he created the International Symposium on Polymers and Advanced Materials, which takes place every other two years in Georgia.

Tamara Tatrishvili, PhD

Assistant Professor, Department of Chemistry, Ivane Javakhishvili Tbilisi State University (TSU); Director of the Institute of Macromolecular Chemistry and Polymeric Materials at TSU, Tbilisi, Georgia; Main Specialist, Office of the Academic Process Management, Ivane Javakhishvili Tbilisi State University (TSU)

Tamara Tatrishvili, PhD, is an Assistant Professor and Main Specialist at the Office of Academic Process Management (Faculty of Exact and Natural Sciences) at Ivane Javakhishvili Tbilisi State University, as well as Director of the Institute of Macromolecular Chemistry and Polymeric Materials at TSU, Tbilisi, Georgia; DAAD alumni, and a member of the Georgian Chemical Society. Her research interests include polymer chemistry, polymeric materials, and chemistry of silicon-organic compounds. Dr. Tatrishvili is the author of more than 190 scientific publications, 12 books, and monographs.

Neha Kanwar Rawat, PhD
Researcher, Department of Materials Science and Engineering, Indian Institute of Science, Bangalore, Karnataka, India

Neha Kanwar Rawat, PhD, is a recipient of a prestigious DS-Kothari Postdoctoral Fellowship and DST Young Scientist Postdoctoral Fellowship and is presently a researcher in the Department of Materials Science and Engineering, Indian Institute of Science, Bengaluru, India. She received her PhD in Chemistry from Jamia Millia Islamia (a central university), India. Her main interests include nanotechnology and nanostructured materials synthesis and characterization, with main focus on green chemistry; novel sustainable chemical processing of nano-conducting polymers/nanocomposites; conducting films, ceramics, silicones, and matrices: epoxies, alkyds, polyurethanes, etc. She is also pursuing her interest in fusing new technology in areas that include electrochemistry and organic-inorganic hybrid nanocomposites for biomedical applications, which also includes protective surface coatings for corrosion inhibition and MW shielding materials. She has published numerous peer-reviewed research articles in journals of high repute. Her contributions have led to many books and chapters in international books published with the Royal Society of Chemistry, Wiley, Elsevier, Apple Academic Press, Nova US, and many others in progress. She is a peer reviewer for many international books and is a member of many groups, including the Royal Society of Chemistry and the American Chemical Society (USA), and a life member of the Asian Polymer Association.

A. K. Haghi, PhD
Professor Emeritus of Engineering Sciences, Former Editor-in-Chief, International Journal of Chemoinformatics and Chemical Engineering; Member, of Canadian Research and Development Center of Sciences and Culture

A. K. Haghi, PhD, has published over 250 academic research-oriented books as well as over 1,000 research papers published in various journals and conference proceedings. He has received several grants, consulted for several major corporations, and is a frequent speaker to national and international audiences. He is the Founder and former Editor-in-Chief of the *International Journal of Chemoinformatics and Chemical Engineering,* published by IGI Global (USA), as well as *Polymers Research Journal,* published by Nova Science Publishers (USA). Professor Haghi has acted as an editorial board member of many international journals. He has served as a member of the Canadian Research and Development Center of Sciences and Cultures (CRDCSC) and the Research Chemistry Centre, Coimbra, Portugal. Dr. Haghi holds a BSc in urban and environmental engineering from the University of North Carolina (USA), an MSc in mechanical engineering from North Carolina A&T State University (USA), and an MSc in applied mechanics, acoustics, and materials from the Université de Technologie de Compiègne (France), and a PhD in engineering sciences from Université de Franche-Comté (France).

Contents

CONTENTS

Contributors

Jimsher Aneli
Institute of Macromolecular Chemistry and Polymeric Materials, Ivane Javakhishvili Tbilisi State University, Tbilisi, Georgia

Shahidul Islam Bhat
Corrosion Research Lab, Department of Chemistry, AMU, Aligarh, Uttar Pradesh, India

Asif Husain
Baba Ghulam Shah Badshah University, Rajouri, Jammu and Kashmir, India;
Jamia Millia Islamia University, New Delhi, India

Mohd Irfan
Archaeological Survey Chemist, Jammu and Kashmir, India

Abdul Wahied Khan
Department of Electrical Engineering, Mewar University, Chittorgarh, Rajasthan, India

Rabia Kouser
Baba Ghulam Shah Badshah University, Rajouri, Jammu and Kashmir, India

Omar Mukbaniani
Department of Macromolecular Chemistry, Ivane Javakhishvili Tbilisi State University, Tbilisi, Georgia;
Institute of Macromolecular Chemistry and Polymeric Materials, Ivane Javakhishvili Tbilisi State University, Tbilisi, Georgia

G. P. Nayaka
Physical and Materials Chemistry Division, CSIR–National Chemistry Laboratory, Pune, Maharashtra, India

G. Santhosh
Department of Mechanical Engineering, NMAM Institute of Technology, Nitte, Karnataka, India

Tamara Tatrishvili
Department of Macromolecular Chemistry, Ivane Javakhishvili Tbilisi State University, Tbilisi, Georgia;
Institute of Macromolecular Chemistry and Polymeric Materials, Ivane Javakhishvili Tbilisi State University, Tbilisi, Georgia

Abbreviations

AChR	acetylcholine receptor
BG	bio-active glass
BPO	benzoyl peroxide
BTCA	1,2,3,4-butane tetracarboxylic dianhydride
CaP	calcium phosphates
CHPO	cyclohexanone peroxide
CNS	central nervous system
CP	calcium phosphate
DCP	dicumyl peroxide
DP	degree of polymerization
ECM	extracellular matrix
ECM-PLGA	ECM-coated polylactic-co-glycolic acid
EDXA	energy dispersive X-ray microanalysis
FTIR	Fourier transition IR spectroscopy
GG	guar gum
GG-g-PAA/APT	guar gum-g-polyacrylic acid-attapulgite
H	hydrogen
HA	hydroxyapatite
HPO	hydrogen peroxide
ISO-HPO	isopropyl hydroperoxide
LCST	lower critical solution temperature
LG	liquid glass
LPO	lauroyl peroxide
MEKP	methyl ethyl ketone peroxide
Mw	molecular weight
N	nitrogen
O	oxygen
PCL	polycaprolactone
PE	polyethylene
PGA	polyglycolic acid
PLA	polylactic acid
PLLA	poly-L lactic acid
PNS	peripheral nervous system
PTFE	polytetrafluoroethylene

SEM	scanning electron microscope
SSA	specific surface area
TBPB	t-butyl hydroperoxide
TCP	tricalcium phosphates
TJRs	total joint replacements
TS	thermal stability
WPC	wood-polymer composites

Preface

In recent years, valuable works have been published to show the challenges related to durable and sustainable products by raising a circular economy in green composite materials. The growing environmental and sustainability awareness has inspired efforts for configuring green composite materials for diverse engineering and medical applications as a new substitute for conventional materials that are sometimes considered non-renewable composites in industrial sectors and daily life. Also to be considered is the fact that bio-composite materials are not always an easy-care substitute and could possibly have some drawbacks such as processing, surface modification, and manufacturing.

This new book provides the most recent studies allied with different aspects of bio-composites. In this new title, the authors of each chapter focus on the overall characteristics, including chemical composition along with engineering and technical properties of the most in-demand green materials employed for fabricating bio-composites. In such engineering and technical approaches, the progress and achievements for enhancing the properties of bio-composites are reviewed in detail.

The present book also reports on the modern techniques along with new developments engaged in the production of bio-composites. In this new volume, after considering applications and trends of bio-composites, the major characterization techniques, along with important environmental effects, are discussed in depth.

This book provides vital information on new green materials and related biocomposites for the first time, in addition to their imminent environmental and medical applications.

Due to the alarming environmental problems and our heavy dependency on finite resources, this book can be broadly considered one of the most important reference books for R&D sectors.

PART I

Biobased Composites for Environmental Applications

CHAPTER 1

Composite Materials on the Basis of Sawdust: An Experimental Approach

OMAR MUKBANIANI,[1,2] JIMSHER ANELI,[2] and TAMARA TATRISHVILI[1,2]

[1]*Department of Macromolecular Chemistry, Ivane Javakhishvili Tbilisi State University, Tbilisi, Georgia*

[2]*Institute of Macromolecular Chemistry and Polymeric Materials, Ivane Javakhishvili Tbilisi State University, Tbilisi, Georgia*

ABSTRACT

In this chapter, composite structures, along with physical properties, such as mechanical strengthening (at bending and impact viscosity), thermal stable properties, and water absorption, were studied in detail. The reflection of technological factors on the thermal stability (TS) of composites is investigated as well.

The value of the water absorption of the composites based on sawdust is in the acceptable range according to the data obtained from mechanical strengthening and TS. The higher the density of composites, the lower their water absorption.

1.1 INTRODUCTION

The high pace of the development of the industry more and more expands the demands on wood materials, however, an increased deficit of natural wood stipulates the scientific-technical research on the theme of analogical materials.

At this time, great attention attracts to the materials obtained as a result of a combination of wood with sub-products formed after its treatment (sawdust, leaves, needless) with the use of different binders. This process takes place

Biocomposites: Environmental and Biomedical Applications. Omar Mukbaniani, Tamara Tatrishvili, Neha Kanwar Rawat, & A. K. Haghi (Eds.)

often as the polymerization of polymer substances (for example oligomers) on the surface of dispersed wood products. Besides there are widespread methods of extrusion and hot pressing in the press molds of high-dispersive thermo-plastic polymers and wood products.

Today composites like sheets are very popular. This material is produced by hot pressing of thermosetting polymers and wood sawdust (Russian abbreviation DSP). This material is widely used in the industry of furniture. The wide spreading of this material is due to its low cost in comparison with poor wood material. However, DSP has many leaks among which are difficult fixing of the nails and screws, high water absorption, and very bad environmental friendliness. This leak is due to phenol (phenol-formaldehyde) contain, which is a very toxic substance and stands out from noted material during a long exploitation period. Therefore, in such a progressive country as the USA, the application of this material in the industry now is restricted partially but soon will be fully restricted. Analogical action is provided in Russia – the materials containing a very harmful fraction of formaldehyde are not produced.

The wood-polymer composites (WPC) are materials of a relatively new generation, in which the role of the binder performs such thermo-plastics polymers as polyethylene (PE), polypropylene, polyvinyl chloride, polystyrene, and others. These materials sometimes are called liquid wood. There is a known rather wide assortment of products made from WPC. Using such methods as extrusion, hot pressing, and rotation formatting one obtained such goods as terraces, floor desks, wall panels, roof coatings, pipes, and so on. WPCs are distinguished from analogs by high stability to atmospheric influences, mechanical, and chemical sustainability, waterproofing, which allows using these materials as coatings of washing rooms, saunas, terraces, and docks, and so on.

The first production of WPC was prepared in the 90th years of the last century. Besides definite achievements without accounting for a lot of restricting factors, it is impossible to suppose about successful perspectives in the sphere of inculcation of the results of the scientific investigations to the industry, the main factor of which is very low environmental friendliness.

The main goal of our project is the obtain new design materials with high technical characteristics on the basis of the industrial and natural wastes (dry leaves, needles, and others) of wood material with the use of clean binder materials.

The main attention during the first quarter was attracted to the purchase of such materials, which were necessary for the fulfillment of the project tasks. It delivered the dry products of wood and bamboo sawdust, straw, wood leaves, and needles dry products. For the first experiments, it was purchased liquid glass (STS).

During the first stage were provided the works taking into account obtaining and the type of wood product. In all cases, the main binder was the liquid glass (LG), although in the separate case, it used high dispersive PE of low density.

The composites were obtained with the use of the following manipulations:

- Weighing of the components with the use of analytical balance;
- Dry mixing of the components;
- Loading of the blend to the standard press forms;
- Heating of blends in the press-forms at definite temperature during fixed time;
- Taking out of the samples from press forms.

There were provided the testing of the obtained samples for the establishment of the following characteristics: density, strengthening on rupture and bending, Yung module, water absorption, and TS (by method Vica). The measurements were fulfilled with the use of standards on the devices of our laboratory. For reliability, the experiments were repeated three times.

During the second quarter, there were provided investigations of the physical mechanical, and thermal properties and water absorption.

Following composites were investigated: the composites consisting of PhES-50(PhES-50), PhES-80(PhES-80), Vin(OEt)$_3$), FES-50 and FES-80 at 3–5 wt.%* and PhES-50(PhES-50), PhES-80(PhES-80), vinylthrietoxisilane (Vin(OEt)$_3$). The weights of FES-50 and FES-80 were 3–5 wt.%*, and vinylthrietoxisilane – 4 wt.%.

As it is known the bulk of the timber is made up of organic substances containing carbon (C), hydrogen (H), oxygen (O), and nitrogen (N). The difference in the content of carbon, hydrogen, and oxygen in a small timber of different breeds: absolutely dry wood contains an average of 49.5% carbon, 6.3% hydrogen, 44.1% oxygen, and 0.1% nitrogen.

The chemical composition of the timber also includes minerals, which form ash during combustion. Depending on the breed of ash wood in wood varies from 0.2 to 1.7%.

Included in the chemical composition of wood carbon, hydrogen, and oxygen – forming complex organic substances, some of which are included in the cell walls, a part – of the cells themselves.

Wood cell walls consist mainly of cellulose, hemicellulose, and lignin, the cell cavity–of tanning and dyeing substances, gums, resins, essential oils, and alkaloids (Table 1.1).

TABLE 1.1 Proportions of the Main Chemical Compound Groups (%) Within Tree Biomass Components*

	Cellulose	Hemicellulose	Lignin	Extractive
	Scots Pine			
Stem wood [14–20, 24, 40]	40.7	26.9	27.0	5.0
Bark [20–25, 40]	22.2	8.1	13.1	25.2
Branches [24–27, 40]	32.0	32.0	21.5	16.6
Needles [16, 24, 25, 40]	29.1	24.9	6.9	39.6
Stump [25]	36.4	28.2	19.5	18.7
Roots [25]	28.6	18.9	29.8	13.3
	Norway Spruce			
Stem wood [16–19, 31, 40]	42.0	27.3	27.4	2.0
Bark [22–25, 30, 32, 33, 40]	26.6	9.2	11.8	32.1
Branches [24, 27, 30, 40]	29.0	30.0	22.8	16.4
Needles [16, 24, 25, 40]	28.2	25.4	8.4	43.3
Stump [25, 34]	42.9	27.9	29.4	3.8
Roots [25]	29.5	19.2	25.5	15.7
	Silver/Downy Birch			
Stem wood [14, 18, 25, 28, 35, 40]	43.9	28.9	20.2	3.8
Bark [24, 25, 36–38, 40]	10.7	11.2	14.7	25.6
Branches [24, 25, 39, 40]	33.3	23.4	20.8	13.5
Leaves [24, 25, 40]	N/A	N/A	11.1	33.0
Stump [25]	29.5	19.4	13.4	4.7
Roots [25]	26.0	17.1	27.1	13.7

*Values presented in the tables are medians of values found in the literature.

1. **Carbohydrates:** The carbohydrate portion of wood comprises cellulose and hemicelluloses. Cellulose content ranges from 40 to 50% of the dry wood weight, and hemicelluloses range from 25 to 35%.
2. **Cellulose:** It is a glucan polymer consisting of linear chains of 1,4-β-bonded anhydroglucose units.

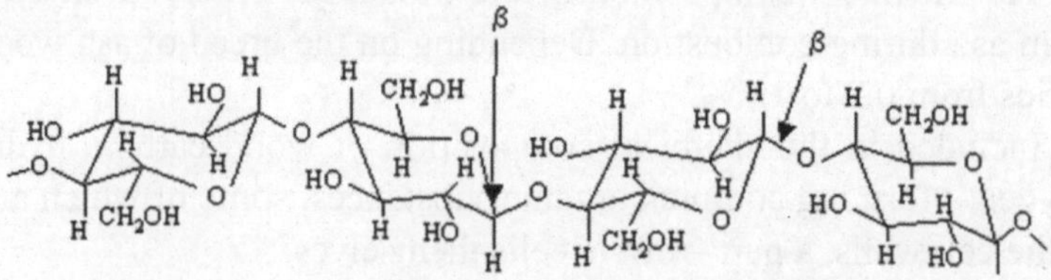

(The notation 1,4-β describes the bond linkage and the configuration of the oxygen atom between adjacent glucose units.) Figure 1.1

shows a structural diagram of a portion of a glucan chain. The number of sugar units in one molecular chain is referred to as the degree of polymerization (DP). Even the most uniform sample has molecular chains with slightly different DP values. The average DP for the molecular chains in a given sample is designated by $\overline{DP}$. Molecular weight (Mw) determinations, done by light-scattering experiments, indicate wood cellulose has a $\overline{DP}$ of at least 9,000–10,000, and possibly as high as 15,000. Cellulose is insoluble in most solvents including strong alkali. It is difficult to isolate from wood in pure form because it is intimately associated with lignin and hemicelluloses. Analytical methods of cellulose preparation are discussed in the section on "Analytical Procedures."

3. **Hemicelluloses:** These are mixtures of polysaccharides synthesized in a wood almost entirely from glucose, mannose, galactose, xylose, arabinose, 4-O methylglucuronic acid, and galacturonic acid residues. Some hardwoods contain trace amounts of rhamnose.

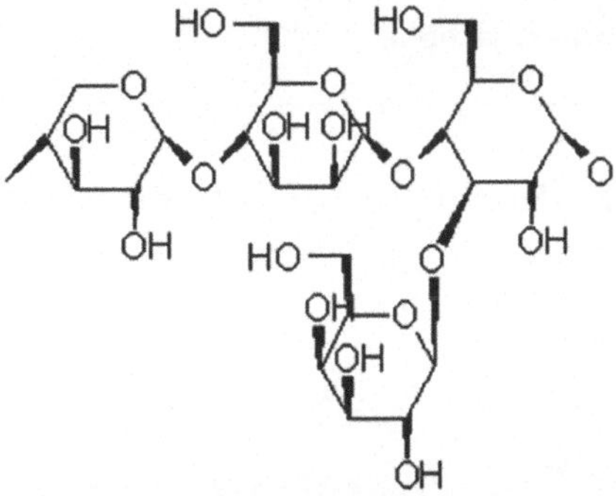

- Xylose - ß(1,4) - Mannose - ß(1,4) - Glucose -
- alpha(1,3) - Galactose

Hemicellulose

Generally, hemicelluloses are of much lower Mw than cellulose and some are branched. They are intimately associated with cellulose and appear to contribute as a structural component in the plant. Some hemicelluloses are present in abnormally large amounts when the plant is under stress, e.g., compression wood has a higher-than-normal galactose content as well as a higher lignin content [11]. Hemicelluloses are soluble in alkali and easily hydrolyzed by acids.

4. **Lignin:** It is a phenolic substance consisting of an irregular array of variously bonded hydroxy- and methoxy-substituted phenylpropane

units. The precursors of lignin biosynthesis are p-coumaryl alcohol (I), coniferyl alcohol (II), and sinapyl alcohol (III). I is a minor precursor of softwood and hardwood lignins; II is the predominant precursor of softwood lignin; and II and III are both precursors of hardwood lignin [15].

Some characteristics of used binders are presented in Table 1.2.

TABLE 1.2 Technical Characteristics of PhES-50 and PhES-80

Appearance	**PhES-50**	**PhES-80**
	Transparent, Colorless Liquid without Mechanical Impurities	
Mass part (%) of hydrogen chloride no more than	0.004	–
Mass part (%) of ethoxy group	27.0–40.0	17.0–23.0
Kinematic viscosity at a temperature (20 ± 2°C), mm²/c	–	–
Flashpoint in an open crucible temperature (°C), no lower	25–150	–
Activity index of aqueous extract hydrogen ions, the pH is not less than.	120	120

The structure of PhES-50 and PhES-80 is presented below:

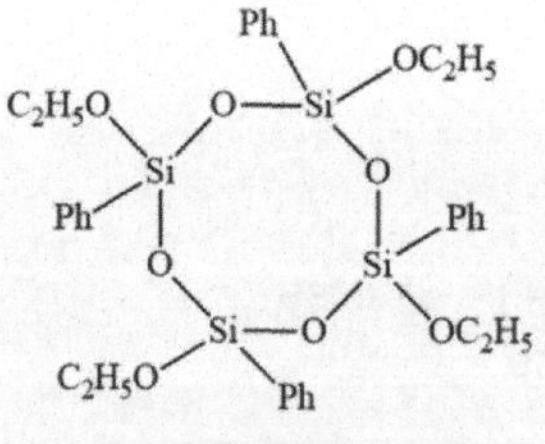

PhES-50

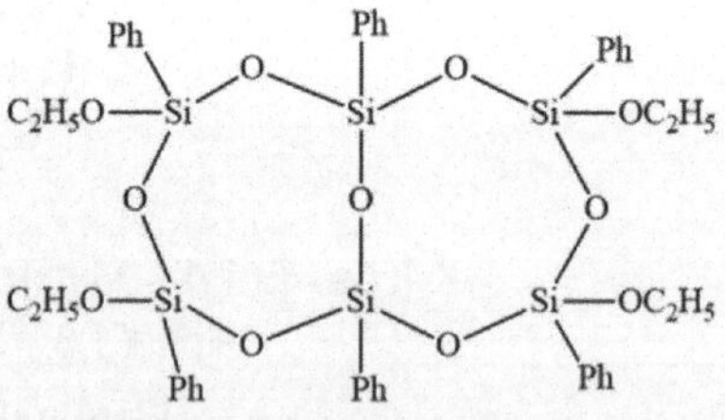

PhES-80

From scientific-technical literature, it is known that the leaves mainly contain sufficient amounts of starch, cellulose, hemicellulose, and pectic substances, lignin, polyphenols. All these above-mentioned compounds may react with proposed binders. For example, phenylethoxysilane (PhES-50 and PhES-80) containing ethoxy group participates in the etherification reaction with a leaf through the macromolecular and intra-molecular reactions. Processes that occur during the curing are complex and varied. A modern look at an overview of the curing LG itself and in the various homogeneous and heterogeneous systems, the most widely encountered in practice, is presented in a number of reviews [41–44].

Acting as an adhesive or binder LG system goes from liquid to solid in many ways and may be divided into three types:

i. The loss of moisture by evaporation at ordinary temperatures;
ii. The loss of moisture from the system, followed by heating above 100°C; and
iii. The transition to the solid state by introducing specific reagents, which are called hardeners. Naturally, these three types are used in combination.

In solution, the DP of silicate anions is known to depend on two factors – the silica modulus and the solution concentration. Each solution has a distribution of degree of the anion polymerization. Distribution is superimposed on the polymer distribution of anions on the charges, which is also determined by these two factors.

It is possible on the intermediate stages the reactions between of ≡Si-OH band contained LG and cellulose molecules, which take place with dehydration reactions. These reactions go through obtaining three-dimension structures. Formation of such structures takes place also with the use of hardener Na_2SiF_6, which accelerates the processes.

Using binders PhES-50 and PhES-80 it is also possible condensation reactions between ethoxy groups with hydroxyl groups according to the scheme: $-OH + C_2H_5O{-}Si\equiv \rightarrow -O{-}Si\equiv + C_2H_5OH$ which may proceed deeper with obtaining three dimensional systems.

The binder colophony derives from pine resin, tall oil, and stump extractives. It is used naturally or in chemically modified forms: hydrogenated, disproportionated, esterified, or polymerized [45]. The colophony is in the structure of different plants' main structure containing the isomeric acid rings. The noted structural ring can introduce to reaction both with PhES-50 and PhES-80 or with LG. In the presence of colophony, it may be realized the donor-acceptor bond with the leaves matrix, which is connected with the formation of additive intermolecular forces and leads to increasing material strengthening.

Natural dry wood sawdust was destructed in the mixer like of coffee mill. The fraction of particles with middle sizes near 50 mcm was used in our experiments. The ingredients dispersive powders PhES, LG, PE, colophony, and wood glue with an amount of 3–20 wt.% were added to the powder of wood sawdust and carefully mixed in the mixer. The obtained blends were placed in the spatial press forms corresponding to the standards and pressed under various pressures and temperatures.

Composites were measured using the following features: (i) Fourier transition IR spectroscopy (FTIR) study; (ii) limit bending strength; (iii) shock viscosity; (iv) thermo-Vikas method; and (v) water absorption coefficient. Results and analysis are provided below.

The microstructure of the samples was studied by an NMM-800RF/TRF type of optical microscope. For composite materials made on the basis of sawdust and different hardeners and additives scanning electron microscopic (SEM) investigations are provided. Besides energy dispersive X-ray examinations have been carried out in parallel with the micro-spectral (EDS) examinations. SEM and EDS analyses were conducted with the use of the microscope Nikon Eclipse LV 150.

From the physical-mechanical characteristics of the composites, the mechanical strength at elongation and shock viscosity was provided with the use of the standard methods and apparatus. TS of the materials was established with the use of the well-known method of Vica. Water absorption was defined by means of weighing samples before and after the exposition of samples in distilled water.

There were measured following characteristics: (i) FTIR spectra; (ii) strengthening on bending; (iii) impact viscosity; (iv) TS by method Vica; and (v) coefficient of the water absorption. The results of these measurements are presented in succeeding sections.

1.2 FTIR SPECTRAL INVESTIGATIONS

For samples, Fourier transforms infrared spectroscopy investigations have been carried out in KBr [48, 49]. The KBr pellets of samples were prepared by mixing (1.5–2.00) mg of samples, finely grounded, with 200 mg KBr (FT-IR grade) in a vibratory ball mixer for 20 s.

In the FTIR spectra of composites one can absorption bands for asymmetric valence oscillation of Si-O-Si bonds with a maximum at 1,066 cm^{-1}, characteristic for siloxane bonds in cyclotetrasiloxane fragments well as for etheric C-O-C and C-O-Si bonds these bands are overlaps. In the spectra, one can see absorption bands at 1267, 1373, 1429, 1515, 1600–1650, 1733, 2800–2950, 3363 characteristics for methyl groups, C–H absorption (–/C–/CH_3), CH_2 cellulose – lignin, C=C aromatic, C=C alkene, (C=O ester), C–H methyl, methylene, and phenyl groups, O–H alcohol accordingly [46, 47] (Figures 1.1–1.4).

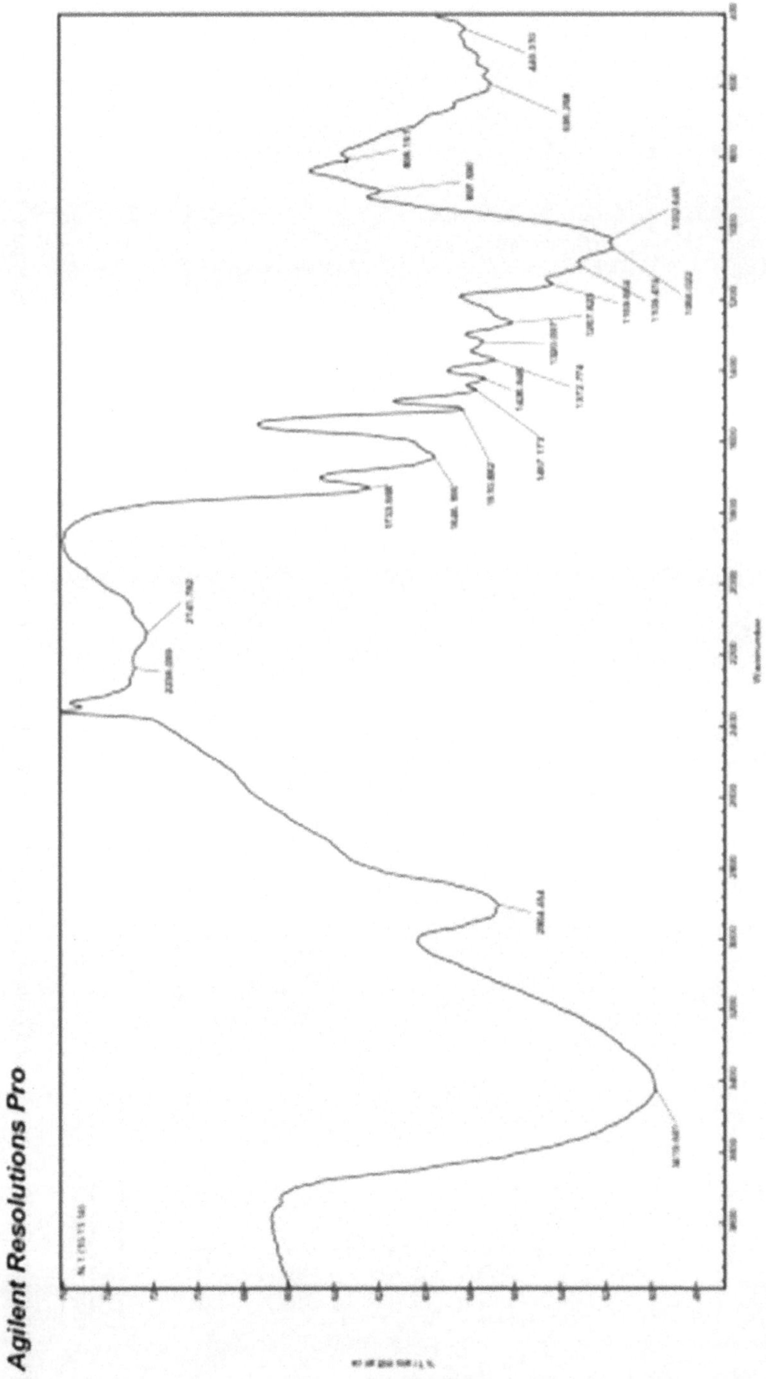

FIGURE 1.1 FTIR spectra of sawdust.

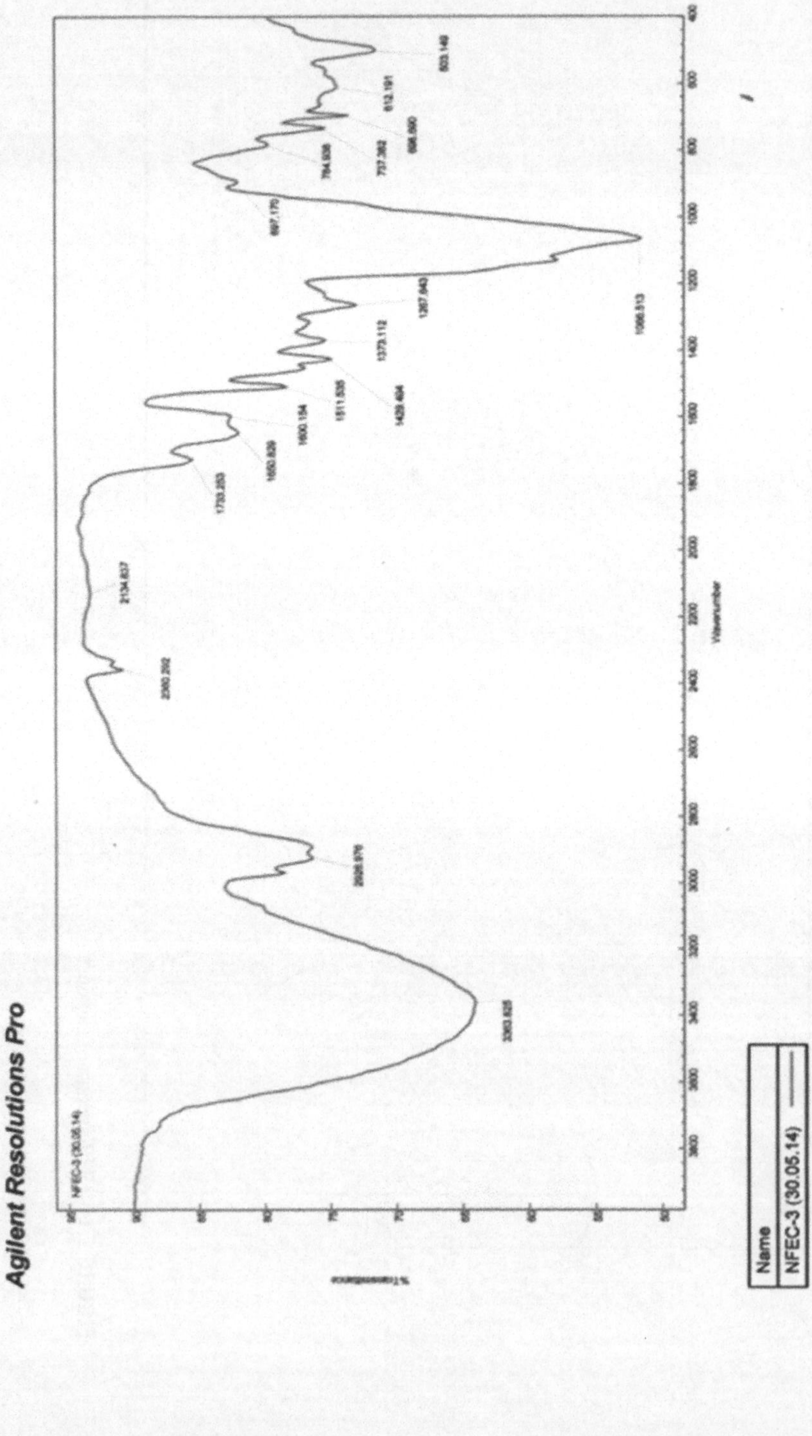

FIGURE 1.2 FTIR spectra of composite 4 sawdust (95%), + (PhES-80 (5%) (120°C) (pressure 170 kg/cm^2, temperature 120°C) (Table 1.9).

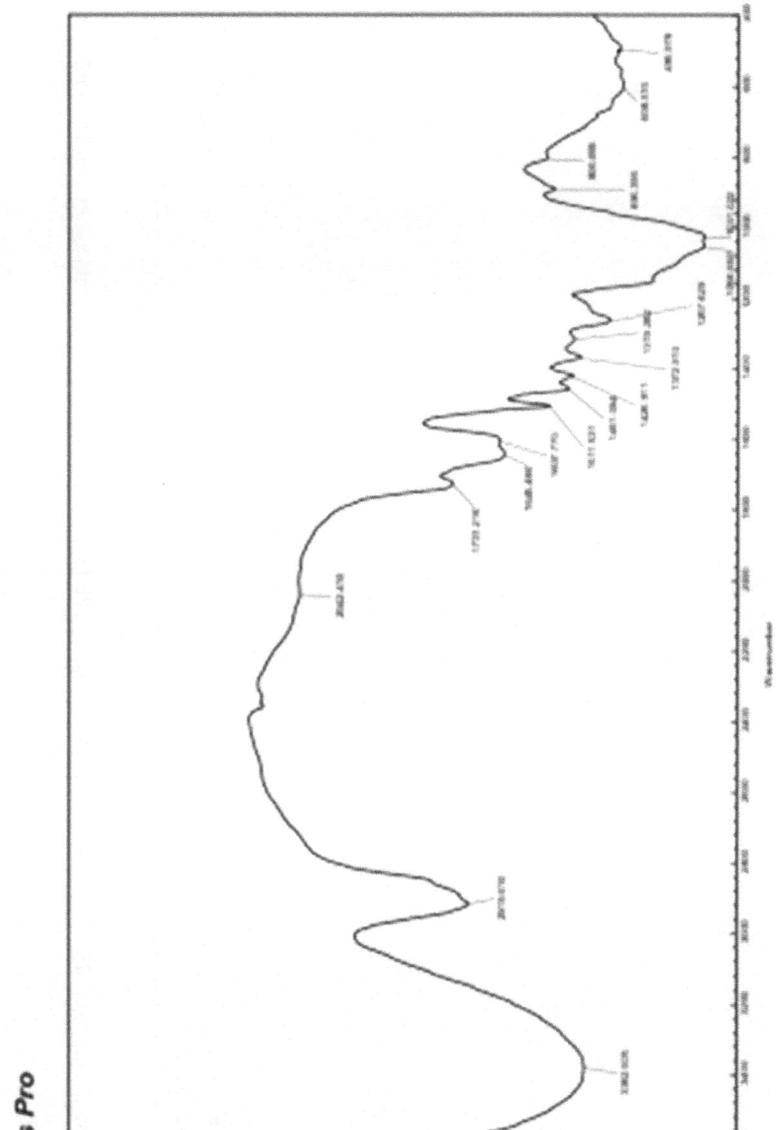

FIGURE 1.3 FTIR spectra of composite 16 PhES-50 (3%) + sawdust (92%) + PE (5%), (pressure 150 kg/cm^2, temperature 110°C) (Table 1.9).

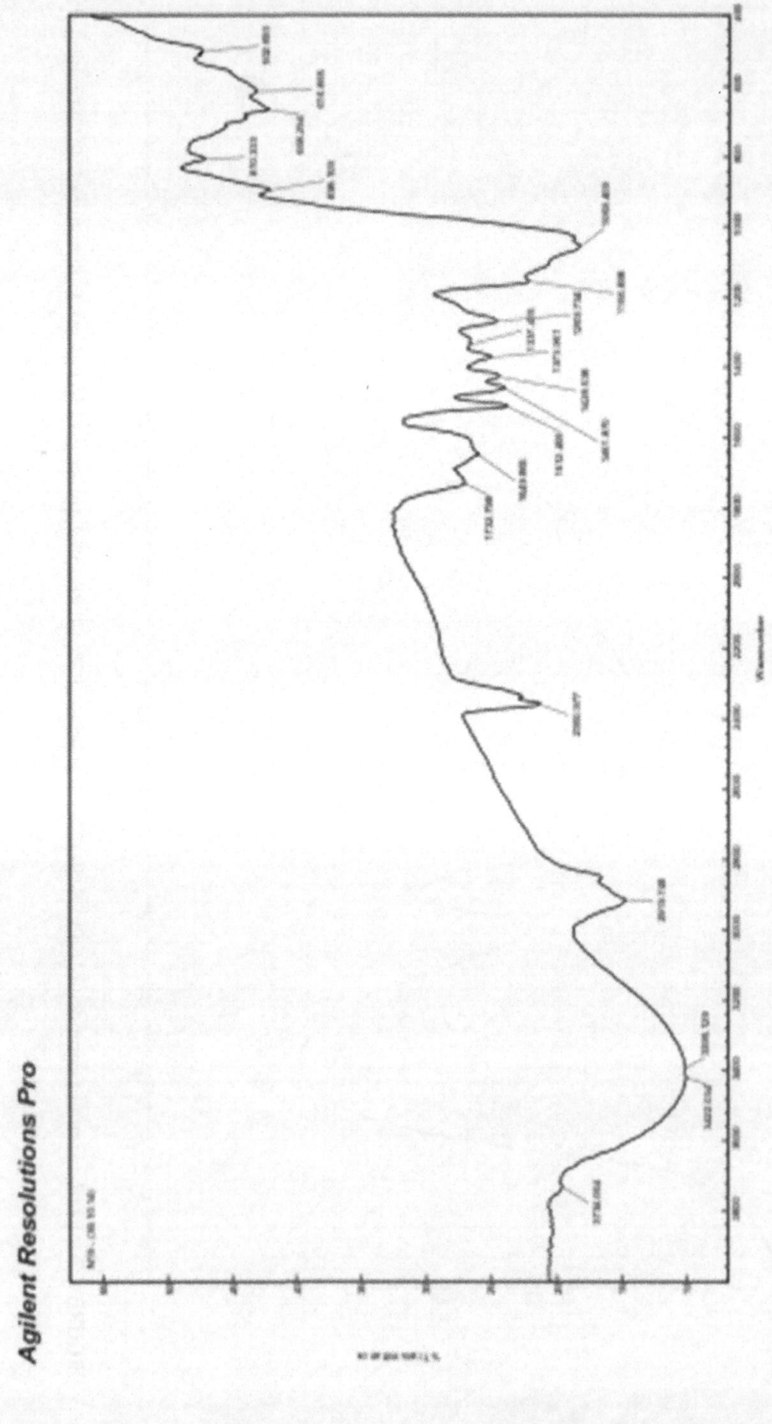

FIGURE 1.4 FTIR spectra of composite 19 PhES-50 (3%) + sawdust (92%) + PE (5%) (pressure 130 kg/cm^2, temperature 150°C) (Table 1.9).

1.3 OPTICAL MICROSCOPE INVESTIGATIONS OF COMPOSITES

The microstructure of wood composites was studied on the NMM-800RF/TRF type optical microscope and SEM.

For investigation of the wood material, the sample with 1 cm longevity was prepared. This sample was polished on the sheet of 25 mcm SiC paper for 1 h and after this sample was displaced to the sending paper with 5 mcm and the polishing was continued for 1 h. After these procedures, the sample was polished additively by means of byazi, after which the sample was displaced to the optical microscope for investigation of different ranges. It investigated the samples prepared under pressure of 170 kg/cm^2 and at different temperatures. The composite with sawdust (97%) +PhES-80 (3%, at temperatures 110, 120, and 125°C and composite with sawdust (90%)+PhES-80 (5% and silylated styrene (5%)). For samples, it was defined as both the transverse surface and longitudinal one or the inner part of the sample, and the inclusions were observed under conditions with increasing 50, 100, 200, and 500 times (*see* Figures 1.5–1.13).

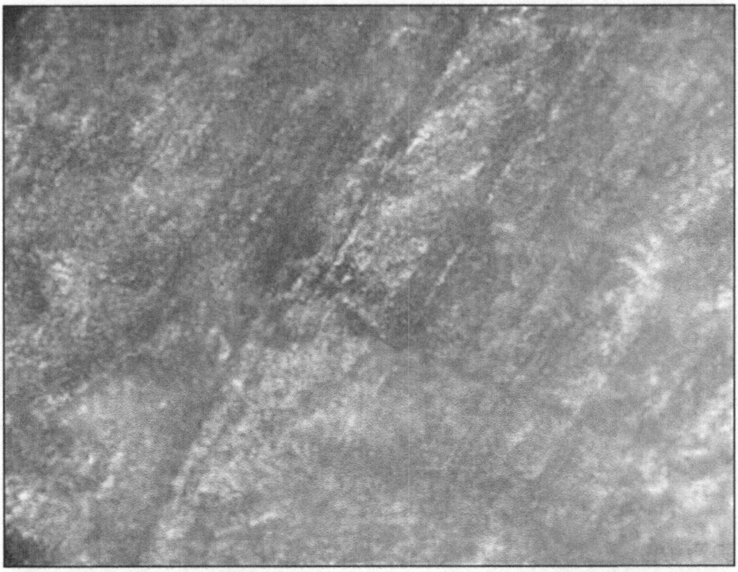

FIGURE 1.5 Optical microscopic data of the transverse surface of composite sawdust (97%) + PhES-80 (3%) obtained at 110°C.

From the optical microscope figures one can see, that the binder rises to surfaces with the change of the inclusions with sizes in the frames 76–200 mcm and cleavage sizes 160–760. The change in temperature does not influence the sizes either inclusions or cleavages.

FIGURE 1.6 Optical microscopic data of the transverse surface of composite sawdust (97%) + PhES-80 (3%) was obtained at 120°C, with cleavage near 600 mcm.

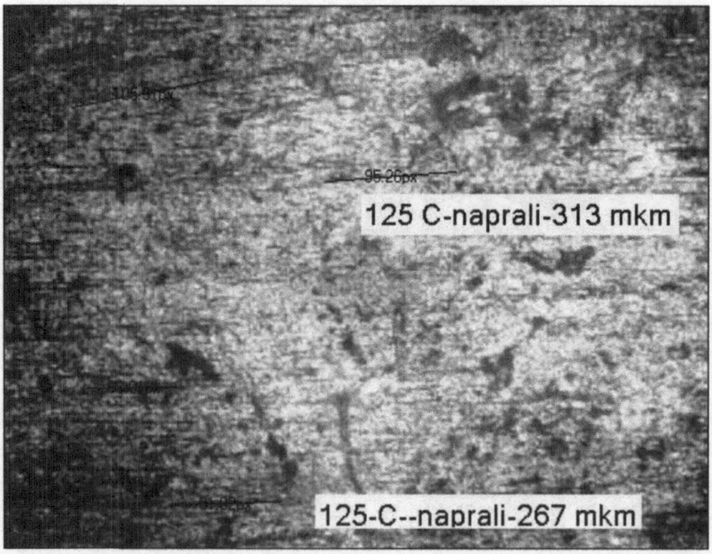

FIGURE 1.7 Optical microscopic data of the transverse surface of composite sawdust (97%) + PhES-80 (3%) was obtained at 125°C, with cleavage near 267–313 mcm.

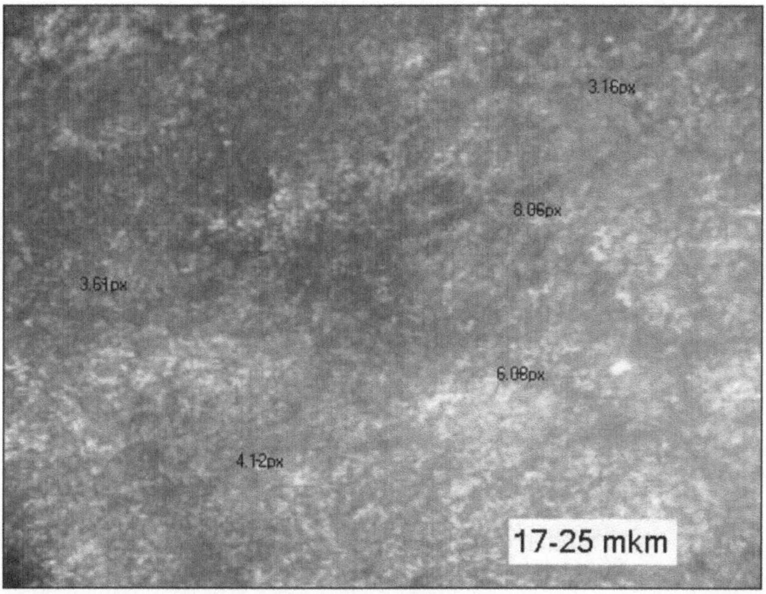

FIGURE 1.8 Optical microscopic data of the longitudinal surface of composite sawdust (97%) + PhES-80 (3%) obtained at 110°C.

FIGURE 1.9 Optical microscopic data of the longitudinal surface of composite sawdust (97%) + PhES-80 (3%) obtained at 110°C, magnification x200, the inclusion size 200 mcm.

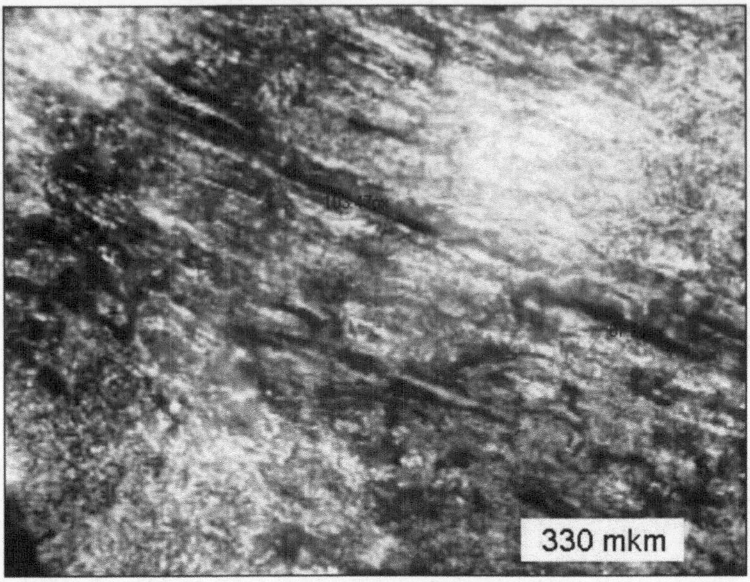

FIGURE 1.10 Optical microscopic data of the longitudinal surface of composite sawdust (97%) + PhES-80 (3%) was obtained at 110°C, the magnification x200, and the cleavage size of about 330 mcm.

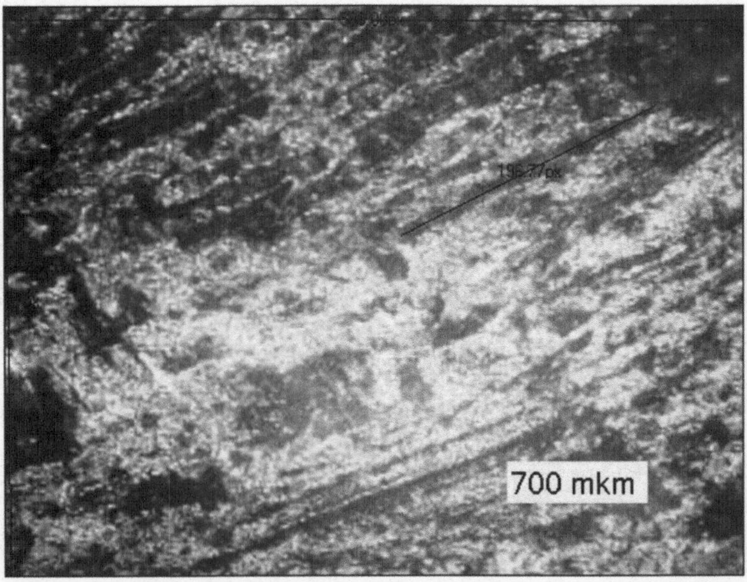

FIGURE 1.11 Optical microscopic data of the longitudinal surface of composite sawdust (97%) + PhES-80 (3%) obtained at 120°C, magnification x100, the inclusion size 700 mcm.

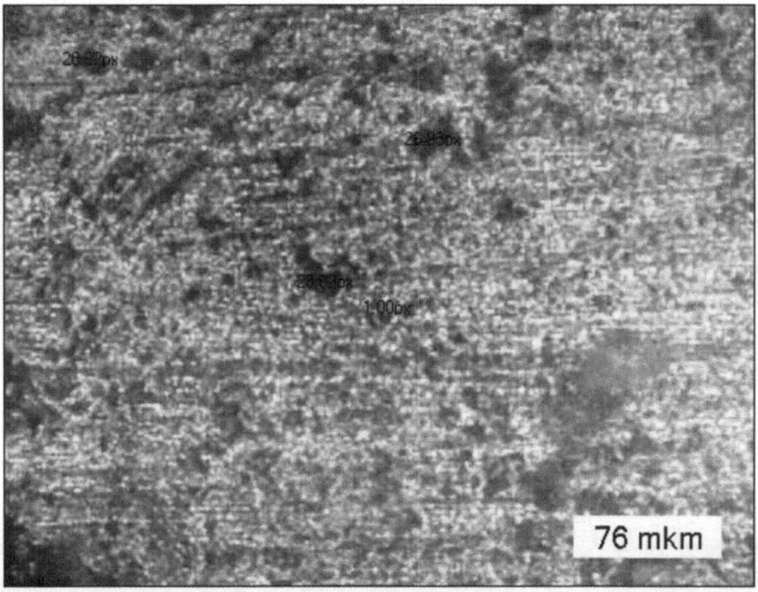

FIGURE 1.12 Optical microscopic data of the longitudinal surface of composite sawdust (97%) + PhES-80 (3%) obtained at 125°C, magnification x100, the cleavage size 76 mcm.

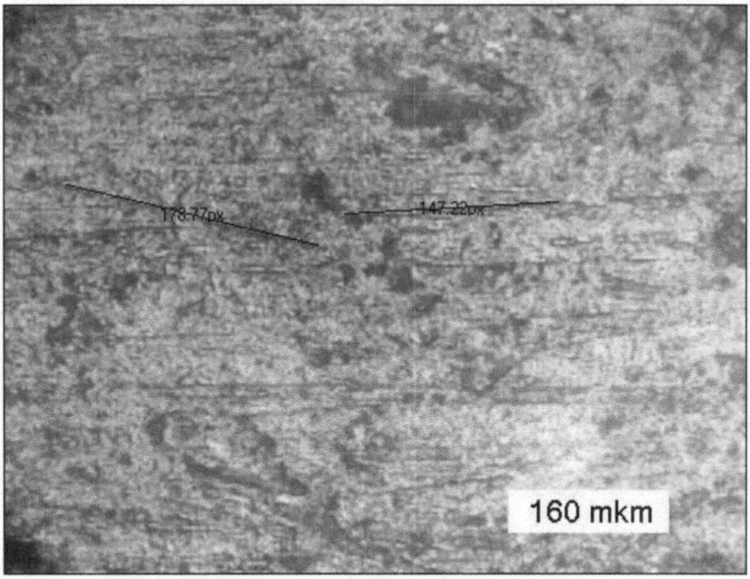

FIGURE 1.13 Optical microscopic data of the longitudinal surface of composite sawdust (97%) + PhES-80 (3%) obtained at 125°C, magnification x100, the inclusion size 160 mcm.

1.4 SCANNING ELECTRON MICROSCOPIC AND ENERGY DISPERSION MICRO-X-RAY ANALYSIS OF COMPOSITE MATERIALS ON THE BASIS OF THE SAWDUST

For composite materials made on the basis of powder like wood sawdust bamboo and different hardeners and additives, SEM investigations are provided. Besides energy dispersive X-ray examinations have been carried out in parallel with the micro-spectral (EDS) examinations. SEM and EDS analyses were conducted with the use of t he microscope Nikon Eclipse LV 150. The micrograms of SEM were obtained at various (x100–x1,000) magnification ratios. Surface analysis, chemical analysis, and visualization of the composites obtained on the basis of leaves were studied by the method of SEM (Figures 1.14–1.40 and Tables 1.3–1.7).

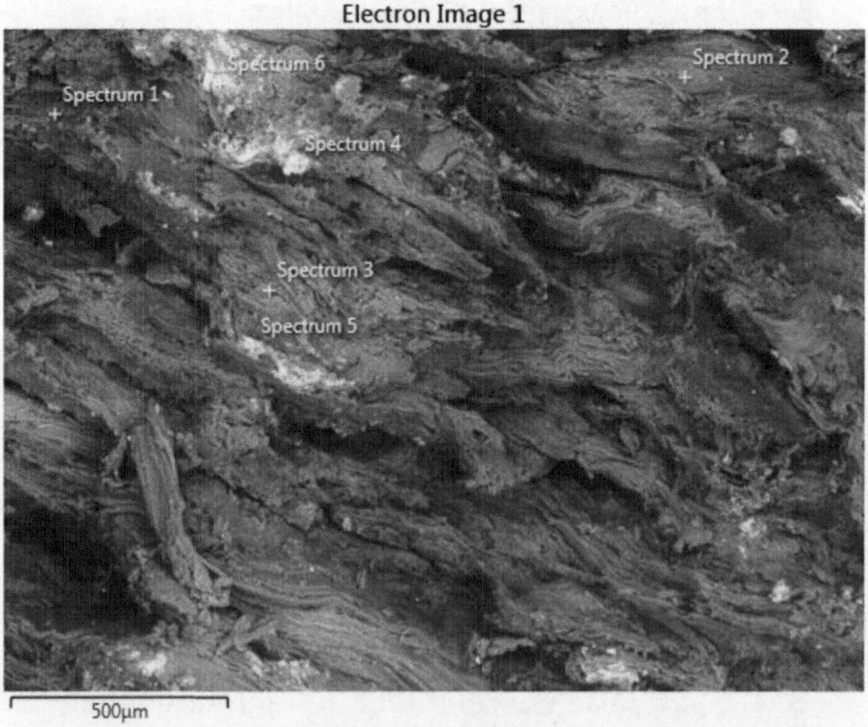

FIGURE 1.14 Scanning electron microscopic micrograms of composite 1.2 (wood sawdust 95% + liquid glass 5%) (Composite 1.2).

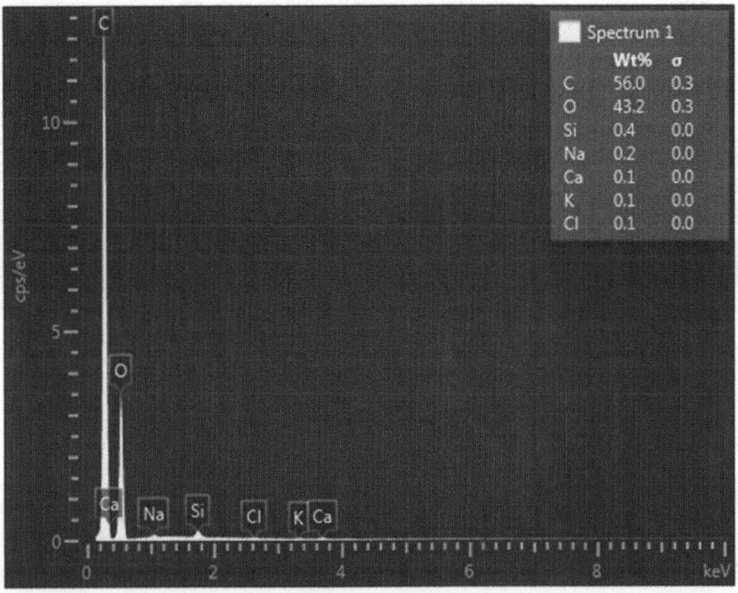

FIGURE 1.15 Energy dispersion micro-X-ray spectral analysis of composite 1.2 (wood sawdust 95% + liquid glass 5%) (Spectrum 1).

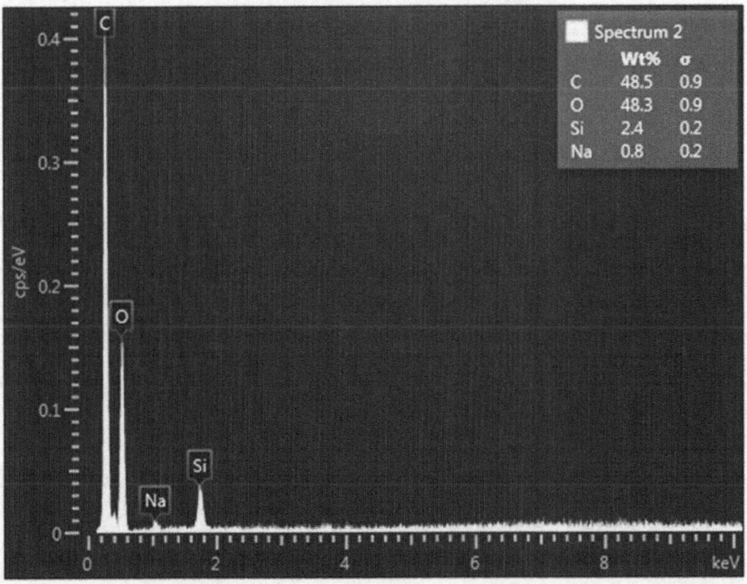

FIGURE 1.16 Energy dispersion micro-X-ray spectral analysis of composite 1.22 (wood sawdust 95% + liquid glass 5%) (Spectrum 2).

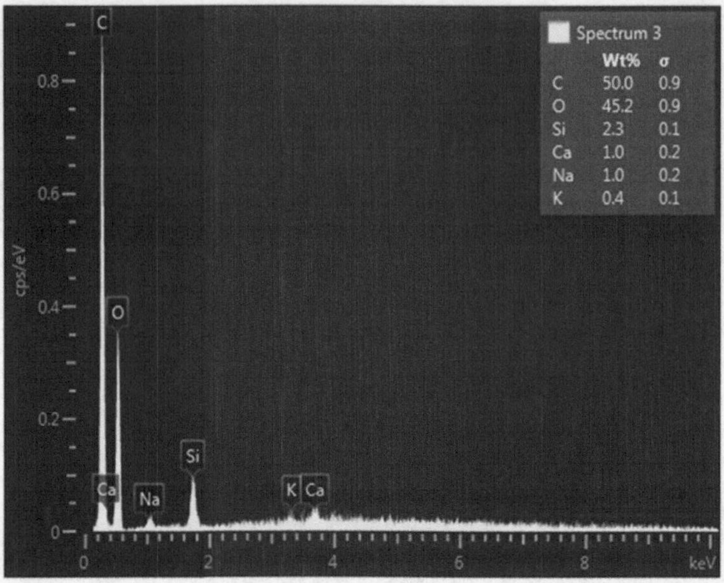

FIGURE 1.17 Energy dispersion micro-X-ray spectral analysis of composite 1.2 (wood sawdust 95% + liquid glass 5%) (Spectrum 3).

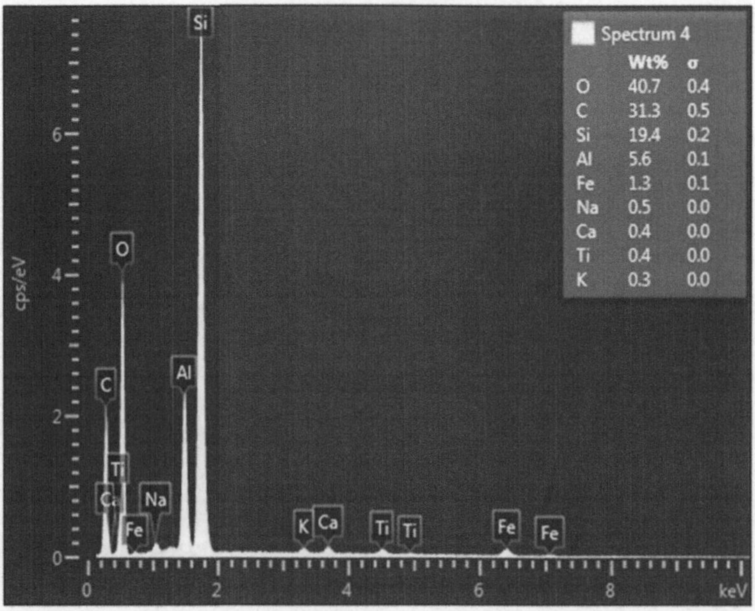

FIGURE 1.18 Energy dispersion micro-X-ray spectral analysis of composite 1.2 (wood sawdust 95% + liquid glass 5%) (Spectrum 4).

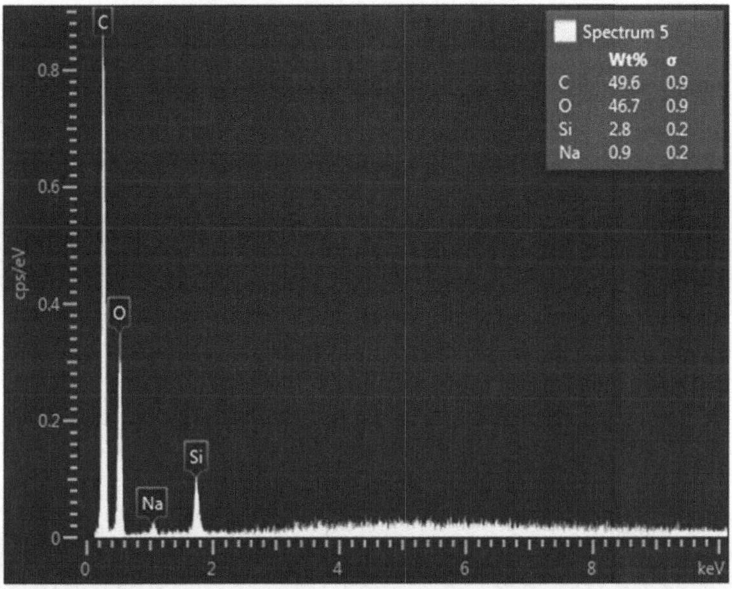

FIGURE 1.19 Energy dispersion micro-X-ray spectral analysis of composite 1.2 (wood sawdust 95% + liquid glass 5%) (Spectrum 5).

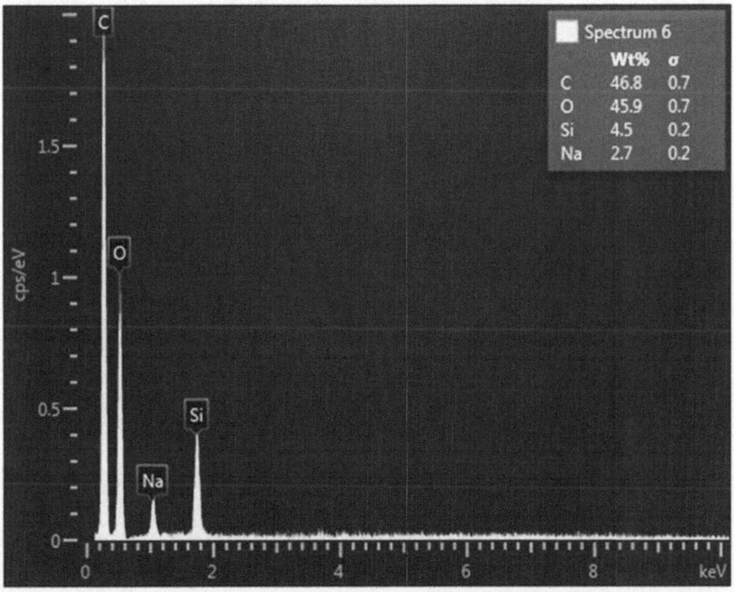

FIGURE 1.20 Energy dispersion micro-X-ray spectral analysis of composite 1.2 (wood sawdust 95% + liquid glass 5%) (Spectrum 6).

TABLE 1.3 Energy Dispersion Micro-X-Ray Spectral Analysis of Composite 1.2 (Wood Sawdust 95% + Liquid Glass 5%)

Result Type	Weight (%)					
Spectrum Label	**Spectrum 6**	**Spectrum 1**	**Spectrum 2**	**Spectrum 3**	**Spectrum 4**	**Spectrum 5**
C	46.83	55.96	48.50	50.00	31.35	49.57
O	45.88	43.16	48.33	45.22	40.65	46.70
Na	2.75	0.23	0.79	0.99	0.48	0.93
Al	–	–	–	–	5.65	–
Si	4.54	0.37	2.38	2.31	19.45	2.80
Cl	–	0.07	–	–	–	–
K	–	0.08	–	0.43	0.28	–
Ca	–	0.13	–	1.04	0.45	–
Ti	–	–	–	–	0.41	–
Fe	–	–	–	–	1.30	–
Total	**100.00**	**100.00**	**100.00**	**100.00**	**100.00**	**100.00**

Statistics	C	O	Na	Al	Si	Cl	K	Ca	Ti	Fe
Max.	55.96	48.33	2.75	5.65	19.45	0.07	0.43	1.04	0.41	1.30
Min.	31.35	40.65	0.23	5.65	0.37	0.07	0.08	0.13	0.41	1.30
Average	47.04	44.99	1.03	–	5.31	–	–	–	–	–
Standard deviation	8.29	2.73	0.89	–	7.05	–	–	–	–	–

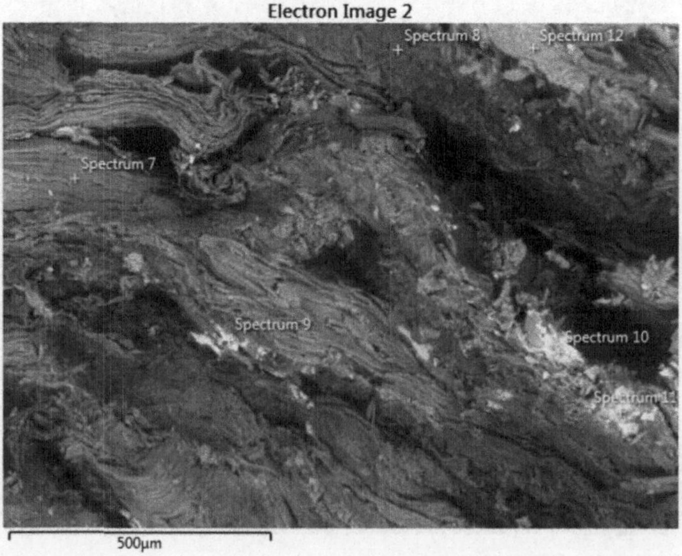

FIGURE 1.21 Scanning electron microscopic micrograms of composite 1.10 (wood sawdust 95% + PhES-80 – 5%).

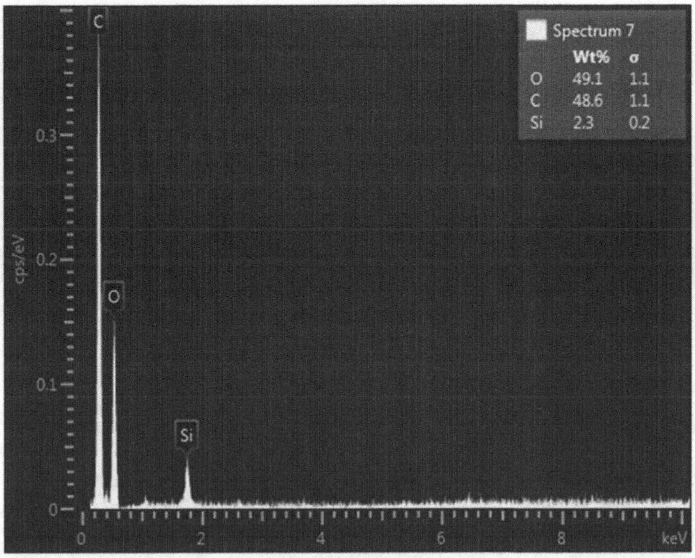

FIGURE 1.22 Energy dispersion micro-X-ray spectral analysis of composite 1.10 (wood sawdust 95% + PhES-80 – 5%) (Spectrum 7).

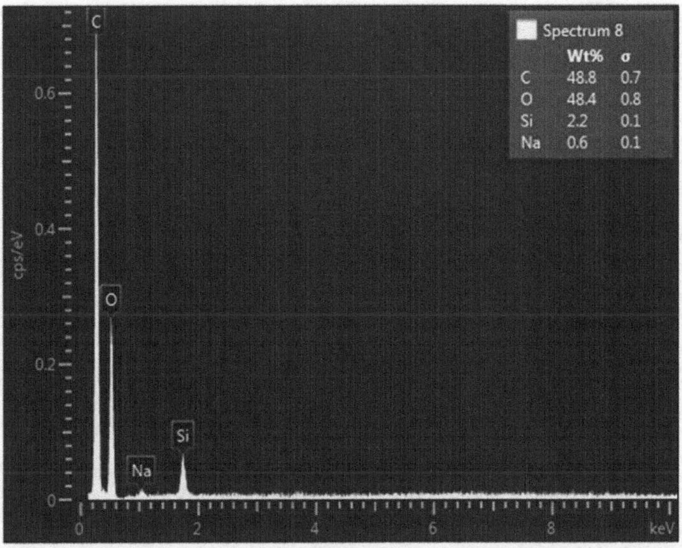

FIGURE 1.23 Energy dispersion micro-X-ray spectral analysis of composite 1.10 (wood sawdust 95% + PhES-80 – 5%) (Spectrum 8).

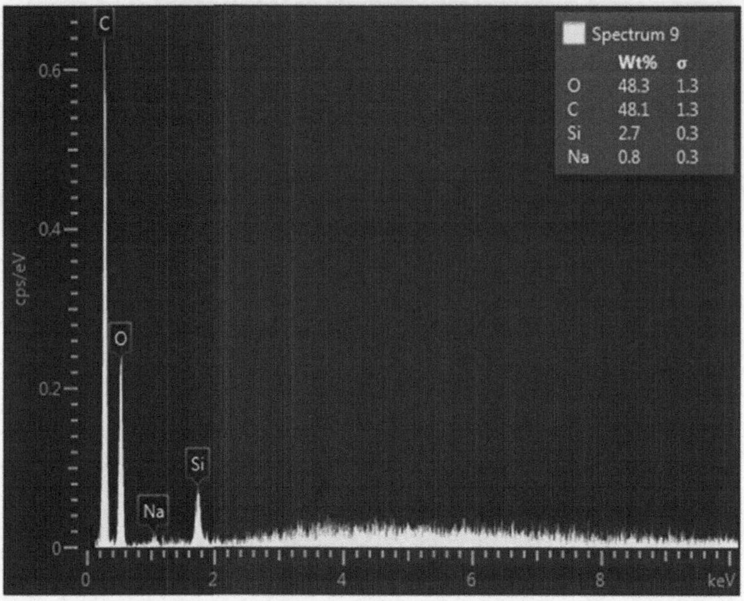

FIGURE 1.24 Energy dispersion micro-X-ray spectral analysis of composite 1.10 (wood sawdust 95% + PhES-80 – 5%) (Spectrum 9).

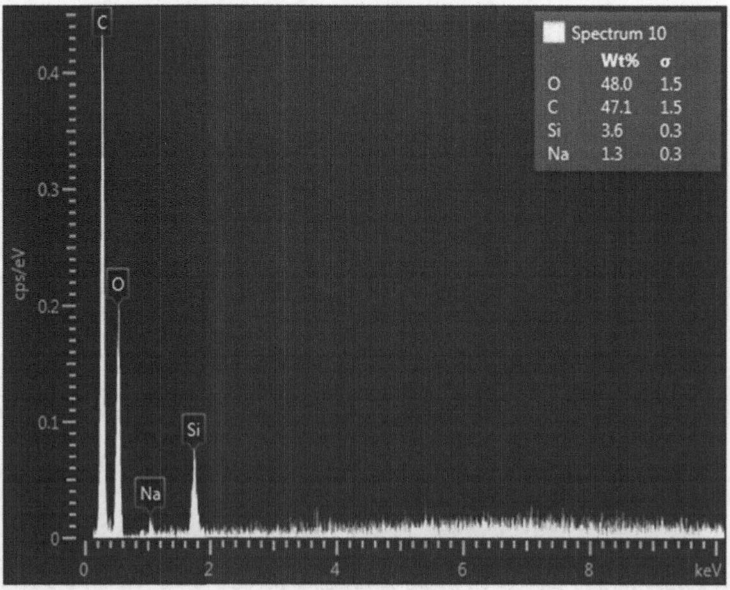

FIGURE 1.25 Energy dispersion micro-X-ray spectral analysis of composite 1.10 (wood sawdust 95% + PhES-80 – 5%) (Spectrum 10).

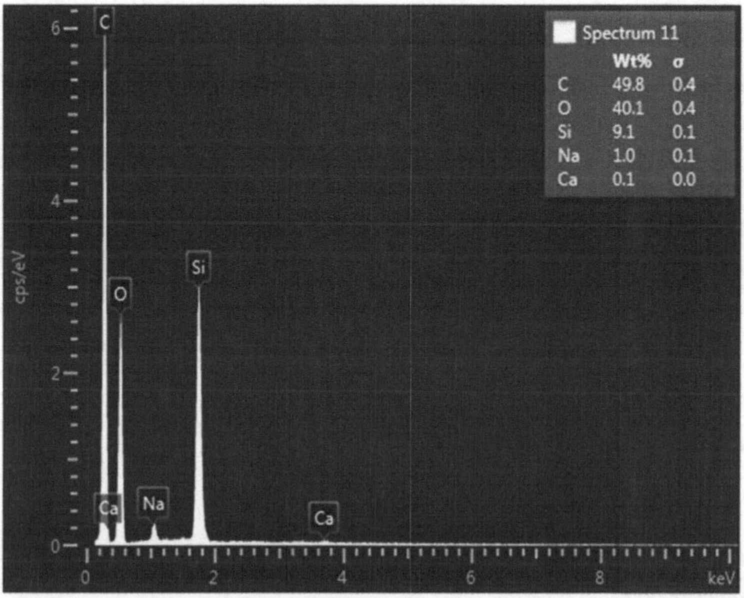

FIGURE 1.26 Energy dispersion micro-X-ray spectral analysis of composite 1.10 (wood sawdust 95% + PhES-80 – 5%) (Spectrum 11).

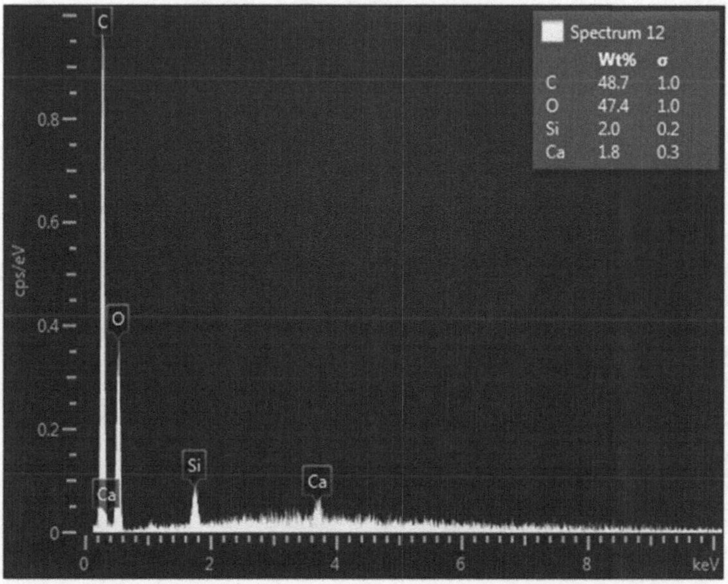

FIGURE 1.27 Energy dispersion micro-X-ray spectral analysis of composite 1.10 (wood sawdust 95% + PhES-80 – 5%) (Spectrum 12).

TABLE 1.4 Energy Dispersion Micro-X-Ray Spectral Analysis of Composite 1.10 (Wood Sawdust 95% + PhES-80 – 5%)

Result Type	Weight (%)					
Spectrum Label	**Spectrum 12**	**Spectrum 7**	**Spectrum 8**	**Spectrum 9**	**Spectrum 10**	**Spectrum 11**
C	48.74	48.64	48.81	48.15	47.07	49.75
O	47.41	49.06	48.41	48.31	48.01	40.05
Na	–	–	0.55	0.85	1.27	1.00
Si	2.03	2.30	2.23	2.70	3.65	9.07
Ca	1.82	–	–	–	–	0.13
Total	**100.00**	**100.00**	**100.00**	**100.00**	**100.00**	**100.00**

Statistics	**C**	**O**	**Na**	**Si**	**Ca**
Max.	49.75	49.06	1.27	9.07	1.82
Min.	47.07	40.05	0.55	2.03	0.13
Average	48.53	46.88	–	3.66	–
Standard deviation	0.88	3.39	–	2.71	–

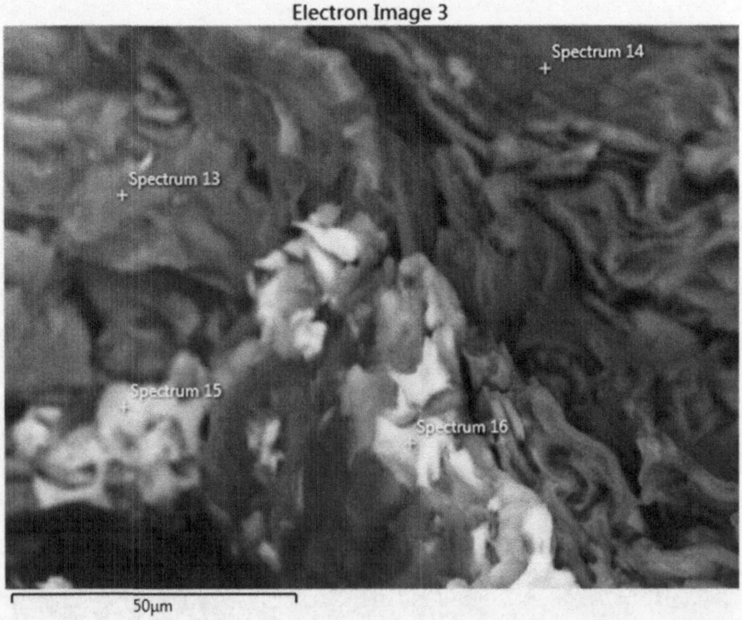

FIGURE 1.28 Scanning electron microscopic micrograms of composite 1.13 (wood sawdust 95% + PE 5%).

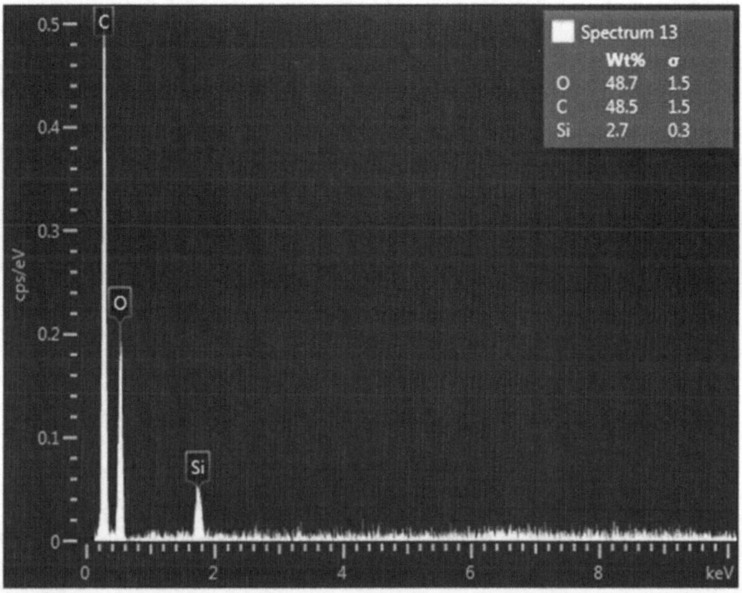

FIGURE 1.29 Energy dispersion micro-X-ray spectral analysis of composite 1.13 (wood sawdust 95% + PE 5%).

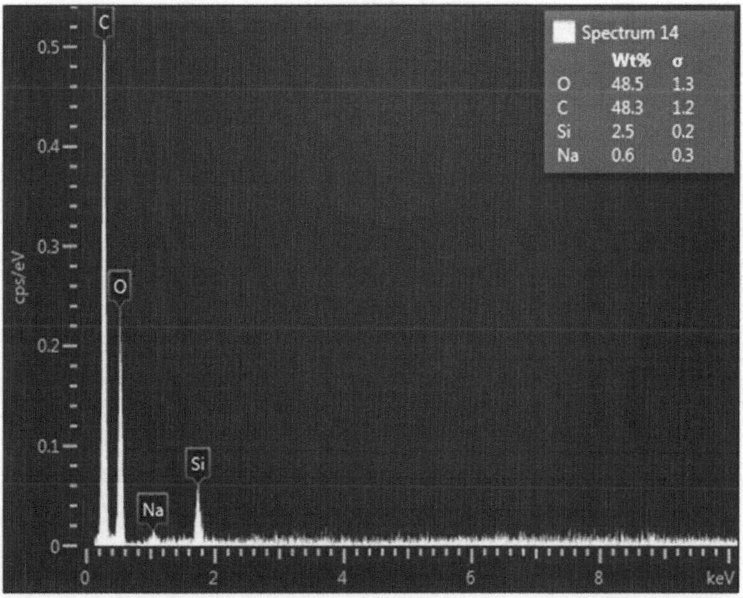

FIGURE 1.30 Energy dispersion micro-X-ray spectral analysis of composite 1.13 (wood sawdust 95% + PE 5%) (Spectrum 14).

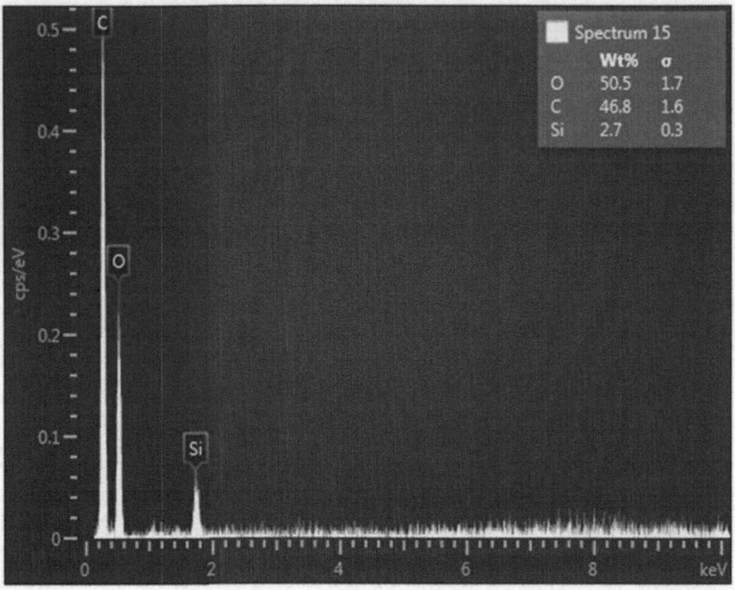

FIGURE 1.31 Energy dispersion micro-X-ray spectral analysis of composite 1.13 (wood sawdust 95% + PE 5%) (Spectrum 15).

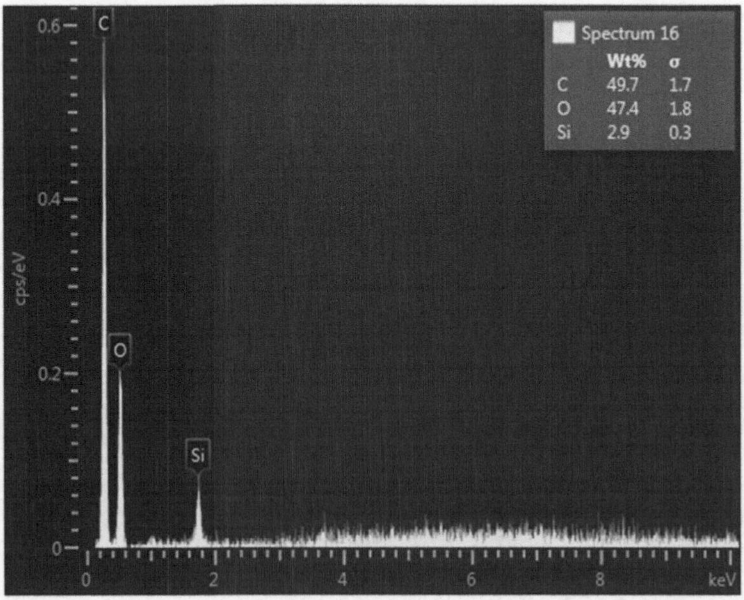

FIGURE 1.32 Energy dispersion micro-X-ray spectral analysis of composite 1.13 (wood sawdust 95% + PE 5%) (Spectrum 16).

TABLE 1.5 Energy Dispersion Micro-X-Ray Spectral Analysis of Composite 1.13 (Wood Sawdust 95% + PE 5%)

Result Type		**Weight (%)**		
Spectrum Label	**Spectrum 16**	**Spectrum 13**	**Spectrum 14**	**Spectrum 15**
C	49.71	48.53	48.35	46.82
O	47.39	48.75	48.55	50.51
Na	–	–	0.60	–
Si	2.90	2.72	2.50	2.66
Total	**100.00**	**100.00**	**100.00**	**100.00**

Statistics	**C**	**O**	**Na**	**Si**
Max.	49.71	50.51	0.60	2.90
Min.	46.82	47.39	0.60	2.50
Average	48.35	48.80	–	2.70
Standard deviation	1.18	1.29	–	0.17

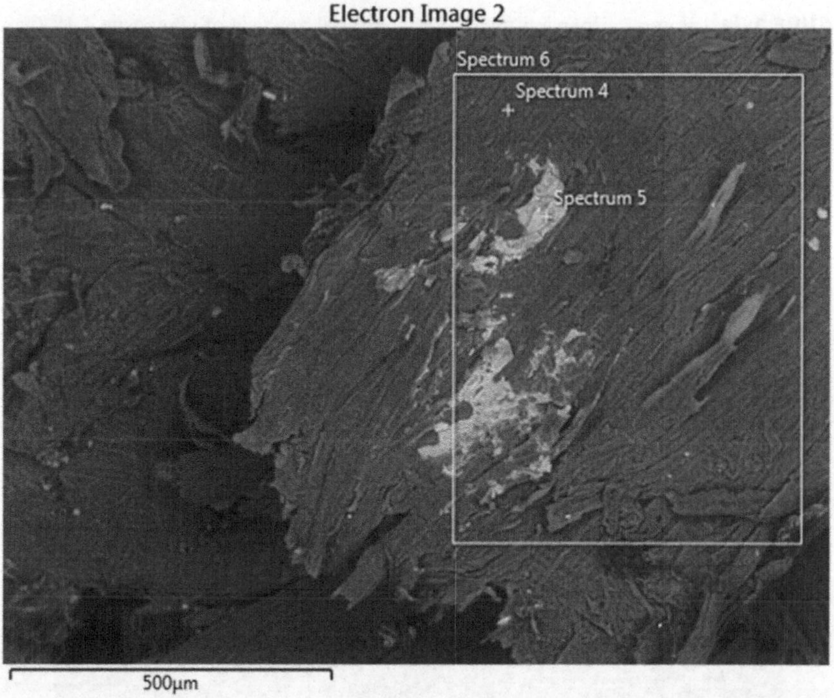

FIGURE 1.33 Scanning electron microscopic micrograms of composite 1.18 (wood sawdust 90% + 10% colophony).

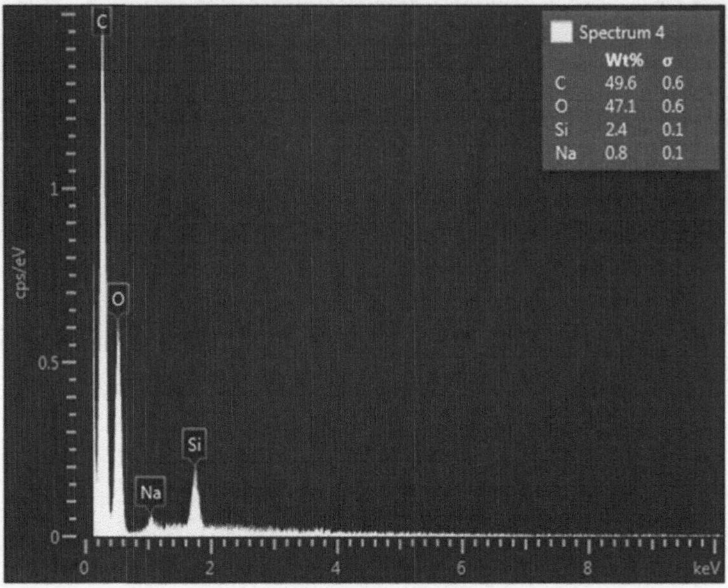

FIGURE 1.34 Energy dispersion micro-X-ray spectral analysis of composite 1.18 (wood sawdust 90% + 10% colophony) (Spectrum 4).

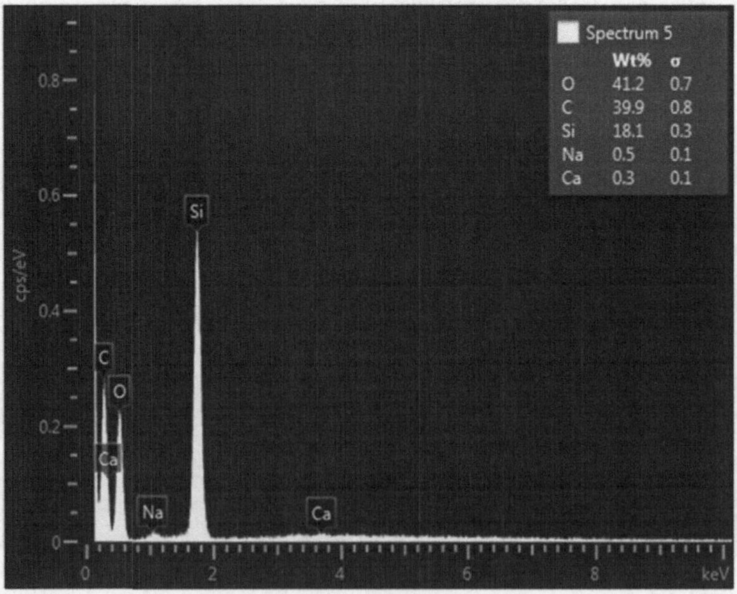

FIGURE 1.35 Energy dispersion micro-X-ray spectral analysis of composite 1.18 (wood sawdust 90% + 10% colophony) (Spectrum 5).

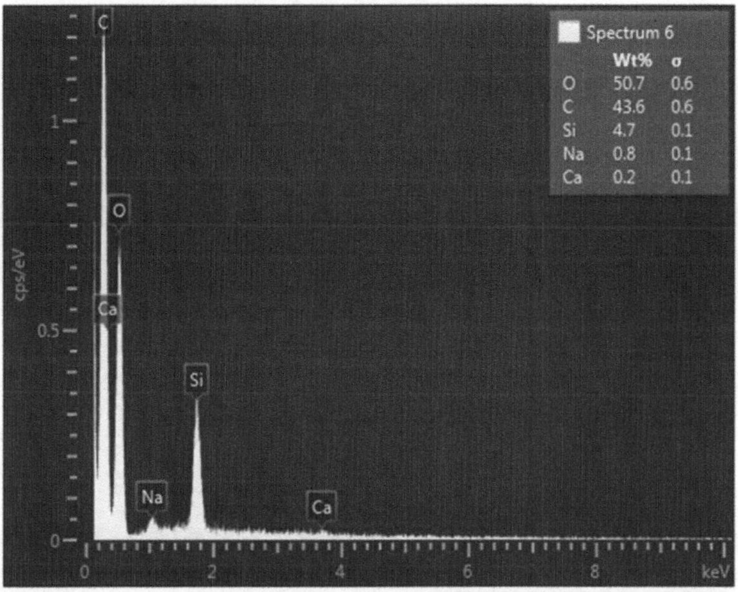

FIGURE 1.36 Energy dispersion micro-X-ray spectral analysis of composite energy dispersion micro-X-ray spectral analysis of composite 1.18 (wood sawdust 90% + 10% colophony) (Spectrum 6).

TABLE 1.6 Energy Dispersion Micro-X-Ray Spectral Analysis of Composite 1.18 (Wood Sawdust 90% + 10% Colophony)

Result Type	**Weight (%)**		
Spectrum Label	**Spectrum 4**	**Spectrum 5**	**Spectrum 6**
C	49.54	38.94	43.61
O	47.20	42.19	50.66
Na	0.85	0.38	0.82
Si	2.41	18.18	4.68
Ca	–	0.31	0.23
Total	100.00	100.00	100.00

Statistics	**C**	**O**	**Na**	**Si**	**Ca**
Max.	48.64	51.66	0.83	17.08	0.30
Min.	40.94	40.19	0.50	3.41	0.24
Average	43.40	45.32	0.70	9.39	–
Standard deviation	5.90	6.79	0.23	7.47	–

Sawdust 85% + PhES-50 – 5% + PE 10%.

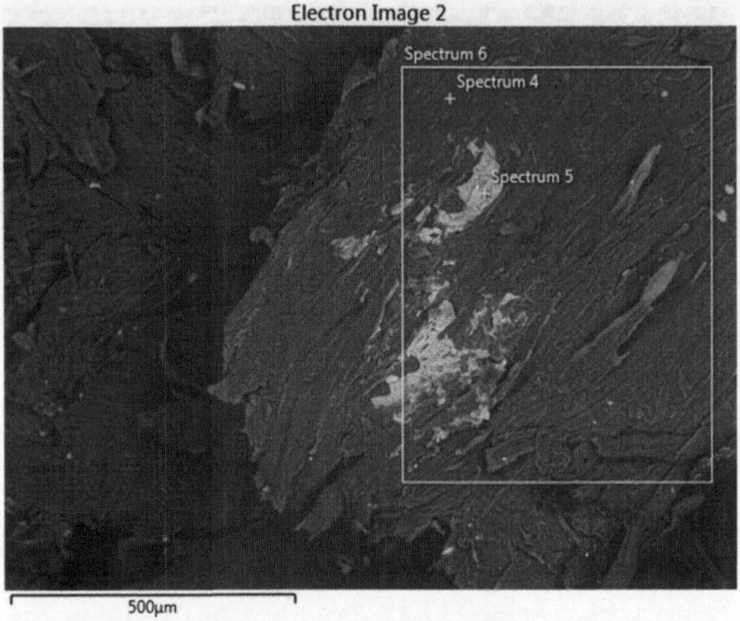

FIGURE 1.37 Scanning electron microscopic micrograms of composite 1.38 (wood sawdust 85% + 10% PhES-50 – 5% + PE 10%). Sawdust 85% + PhES-50 – 5% +PE 10%.

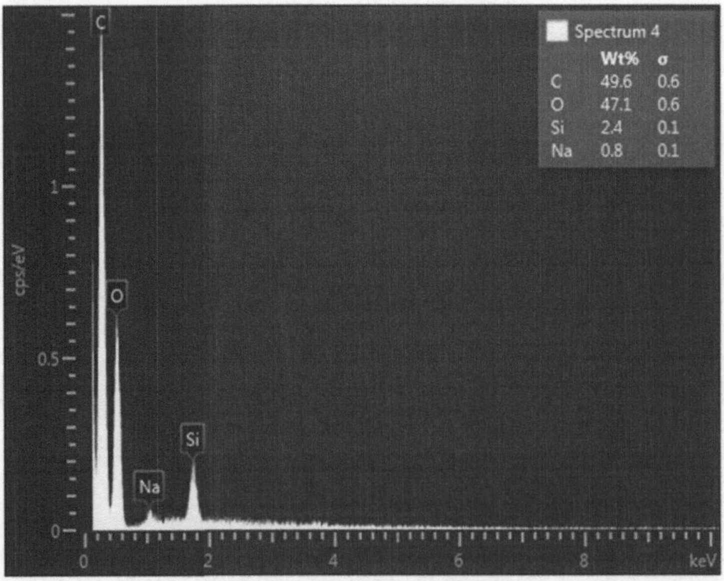

FIGURE 1.38 Energy dispersion micro-X-ray spectral analysis of composite 1.38 (wood sawdust 85% + 10% PhES-50 – 5% + PE 10%) (Spectrum 4).

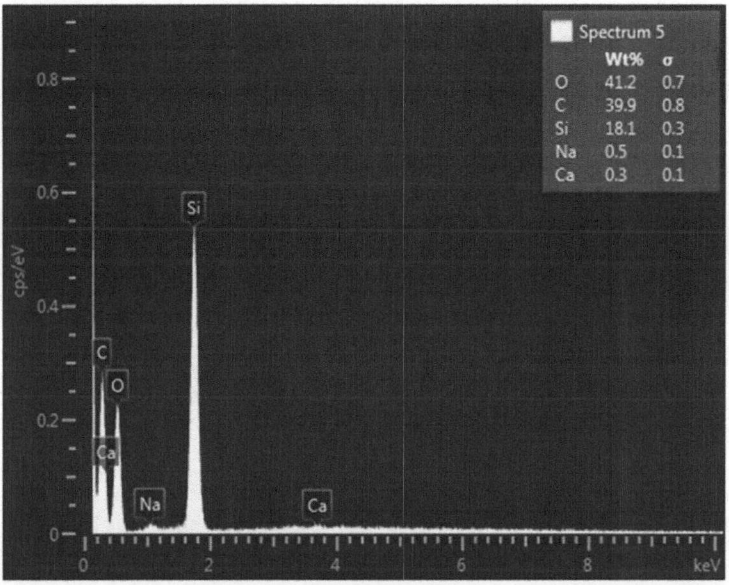

FIGURE 1.39 Energy dispersion micro-X-ray spectral analysis of composite 1.38 (wood sawdust 85% + 10% PhES-50 – 5% + PE 10%) (Spectrum 5).

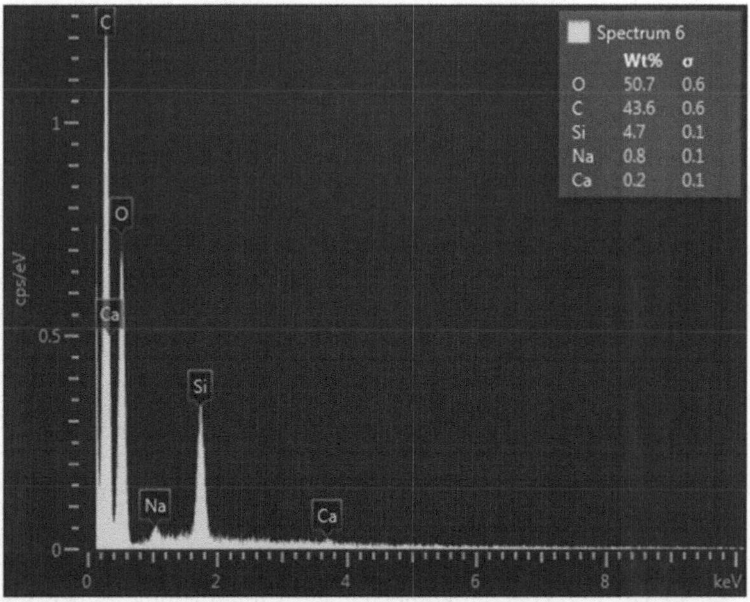

FIGURE 1.40 Energy dispersion micro-X-ray spectral analysis of composite 1.38 (wood sawdust 85% + 10% PhES-50 – 5% + PE 10%) (Spectrum 6).

TABLE 1.7 Energy Dispersion Micro-X-Ray Spectral Analysis of Composite 1.38 (Wood Sawdust 85% + 10% PhES-50 – 5% + PE 10%)

Result Type	Weight (%)		
Spectrum Label	**Spectrum 4**	**Spectrum 5**	**Spectrum 6**
C	49.64	39.94	43.61
O	47.10	41.19	50.66
Na	0.85	0.48	0.82
Si	2.41	18.08	4.68
Ca	–	0.31	0.23
Total	100.00	100.00	100.00

Statistics	C	O	Na	Si	Ca
Max.	49.64	50.66	0.85	18.08	0.31
Min.	39.94	41.19	0.48	2.41	0.23
Average	44.40	46.32	0.72	8.39	–
Standard deviation	4.90	4.79	0.21	8.47	–

1.5 GENERAL PROPERTIES OF COMPOSITES

Following composites on the basis of wood sawdust were obtained and investigated in the project frames: (a) composites with two ingredients (sawdust + binder 1); (b) composites with three ingredients (sawdust + binder 1 + binder 2). The numerical data of the noted parameters for these composites are tabulated in Table 1.8.

The table data allow us to make the following conclusions:

1. The strengthening of the composite containing bamboo powder is 4 times better than the analog, which contains bamboo less than 5 wt.%. Although here probably this difference is due to the difference in the percent amount of liquid glass (LG), i.e., at more high content of last leads to worse results because of "excessive" wetting of the filler; the composite with highly dispersed filler exposes relatively less strengthening in comparison with analog containing more big particles of the same filler, which may be due to the armoring function of last filler.
2. The composites containing wood sawdust and additively polyethylene (PE) powder show mechanical strengthening higher than that for analog composites without PE. This fact naturally is described by the additive amplifier role of the polymer filler.

TABLE 1.8 Some Characteristics of the Composites Obtained on the Basis of the Renewable Raw Materials

SL. No.	Composition (wt.%)	Density (g/cm3)	Strengthening at Stretching (MPa)	Strengthening at Bending (MPa)	Young Module (MPa)	Thermal Stability (by method Vica) (°C)	Water Absorption (%)
1.	Bamboo (90%)+ LG (10%)	0.9	0.6	4	100	>180	4
2.	Bamboo (95%) + LG (5%)	0.9	2.6	4	52	>180	7
3.	Fine dispersed bamboo (95%)+ LG (5%)	0.9	1.6	4	357	>180	6
4.	Wood sawdust + (95%) + LG (5%)	0.9	1.4	4	254	>180	20
5.	Fine dispersed sawdust (95%) + LG (5%)	–	1.9	4	–	>180	19
6.	Wood sawdust (90%) + LG (5%) + Polyethylene (5%)	0.9	5.0	6	913	>180	11
7.	Fine dispersed sawdust (90%) + PE (5%) + LG (5%)	0.9	3.8	6	865	>180	9
8.	Wood leaf (95%) + LG (5%)	0.9	1.0	4	–	>180	16
9.	Straw (95%) + LG (5%)	0.9	4.5	4	–	–	20
10.	Pine needles (95%) + LG (5%)	0.9	4	4	832	>150	14
11.	Bamboo (70%) + PhES-80 (30%)	0.7	0.4	0.8	–	–	55
12.	Bamboo (75%) + PhES-80 (25%)	0.7	0.5	1.1	–	–	50
13.	Bamboo (85%) + PhES-80 (15%)	0.8	0.7	2	–	–	38
14.	Cheap board (DSP)	–	2.7	–	–	–	55

3. Comparison of the mechanical properties of the composites containing woods leaves, pine needles, and hey shows that composites with the first filler are less durable than the ones with the other two materials; it is clear that in these cases the needles and hay have thread-like structure and consequently possess the armoring properties.
4. Comparison of the experimental results obtained by us show that the composites on the basis of wood materials waste products show some important mechanical properties and water prove better than ones for widespread construction material-DSP and, what is more important, our composites are pollution-free materials.
5. According to the table, apparently, in this case, the binder, as in the case of composite number 1, "excessive wetting" wood filler impairs the mechanical properties of the composite. Preliminary experiments on the composites containing a binder of the type PhES-80 and bamboo have shown that these materials require careful selection of the proportions of the components of the composite material as the above relations are not presented to them provide requirements for important performance characteristics. With this purpose in the future, we will carry out the optimization of composites with ecologically cleaner binder PhES-80.

1.5.1 INVESTIGATION OF PHYSICAL–MECHANICAL PROPERTIES

Following composites on the basis of wood sawdust were obtained and investigated in the project frame s: (a) composites with two ingredients (sawdust + binder 1); (b) composites with three ingredients (sawdust + binder 1 + binder 2). The numerical data of the noted parameters for these composites are tabulated in Tables 1.6 and 1.7.

The noted properties of the investigated materials were tested in two directions: (i) bending strengthening; and (ii) impact viscosity.

1.5.2 EFFECT OF TECHNOLOGICAL FACTORS ON THE IMPACT VISCOSITY OF THE COMPOSITES BASED ON SAWDUST

It was interesting to define some properties in dependence on some technological factors (temperature, pressure). There were tested The composites obtained above on the impact viscosity. The results of the experimental measurements are presented in Table 1.9.

TABLE 1.9 Dependence of the Value of Impact Viscosity of Composites Containing Sawdust on the Conditions (Temperature, Pressure) of Obtaining

SL. No.	Composite (Mass%)	Technological Parameters		Samples Cross-Section (10^{-5} M^2)	Impact Viscosity[3] (kJ/m^2)
		Pressure (MPa)	Temperature (°C)		
1.	Sawdust (95%) + PhES-80 (5%)	17	90	12	16.3
2.	Sawdust (95%) + PhES-80 (5%)	17	100	12	17.0
3.	Sawdust (95%) + PhES-80 (5%)	17	110	12	19.4
4.	Sawdust (95%) + PhES-80 (5%)	17	120	12	20.8
5.	Sawdust (95%) + PhES-80 (5%)	8	110	11	18.1
6.	Sawdust (95%) + PhES-80 (5%)	10	110	11	17.2
7.	Sawdust (95%) + PhES-80 (5%)	12	110	11	18.8
8.	Sawdust (95%) + PhES-80 (5%)	15	110	12	20.7
9.	Sawdust (95%) + PhES-50 (5%)	8	110	12	16.3
10.	Sawdust (95%) + PhES-50 (5%)	10	110	12	18.9
11.	Sawdust (95%) + PhES-50 (5%)	12	110	9	20.1
12.	PhES-50 (5%) + Sawdust (95%)	15	110	11	22
13.	Sawdust (92%) + PhES-50 (3%) + PE (5%)	8	110	12	21.8
14.	Sawdust (92%) + PhES-50 (3%) + PE (5%)	10	110	13	17.0
15.	Sawdust (92%) + PhES-50 (3%) + PE (5%)	12	110	13	18.3
16.	Sawdust (92%) + PhES-50 (3%) + PE (5%)	15	110	10	13.6
17.	Sawdust (90%) + PhES-50 (5%) + PE (5%)	13	120	–	–
18.	Sawdust (92%) + PhES-50 (5%) + PE (5%)	13	130	13	17.7

TABLE 1.9 *(Continued)*

SL. No.	Composite (Mass%)	Technological Parameters		Samples Cross-Section (10^{-5} M^2)	Impact Viscosity[3] (kJ/m^2)
		Pressure (MPa)	Temperature (°C)		
19.	Sawdust (92%) + PhES-50 (5%) + PE (5%)	13	150	12	17.7
20.	Sawdust (92%) + PhES-50 (5%) + PE (5%)	12	120	13	16.9
21.	Sawdust[1] (92%) + PhES-80 (5%) + PE (3%)	12	120	16	19.8
22.	Sawdust[2] (90%) + PhES-50 (3%)[2] + PE (3%) + Vin$(OEt)_3$ (4%)	12	120	14	22.4

[1] Dispersity <0.5 mm;

[2] Dispersity up to 0.1 mm;

[3] Hummer mass 0.1 kg.

On the basis of Table 1.9, it may be made the following conclusions. The impact viscosity of the composite based on sawdust and PhES-80 depends on the temperature at pressing (pressure is constant). Namely the increasing temperature in the range of 90–120°C leads to increasing of the impact viscosity in the range of 16–21 kJ/m, which may be connected to enhancing of the caking at increased temperature (samples 1–4). The analogical process takes place at increasing technological pressure in the range of 8–15 MPa at constant temperature (samples 5–8).

Although this increase in comparison with the first process is lower to some extent. In this case, increasing pressure leads to compressing of ingredients and consequently to increasing the interaction between them and enhancing the physical and chemical contacts. In the same conditions the composites with PhES-50 the mechanical strengthening increases in the interval 16–22 kJ/M^2.

It must be noted that the variation of temperature and pressure in the case of PE does not lead to any essential change in mechanical strengthening. This fact indicates the existence of some additive factors. Here it may be supposed that at temperatures higher than 150C the softening of PE reaches to level when the part of PE is leaked from press forms, which is equal to a decrease in the concentration of PE.

Comparison of the samples 21 and 22 show that the higher is dispersive the sawdust the higher is its mechanical properties. This fact is described in terms of increasing the contact surface between ingredients.

The bending strength of the tested compositions was taken in two modes: during increasing the temperature at fixed pressure by using working molds with appropriate standards, in a wide temperature range. The temperature in the forms varied in the range of 80–170°C, while the pressure was at intervals of 10–15 MPa. Samples were prepared for 10 minutes.

Experimental data on the bending strength is given in Table 1.10. Data from Table 1.10 allow us to make the following conclusions.

The bending strength of the composite, which was obtained from 95% wood Sawdust (average grain size is less than 1 mm) and 5% binder PhES-80, generally depending on the technological parameters – temperature and pressure. The strength of the composites, which are derived in the temperature range of 90–120°C and at a fixed pressure of 17 MPa monotonously increases between 4.5 and 21 MPa within. The bending strength composites that are obtained at the pressure interval 8–15 MPa at the constant temperature (110°C) conditions, monotonously decrease with an increase of technological pressure (8 MPa–22 MPa). Clearly, monolithic high-strength material is obtained at optimal values of the parameters of the technological conditions. According to our data, the best results are obtained in the case when the temperature of the composite is 120°C and a pressure of 17 MPa, or when the temperature is 110°C and a pressure of 8 MPa.

In composites, which are received in the same mode, where PhES-50 is used instead of PhES-80, the influence of the technological regime on the impact bending strength is less noticeable. This option will vary in the range of 12–15 MPa (when the composite-making regime is 110°C temperature and 8–15 MPa pressure) and 14–19 MPa (a composite-making regime is the same, but 3% of the amount taken PhES-50).

Composites containing PE (5%) generally exhibit higher mechanical properties and depending on the mode of getting less than its not-containing analogs. It is obvious that in this case, the PE as a binder performs an additional role. It should be noted that a composite containing 3% of the PE gives better properties to composites than 5% by weight, which indicates that the elastic phase which is obtained by PE with the percentage reduction is reduced, and thus the strength of the composite increases. For example, wood composite, which includes PhES-80 (3%), and PE (3%) demonstrated a 33 MPa equal bending strength limit. Also distinguished by high strength composite, which additionally contains a binder of vinyltriethoxysilanes (29 MPa) and dispersed sawdust. This result can be explained by the last phase separator surface (sawdust-binder) and raising the heterogeneous bonds with increasing concentration. The experiments show that the composites making technological conditions significantly affect the microstructure of the composites.

Further, it provided a systematic investigation of the impact viscosity for composites based on sawdust and different binders at their separate and combined application in composites.

The experimental data are presented in Table 1.10.

1. **Strengthening on Bending:** The samples 1.9–1.12 present the composites with PhES-80 and sawdust (particle sizes up to 1 mm), the impact viscosity of which increases with increasing temperature (90–120°C) at constant pressure (17 MPa). This fact shows that these conditions cause rather high mechanical properties and give the possibility to suggest that here it takes place some heterogenic reactions between active groups of the ingredients (PhES-80 and Sawdust), which lead to increasing the chemical bonds' concentration and consequently–mechanical properties. Although this parameter is at 110°C and 8–15 MPa this parameter for obtained composites changes only up to 15 MPa.

TABLE 1.10 Mechanical Properties of the Composites Based on Sawdust and Different Binder

SL. No.	Composite (Mass %)	Pressure (MPa)	Temperature (T, °C)	Bending strength, Kg/cm^2	Impact Viscosity (kJ/m^2)
1.	Sawdust 97% + LG 3%	150	130	3.6	22.0
2.	Sawdust 95% + LG 5%	150	130	6.2	25.0
3.	Sawdust 90% + LG 10%	150	130	4.0	19.0
4.	Sawdust 85% + LG 15%	150	130	7.1	27.0
5.	Sawdust 97% + PhES-50 – 3%	250	130	12.2	23.3
6.	Sawdust 95% + PhES-50 – 5%	250	130	3.6	2.1
7.	Sawdust 90% + PhES-50 – 10%	150	130	9.5	30.0
8.	Sawdust 85% + PhES-50 – 15%	150	130	9.2	20.0
9.	Sawdust 97% + PhES-80 – 3%	150	130	11.4	32.0
10.	Sawdust 95% + PhES-80 – 5%	150	130	8.0	25.0
11.	Sawdust 90% + PhES-80 – 10%	150	130	13.4	35.2
12.	Sawdust 85% + PhES-80 – 15%	150	130	9.0	25.0
13.	Sawdust 90% + PE 10%	150	130	10.0	35.0
14.	Sawdust 95% + PE 5%	150	130	16.3	39.03
15.	Sawdust 85% + PE 15%	150	130	17.0	32.5
16.	Sawdust 80% + PE 20%	150	130	15.6	49.6
17.	Sawdust 95% + Colophony 5%	150	130	19.9	52.4
18.	Sawdust 90% + Colophony 10%	150	130	11.8	21.0
19.	Sawdust 85% + Colophony 15%	150	130	11.0	30.0
20.	Sawdust 80% + Colophony 20%	150	130	23.3	46.3
21.	Sawdust 95% + Wood glue 5%	150	130	16.5	36.2

TABLE 1.10 *(Continued)*

SL. No.	Composite (Mass %)	Pressure (MPa)	Temperature (T, °C)	Bending strength, Kg/cm^2	Impact Viscosity (kJ/m^2)
22.	Sawdust 90% + Wood glue 10%	150	130	20.4	35
23.	Sawdust 85% + Wood glue 15%	150	130	40.0	46.0
24.	Sawdust 80% + Wood glue 20%	150	130	23.9	63.5
25.	Sawdust 94% + PhES-50 – 3% + LG 3% + Hardener	150	130	16.7	33.3
26.	Sawdust 90% + PhES-50 – 5% + LG 5% + Hardener	150	130	17.2	26.0
27.	Sawdust 80% + PhES-50 – 10% + LG 10%	150	130	18.6	40.7
28.	Sawdust 80% + PhES-50 – 15% + LG 5%	150	130	18.6	42.6
29.	Sawdust 80% + PhES-50 – 5% + LG 15%	150	130	16.9	17.3
30.	Sawdust 94% + PhES-80 – 3% + LG 3% + Hardener	150	130	17.1	41.2
31.	Sawdust 90% + PhES-80 – 5% + LG 5%	150	130	20.2	40.2
32.	Sawdust 91% + PhES-50 – 3% + LG 3% + PE 3%	150	130	24.5	37.4
33.	Sawdust 85% + PhES-50 – 5% + LG 5% + PE 5%	150	130	15.0	38.6
34.	Sawdust 91% + PhES-80 – 3% + LG 3% + PE 3%	150	130	15.3	37.3
35.	Sawdust 85% + PhES-80 – 5% + LG 5% + PE 5%	150	130	18.3	3.08
36.	Sawdust 94% + PhES-50 – 3% + PE 3%	150	130	16.7	43.0
37.	Sawdust 90% + PhES-50 – 5% + PE 5%	150	130	14.0	41.5
38.	Sawdust 85% + PhES-50 – 5% + PE 10%	150	130	20.0	40.0
39.	Sawdust 94% + PhES-80 – 3% + PE 3%	150	130	17.3	44.0
40.	Sawdust 90% + PhES-80 – 5% + PE 5%	150	130	13.0	46.0
41.	Sawdust 85% + PhES-80 – 5% + PE 10%	150	130	18.0	41.6
42.	Sawdust 85% + PhES-50 – 5% + PVA 10%	150	130	17.3	38.0

TABLE 1.10 *(Continued)*

SL. No.	Composite (Mass %)	Pressure (MPa)	Temperature (T, °C)	Bending strength, Kg/cm^2	Impact Viscosity (kJ/m^2)
43.	Sawdust 88% + PhES-50 – 3% + LG 3% + PE 3% + $VinSi(OEt)_3$ – 3%	150	130	15.0	45.0
44.	Sawdust 80% + PhES-50 – 5% + LG 5% + PE 5% + $VinSi(OEt)_3$ – 5%	150	130	19.0	37.5
45.	Sawdust 88% + PhES-80 – 3% + LG 3% + PE 3% + $VinSi(OEt)_3$ – 3%	150	130	13.8	35.0
46.	Sawdust 80% + PhES-80 – 5% + LG 5% + PE 5% + $VinSi(OEt)_3$ – 5%	150	130	16.7	46.0
47.	Sawdust 90% + PhES-80 – 5% + PE 5% Colophony	150	130	13.0	42.0
48.	Sawdust 94% + LG 3% + PE 3%	150	130	17.1	50.0
49.	Sawdust 90% + LG 5% + PE 5%	150	130	23.0	45.0
50.	Sawdust 94% + LG 4% + PhES-80 – 2%	150	130	9.9	24.4
51.	Sawdust 94% + LG 2% + PhES-80 – 4%	150	130	7.6	34.2
52.	Sawdust 85% + LG 5% + PE 10%	150	130	15.6	54.0
53.	Sawdust 85% + LG 10% + PE 5%	150	130	15.2	47.9
54.	Sawdust 85% + LG 7.5% + PE 7.5%	150	130	14.4	57.6
55.	Sawdust 85% + PhES-50 – 7.5% + PE 7.5%	150	130	8.8	44.4
56.	Sawdust 85% + PhES-50 – 10% + PE 5%	150	130	9.2	36.7

The strengthening for composite obtained on the basis of a blend of sawdust (95%) with particle sizes less than 1 mm and 5% PhES-80 generally depends on the technological parameters–temperature and pressure. Namely the strengthening of composites obtained at temperatures 90–120°C and pressure 17 MPa monotonically increases in the range 4.5–21 MPa. The strengthening of composites obtained at 8–15 MPa and constant temperature (110°C) monotonically decreases with increasing technological pressure. It is clear that the monolithic material may be obtained only at optimal technological parameters. In our case best results are obtained under conditions of 120°C and pressure of 17 MPa or when the temperature is equal to 110°C and pressure – 8 MPa.

In the composites obtained in the analogical regime, where PhES-50 is used instead of PhES-80 the effect of the technological conditions does not observe. This parameter changes in the range 12–15 MPa (when the sample is obtained at 110°C and pressure 8–15 MPa) and 14–19 (when the regime of obtaining composites is the same and the PhES-50 amount is 3%).

Composites that contain PE (5%) in general reveal high mechanical properties and depend on the regime less essential than the analogies without PE. It is clear that in this case, PE plays the role of additive binder. It is very unexpected that the composite containing 3% PE is better than that connected with 5%. This fact may be explained in terms of the creation of the self-phase of PE by increasing the amount of this binder. Higher results (33 MPa at bending) are obtained for the composites containing PhES-80 (3%) and 3% PE together. With high data are characterized the composite containing additively vinyltriethoxysilane (29 MPa). This result shows the increase of the inter-phase surfaces and consequently the interactions between phases.

The experiments provided by us show that the selection of the components and the technological conditions are the factors that defined the best mechanical properties of the composites.

2. **Impact Viscosity:** First of all, the comparison of the dependence of the mechanical properties of these samples on the concentration of the ingredients shows that it has extreme character. So, the impact viscosity of composites with LG reaches maximum at concentration LG 10 wt.% (27 kJ/m^2 for sample 1.3). The composites containing PhES-50 (N1.5–1.8) and separately PhES-80 (1.9–1.12) are characterized with

maximums of this parameter (30 and 35) at the same concentrations. However, for composites with PE (1.13–1.16) maximum of this parameter corresponds to 20 wt.% (49 kJ/m^2), while composites with colophony with a maximum (52 kJ/m^2) of this parameter are characterized as composites containing 15 wt.% of this binder (1.19). More high values show the composites containing wood glue – 63 kJ/m^2 (1.24). The finding of the composites with higher mechanical properties by use of two, three, and even four binders, unfortunately, didn't give us positive results. The main reason for such behavior of the values of impact viscosity of the considered materials is the compatibility of the ingredients in the frames of the composite structure. The second factor is the wettability of the composite components. The components with high wettability between them essentially increase the mechanical properties of considered composites. In this case effect of wettability is directly connected with the degree of the physical and chemical (formation of chemical bonds) interactions between ingredients.

1.6 INVESTIGATION OF THE THERMAL STABILITY (TS) OF THE COMPOSITES BASED ON WOOD SAWDUST

There were fulfilled the investigations with the use of the method Vica. The experimental results are presented in Figures 1.41–1.43.

On the basis of data from Figures 1.41–1.43 it may be made following conclusions:

1. TS for composites based on sawdust and PhES-80 increases when the external pressure increases to a definite level (Figure 1.41). The higher is pressure the higher the TS. This result is connected with two factors: (i) The sum volume of micro-empties at increasing technological pressure decreases and the composite becomes more monolithic; (ii) PhES-80 actively participates in the heterogenic reactions in the mass of composite, as a result of which the concentration of new chemical bonds increases;
2. The composites based on sawdust and PhES-50 in comparison with ones based on PhES-80 are less stable (Figure 1.42). In this case, it may be suggested that the composites based on PhES-50 have a greater number of micro-empties and low heterogenic bonds;

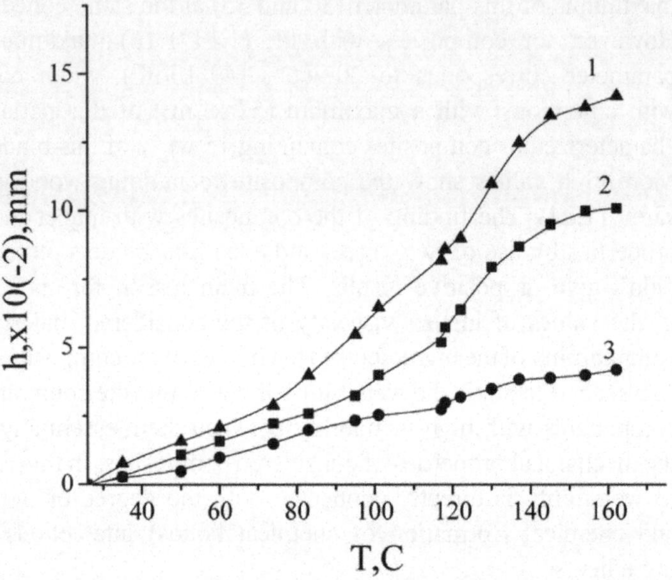

FIGURE 1.41 Dependence of the softening on temperature for composites based on sawdust and PhES-80 (5%), obtained under pressure 10 (1), 12 (2), and 15 (3) MPa at constant (110°C) temperature.

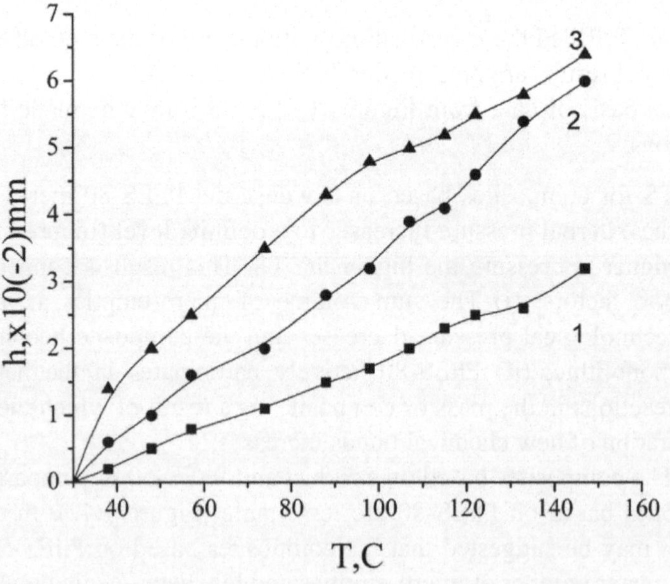

FIGURE 1.42 (5%), obtained under pressure 10 (1), 12 (2), and 15 (3) MPa at constant (110°C) temperature.

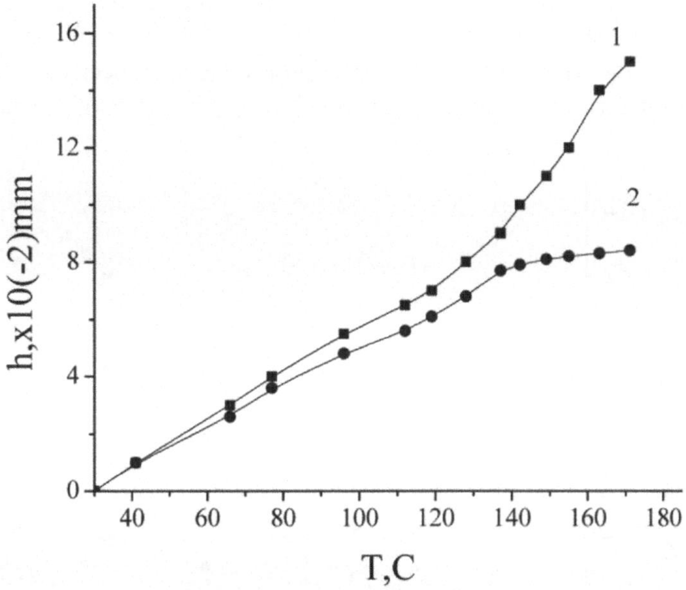

FIGURE 1.43 ependence of the softening on temperature for composites based on sawdust and PhES-80 (3%) with particle sizes up to 1 mm + PE (3%) + Vin(OEt)$_3$ (4 wt.%) – 1 and PhES-80 (3%) + sawdust (95%) with particles sizes up to 0.5 mm + PE (5%) – 2. The samples are produced at 120°C and 12 MPa.

3. Effect of the sawdust particle sizes in TS of the composites reflects in Figure 1.43. The composites containing particles with sizes up to 1 mm are less stable in comparison with ones containing sawdust particles with sizes up to 0.5 mm. These results are described for the same reason as was expressed above (i.e., the square of the sum of surfaces of contacting phases in the first case is less than in the second one and consequently the possible heterogenic reactions density in the first case is less than in the second one).

There were tested also on TS many composites with one, two, and three binders. The experimental results are presented in Figures 1.44–1.52.

The curves presented in Figure 1.44 show that the variation of the pressure effects on the TS of composites is rather essential. Partially, the decreasing of pressure in the range of 10–15 MPa leads to increasing in stability of composites. It is depended on the density of the bonds formed between ingredients as well as the mechanical properties of these composites increase also). One of the reasons may be suggested so-called cell effect of Frank and Rabinovich. This effect is explained in the following manner:

the volumes of the reaction volumes decrease at increasing pressure, which leads to a decrease in the movement of reagent particles, and consequently, the reaction rate decreases, or the reaction products, respectively decrease. This fact reflects on the mechanical and thermal properties of the materials.

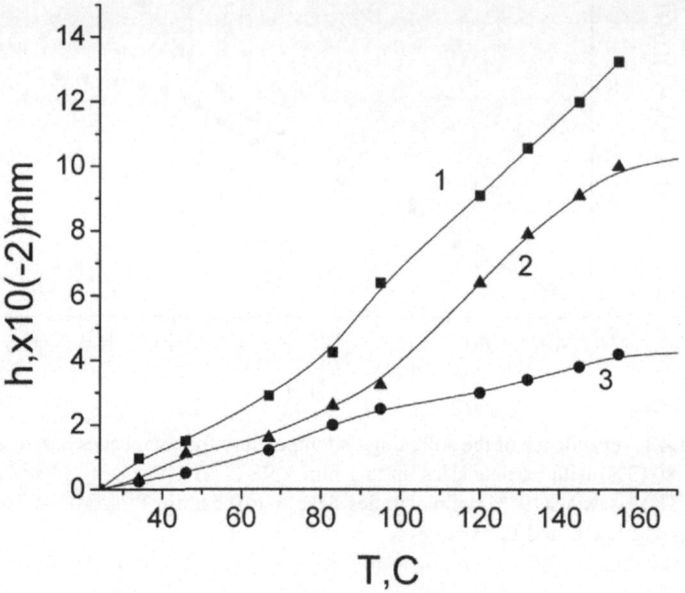

FIGURE 1.44 Dependence of the softening on temperature for composite based on sawdust and PhES-80 (5%), obtained at constant temperature 110°C and different pressures: 15 (1), 12 (2), and 10 (3) MPa.

The curves presented in Figure 1.45 show the difference in the thermostable properties of composites based on two types of PhES. Apparently, 3% of the binder more effectively influences the TS of the composite than that containing 5% of the binder (in this case the volume of the soft phase is higher than in the second composite). Analogical curves were obtained after the aging of composite sawdust (95%)+ LG (5%).

It was interesting to define – how effects the thermal aging of composites on their TS. In Figure 1.46, there are presented the curves corresponding to composite Sawdust (95%)+ LG (5%) before (2) and after (1) aging. One can see that after aging TS of the composite decrease to some extent because of the formation of some structural defects (caverns, empties) and the removal of the light fraction of the composite as a result of prolonged heating.

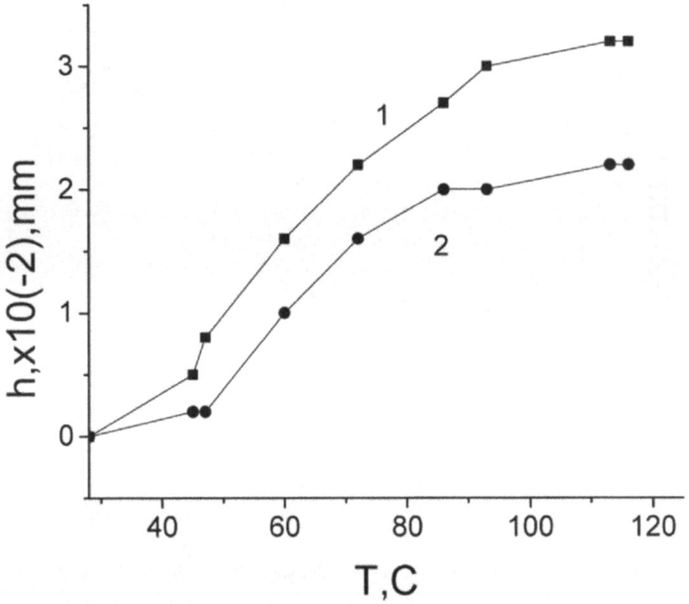

FIGURE 1.45 Thermal stability of composites on the basis of sawdust (95%) + PhES-80 (3%) (1) and sawdust (97%) + PhES-80 (5%) (2).

However, in the case of composite Sawdust (90%)+ LG (5%) + PhES-80 (5%) effect of aging is the opposite – the TS of the initial composite is lower than that after aging (Figure 1.47). Obviously in this case some additive reactions with the participation of LG and PhES-80 take place, as a result of which some more stable products form in composite.

This dependence is characteristic of such composites, obtained in the noted conditions. In the next investigations, we attempt to obtain the composites with more TS by way of inserting 2 or even 3 binders. However, after these experiments only in some cases, we obtained more or less stable composites. For example, with rather high TS is characterized by the composite Sawdust (95%) +LG (5%) (Figure 1.48), while in the case of composite Sawdust (90%) +PE (3%)+ PhES-80 (3%) + vinyltrietoxisilane (4%) (Figure 1.49) is sufficiently low, which indicates on bed wettability of the components in this composite and increasing of "soft phase."

In the case of composites: 1 – Sawdust (85%) + LG (5%) + PE (5%) + PhES-80 (5%), 2 – PE (5%) + PhES-50 (5%), and 3 – Sawdust (85%) + LG (5%) + PhES-50 (5%) TS little differs one from others. Probably the microstructure of these composites differs little from each other.

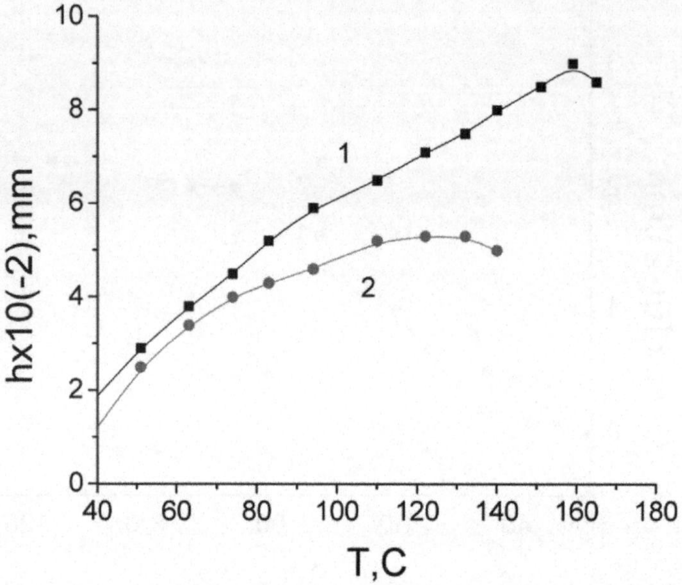

FIGURE 1.46 The thermal stability of the composite sawdust (90%) + LG (5%) + PhES-80 (5%) was obtained as a result of hot pressing at 130°C and 13 MPa before (1) and after aging (2).

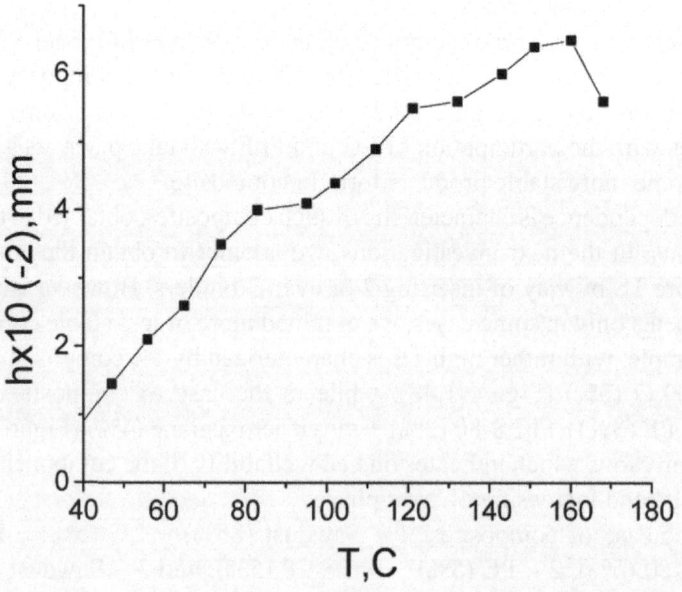

FIGURE 1.47 Thermal stability of the composite sawdust (95%) + LG (5%) was obtained at 130°C and 13 MPa.

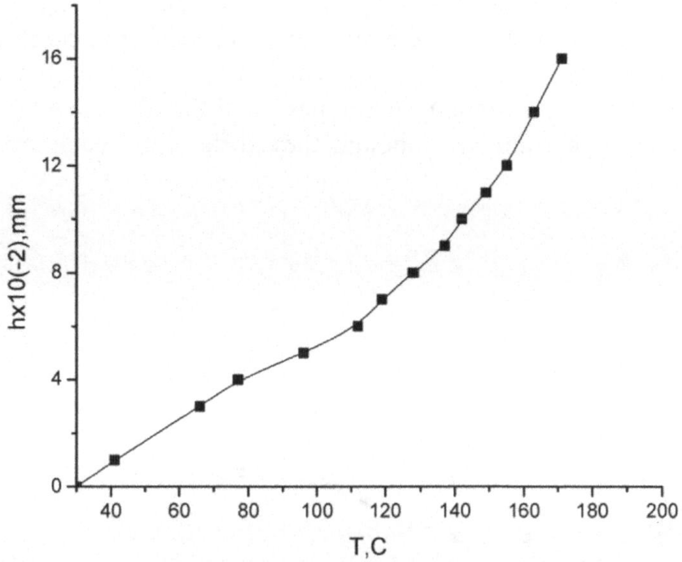

FIGURE 1.48 Thermal stability of composite sawdust (90%) + PE (3%) + PhES-80 (3%) + vinyltriethoxysilane (4%).

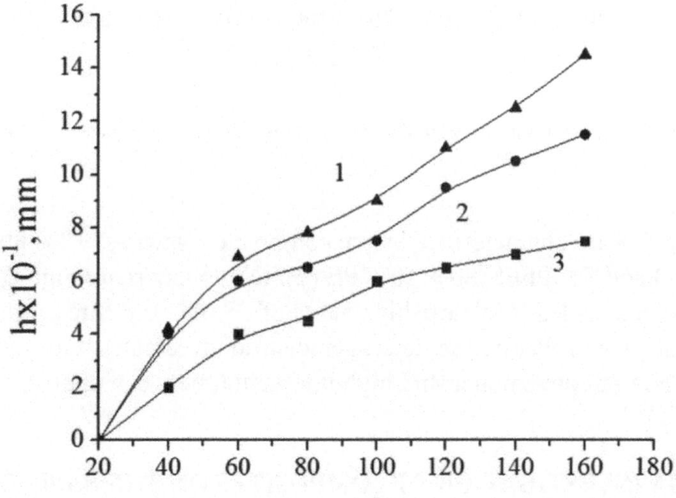

FIGURE 1.49 Temperature dependence of softening of composites based on sawdust with 3 (1), 5 (2), and 10% (3) of liquid glass.

The composites containing only one type of binder – liquid glass at different contents show the increasing TS at increasing concentrations of this binder (Figure 1.50).

The composites containing two types of PhES (fractions 50 and 80) show rather high stability, although they differ little each from other (Figure 1.50).

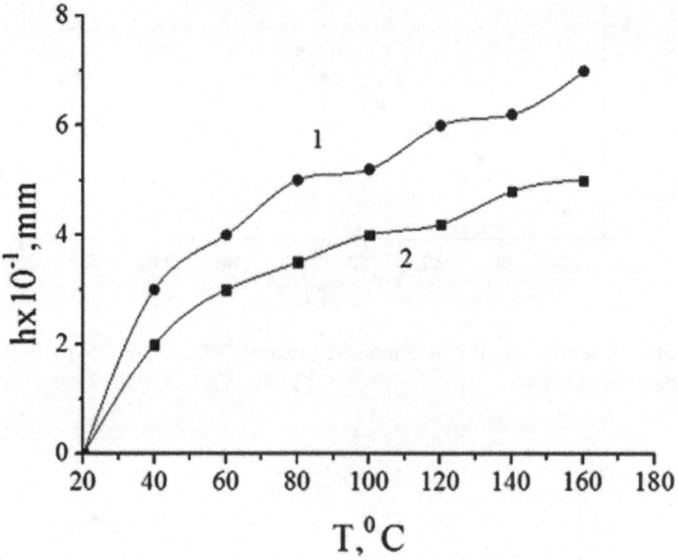

FIGURE 1.50 Temperature dependence of softening of composites based on sawdust with PhES-50 (1) and PhES-80 (2).

Higher TS are characterized by the composites: sawdust +3% PhES-80 + 3% PE (1) and 5% PhES-80 + 10% PE (2). Obviously more high content of PE in composites leads to more high value of TS of composite (Figure 1.51). In the last case probably, the heterogenic structure creates with rather high combination of convenience and high interaction between phases.

1.7 TGA INVESTIGATION OF COMPOSITES ON THE BASIS OF WOOD SAWDUST

Thermogravimetric investigations were carried out on a "Paulic-Pauluc-Erday" derivatograph (model MOM-102). The test conditions were a temperature rise rate of ≈10 deg/min in an open area. Figure 1.52, there are presented the TG curves of composites. From this, it is evident that with an increase in temperature, the mass losses of composites rise. As it is seen from thermogravimetric curves about 10% mass losses one can observe in the temperature range up to 220–230°C in this temperature interval condensation processes may proceed with unreacted hydroxyl groups as well as with hydroxyl and ethoxy groups. The main destruction process takes place from 230–280°C temperature and after 500–600°C the full destruction of composite materials takes place. It has been found that the used bindery does not affect the thermal oxidative stability of the composites.

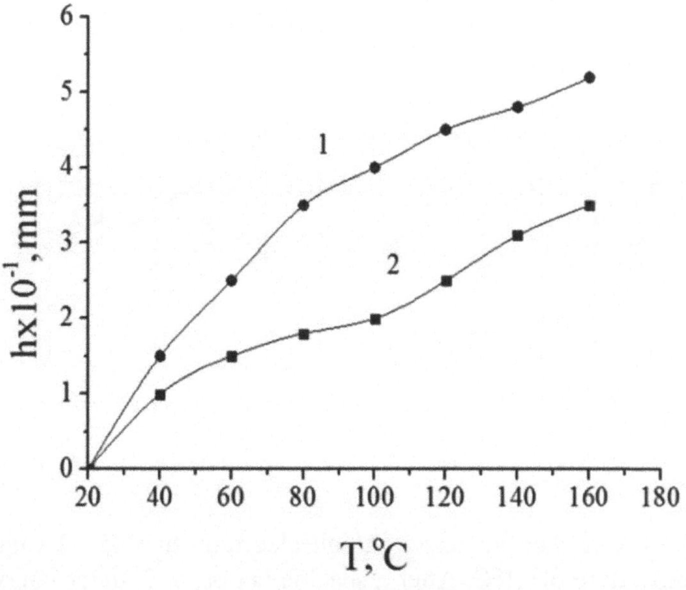

FIGURE 1.51 Temperature dependence of softening of composites based on sawdust with 3% of PhES-80 + 3% PE (1) and 5% PhES-80 + 10% PE (2).

1.8 INVESTIGATIONS OF THE WATER ABSORPTION OF COMPOSITES BASED ON SAWDUST

Water absorption is one of the most important exploitation characteristics for wood composites because they are used usually under conditions of a wet atmosphere. The composites based on sawdust and different binders were tested on water absorption.

Water absorption of samples was prepared as the known standard method. The initial weight of each specimen was taken before immersing into the water by electronic balance with an accuracy of 10^{-4}.

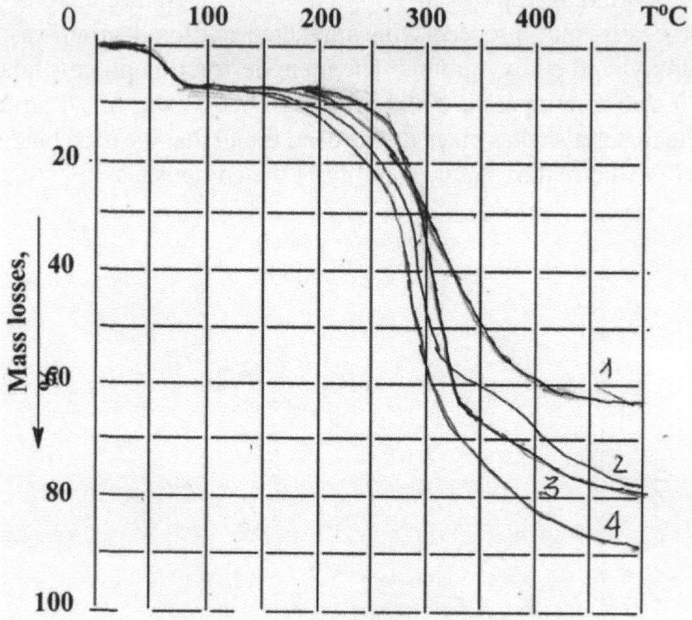

FIGURE 1.52 TGA curves of composites, in an open area, temperature rise rate 10°C/min. Where curve 1 corresponds to composite: wood sawdust (95%) + PhES-80 (5%); 2 – sawdust (95%) + PhES-50 (5%); 3 – sawdust (95%) + PE (5%); and 4 – sawdust (95%)+ LG (5%).

Samples were then immersed in collected rainwater of pH value 5.6 at room temperature of 31°C. After a specific period of 24-hour intervals, the samples were taken out and wiped with tissue paper to remove the moisture. Then the weight of the specimens was taken and tabulated. The readings were taken for 21 days for an interval of 24 hours. The (water absorption) moisture content, M(t) absorbed by each specimen is calculated from its weight before, w(0) and after, w(t) absorption by using the following equation:

$$M(t) = 100\left[\frac{W_t - W_0}{W_0}\right]$$

The results of the corresponding experiments are presented in Table 1.12.

Table 1.12 suggest that water absorption greatly is dependent on the composition of compounds, as well as making its technological conditions.

For example, water absorption of composite, consisting of PhES-50 (5%)+ PE (5%) + Sawdust (90%) is only 0.06%, in other at that time for other composites, water absorption is about 1–2%. This fact shows that the composite contains relatively low concentrations of micro-emptiness spaces in the composite mass. It is expected existence of various separate emptiness in spaces.

It should be noted that the value of the water absorption of composite differs not only in its absolute sense but rather a sign. Negative values correspond to composites, from which water has driven the reaction products. The positive water absorption characterizes composites, which are relatively less washed-up these products at the same time production of material diffused water is completely not driven out of the composite, which depends on the micro-emptiness number, their size, and the Channels among them. The bending strength (compression) test showed that the positive water absorption of samples has a high resistance compared with samples with negative water absorption. It shows that in the composites with negative water absorptions more excreted substances that could participate in the realization of internal bonds in the material.

Thus, in the second quarter of the plan obtained by the analysis of the results of the testing of the materials can be made to the basic conclusion that Sawdust's wood composites based on the physical properties of these composites are the development of recipes and technological conditions for their acceptance. We hope that the study materials as a result of a systematic study of the structural analysis of some additional methods will allow a deeper understanding of the nature of the differences in the properties of materials.

Thus, the results of the analysis of the testing materials can be made to the basic conclusion that the wood sawdust composites' physical-mechanical, as well as water absorption, mainly depends on the composition (recipes) of materials, and technological conditions (temperature, pressure) for their acceptance. It must be noted that not all composites appear stable in the medium of water. Some of them were destructed after the exposition of them in water for 24 hours (in accordance with standard). Such composites are based on liquid glass (1.1–1.3) and PhES-50 (1.5, 1.8) (Table 1.11).

It is characteristic, that the composites containing relatively low concentrations of the water are not stable in water medium. However, the same composite with 15% LG (1.4) is rather stable in comparison with analogs with less content of LG. Don't destructed but have rather low water stability in the composites based on sawdust and PhES-es. Probably these microstructures contained micro-empties and cavities, which are communicated each

TABLE 1.11 Water Absorption of Samples at Various Temperatures and Pressure

SL. No.	Composite	Mass (g)	Volume (cm^3)	Density (g/cm^3)	Mass in Water, 3 h Exposition (g)	Water Absorption, 3 h Exposition (%)	Mass in Water, 24 h exposition (g)	Water Absorption, 24 h exposition (%)
1.	PhES-80 (5%) + Sawdust (90%), P = 120 kg/cm^2 MPa, T = 120°C, Exposition 10 min.	6,425	4,768	1,348	6,553	1.99	6,393	–0.50
2.	PhES-80 (3%) + PE (3%) + Sawdust (90%) + VinSi(OEt) (4%) P = 12 MPa T = 120°C, Exp. 10 min.	7.443	5.475	1.359	7.535	1.24	7.348	–1.28
3.	PhES-80 (5%) + Sawdust (95%), T = 100°C, P = 17 MPa, Exp. 10 min.	4,757	3,532	1,347	4,935	Destroyed partially	–	–
4.	PhES-50 (3%) + Sawdust (92%) + PE (3%), T = 120°C, P = 12 MPa, Exp. 10 min.	5,267	3,797	1,387	6,864	Destroyed	–	–
5.	PhES-50 (3%) + Sawdust (92%) + PE (5%), T = 120°C, P = 12 MPa, Exp. 10 min.	5,469	3,885	1,408	7,220	Destroyed	–	–
6.	PhES-50 (3%) + Sawdust (92%) + PE (5%), T = 110°C, P = 15 MPa, Exp. 10 min.	4.234	3,179	1,332	5,173	22.17	4,309	–1.77
7.	PhES-50 (5%) + PE (5%) + Sawdust (90%), T = 120°C, P = 10 MPa, Exp. 10 min.	5,096	3,885	1,312	5,260	3.22	5,099	0.058
8.	PhES-50 (5%) + PE (5%), Sawdust (90%), T = 120°C, P = 12 MPa, Exposition 10 min.	4,770	3,709	1,286	4,893	2,570	4,721	–1.02

TABLE 1.11 *(Continued)*

SL. No.	Composite	Mass (g)	Volume (cm^3)	Density (g/cm^3)	Mass in Water, 3 h Exposition (g)	Water Absorption, 3 h Exposition (%)	Mass in Water, 24 h exposition (g)	Water Absorption, 24 h exposition (%)
9.	PhES-50 (5%) + PE (5%) + Sawdust (90%), T = 120°C, P = 13 MPa, Exp. 10 min.	5,304	4,238	1,252	5,517	4.02	5,270	–0.64
10.	PhES-50 (5%) + Sawdust (95%), T = 120°C, P = 15 MPa, Exp. 10 min.	4.980	3,885	1,282	5,266	5.74	5,036	1,124
11.	PhES-50 (3%) + PE (3%) + Sawdust (90%), $VinSi(OEt)_3$ (4%), T = 120°C, P = 12 MPa, Exposition 10 min.	5.098	3.797	1.343	5.240	14.20	5.027	–1.39

TABLE 1.12 Water Absorption of Composites on the Basis of Wood Sawdust

SL. No.	Composite (Mass%)	Mass (g)	Volume (cm^3)	Density (g/cm^3)	Mass in Water, 3 h Exposition (g)	Mass in Water, 24 h Exposition (g)	Water Absorption, 24 h exposition (%)
1.	Sawdust 97% + LG 3%	4.47	3.18	1.41	–	–	Destroyed
2.	Sawdust 95% + LG 5%	4.72	3.52	1.34	–	–	Destroyed
3.	Sawdust 90% + LG 10%	3.76	2.81	1.31	–	–	Destroyed
4.	Sawdust 85% + LG 15%	3.78	2.81	1.35	4.13	4.45	17.72
5.	Sawdust 97% + PhES-50 – 3%	3.84	2.81	1.35	–	–	Destroyed
6.	Sawdust 95% + PhES-50 – 5%	3.83	2.81	1.36	5.21	5.49	43.34
7.	Sawdust 90% + PhES-50 – 10%	3.64	2.81	1.29	5.86	6.51	78.84
8.	Sawdust 85% + PhES-50 – 15%	3.46	2.46	1.40	5.98	–	Destroyed
9.	Sawdust 97% + PhES-80 – 3%	3.30	2.29	1.44	4.45	4.68	41.81
10.	Sawdust 95% + PhES-80 – 5%	3.83	2.81	1.36	5.75	6.06	58.22
11.	Sawdust 90% + PhES-80 – 10%	4.33	3.18	1.36	4.46	5.61	29.50
12.	Sawdust 85% + PhES-80 – 15%	4.24	3.18	1.33	4.24	4.47	50.00
13.	Sawdust 95% + PE 5%	3.48	2.64	1.32	3.53	3.65	4.89
14.	Sawdust 90% + PE 10%	3.99	2.99	1.33	4.37	4.70	17.8
15.	Sawdust 85% + PE 15%	3.23	2.64	1.22	3.40	3.53	9.30
16.	Sawdust 80% + PE 20%	3.68	2.64	1.33	3.53	3.60	2.86
17.	Sawdust 95% + Colophony 5%	3.82	2.64	1.45	3.86	3.97	3.93
18.	Sawdust 90% + Colophony 10%	3.68	2.64	1.39	3.74	3.82	3.80
19.	Sawdust 85% + Colophony 15%	3.64	2.64	1.37	3.83	4.50	23.50
20.	Sawdust 80% + Colophony 20%	3.44	2.46	1.4	3.44	3.48	8.14

TABLE 1.12 *(Continued)*

SL. No.	Composite (Mass%)	Mass (g)	Volume (cm^3)	Density (g/cm^3)	Mass in Water, 3 h Exposition (g)	Mass in Water, 24 h Exposition (g)	Water Absorption, 24 h exposition (%)
21.	Sawdust 95% + 5% Wood glue	3.40	2.46	1.38	3.49	3.87	13.82
22.	Sawdust 90% + 10% Wood glue	3.84	2.64	1.45	3.95	4.13	7.55
23.	Sawdust 85% + 15% Wood glue	3.80	2.64	1.43	3.98	4.34	14.2
24.	Sawdust 80% + 20% Wood glue	3.23	2.29	1.41	3.27	3.47	7.43
25.	Sawdust 94% + PhES-50 – 3% + LG 3% Hardener	3.92	2.64	1.48	4.03	4.33	10.2
26.	Sawdust 90% + PhES-50 – 5% + LG 5% + Hardener	3.87	2.64	1.47	4.54	5.24	33.3
27.	Sawdust 80% + PhES-50 – 10% + LG 10%	3.68	2.64	1.39	3.80	4.00	8.1
28.	Sawdust 80% + PhES-50 – 15% + LG 5%	3.18	2.29	1.39	3.29	3.68	15.72
29.	Sawdust 80% + PhES-50 – 5% + LG 15%	3.41	2.29	1.39	3.49	3.65	7.04
30.	Sawdust 94% + PhES-80 – 3% + LG 3% + Hardener	3.87	2.81	1.38	4.02	4.99	28.94
31.	Sawdust 90% + PhES-80 – 5% + LG 5%	3.89	2.81	1.38	4.00	4.36	12.08
32.	Sawdust 91% + PhES-50 – 3% + LG 3% + PE 3%	3.89	2.81	1.38	4.00	4.38	12.88
33.	Sawdust 85% + PhES-50 – 5% + LG 5% + PE 5%	3.81	2.81	1.35	3.90	4.36	14.43
34.	Sawdust 91% + PhES-80 – 3% + LG 3% + PE 3%	3.98	2.81	1.42	4.08	4.56	14.57
35.	Sawdust 85% + PhES-80 – 5% + LG 5% + PE 5%	3.72	2.81	1.32	3.82	4.26	14.51
36.	Sawdust 94% + PhES-50 – 3% + PE 3%	3.92	2.81	1.37	3.92	4.95	28.90
37.	Sawdust 90% + PhES-50 – 5% + PE 5%	3.76	2.81	1.33	3.82	4.17	10.90
38.	Sawdust 85% + PhES-50 – 5% + PE 10%	3.77	2.81	1.34	3.84	4.51	19.62
39.	Sawdust 94% + PhES-80 – 3% + PE 3%	3.67	2.81	1.31	3.76	4.43	20.70
40.	Sawdust 90% + PhES-80 – 5% + PE 5%	3.86	2.81	1.37	3.94	4.76	23.31

TABLE 1.12 *(Continued)*

SL. No.	Composite (Mass%)	Mass (g)	Volume (cm^3)	Density (g/cm^3)	Mass in Water, 3 h Exposition (g)	Mass in Water, 24 h Exposition (g)	Water Absorption, 24 h exposition (%)
41.	Sawdust 85% + PhES-80 – 5% + PE 10%	3.89	2.99	1.30	3.94	4.11	5.65
42.	Sawdust 85% + PhES-50 – 5% + PVA 10%	3.56	2.81	1.27	3.70	3.88	8.98
43.	Sawdust 88% + PhES-50 – 3% + LG 3% + PE 3% + $VinSi(OEt)_3$ – 3%	3.84	2.81	1.37	3.93	4.32	12.5
44.	Sawdust 80% + PhES-50 – 5% + LG 5% + PE 5% + $VinSi(OEt)_3$ – 5%	3.61	2.81	1.28	3.71	3.95	9.41
45.	Sawdust 88% + PhES-80 – 3% + LG 3% + PE 3% + $VinSi(OEt)_3$ – 3%	3.90	2.81	1.39	4.02	4.36	11.79
46.	Sawdust 80% + PhES-80 – 5% + LG 5% + PE 5% + $VinSi(OEt)_3$ – 5%	3.57	2.64	1.35	3.66	3.88	8.68
47.	Sawdust 90% + LG 5% + PE 5%	3.81	2.81	1.36	4.09	4.34	13.91
48.	Sawdust 94% + LG 3% + PE 3%	3.93	2.81	1.40	4.22	4.88	24.17
49.	Sawdust 85% + LG 5% + PE 5% + Colophony 5%	3.53	2.64	1.34	3.57	3.70	4.81
50.	Sawdust 94% + LG 4% + PhES-80 – 2%	3.86	2.64	1.46	4.09	5.95	53.62
51.	Sawdust 94% + LG 2% + PhES-80 – 4%	3.88	2.81	1.38	4.05	5.14	32.47
52.	Sawdust 85% + LG 5% + PE 10%	3.70	2.64	1.40	3.76	4.50	21.62
53.	Sawdust 85% + LG 10% + PE 5%	3.59	2.64	1.35	3.68	4.21	17.27
54.	Sawdust 85% + LG 7.5% + PE 7.5%	3.70	2.64	1.40	3.77	4.46	20.54
55.	Sawdust 85% + PhES-50 – 7.5% + PE 7.5%	3.78	2.61	1.35	4.24	6.20	64.02
56.	Sawdust 85% + PhES-50 – 10% + PE 5%	3.55	2.64	1.34	4.30	5.92	66.76

to other. The water diffusion through these channels leads to effective water sorption by these composites. Essentially from composites noted above the ones contenting the PE distinguish very essentially (compare several percentages for composites contenting PE (1.13, 1.15, 1.16) with 43–78% for composites with PhES-50 or PhES-80 (1.6, 1.7). Obviously, the low water absorption of composites with PE is due to the high hydrophobicity of the composites contenting PE because of the good distribution of PE between particles of sawdust. High hydrophobicity is characterized by the composites contenting colophony and wood glue. The level of water absorption of composites containing the last three binders has nearly one and the same order, because of all they are good hydrophobic substances. To some extent, good hydrophobicity shows also the composites containing vinyltriethoxysilane and some composites with a combination of PhES and PE 1.41, 1.46, 1.55, 1.56). Optimization of the recapture and technological conditions – the most real ways to reach high hydrophobicity of wood composition materials.

1.9 CONCLUSIONS OF COMPOSITES ON THE BASIS OF SAWDUST

Here is presented a new composite material on the basis of wood sawdust and such binders as liquid glass, phenylethoxysisilane of two types PhES-50 and PhES-80 (first molecular mass is less than for second one) in directions of obtaining and investigation of composite structure and such physical properties, as mechanical strengthening (at bending and impact viscosity), thermal stable properties and water absorption.

The structural investigations were provided with the use of optical and scanning electron microscope (SEM) to define the morphology of the composites and the character of distribution of the ingredients in the composite body with the use of the figure of optical and scanning electron microscopes. Energy dispersive X-ray micro-analysis was used as the method for establishing the elemental analysis of the researched materials. From mechanical properties, there were studied the strengthening at bending and the impact viscosity. The methods of Vica and thermal gravimetry were used for the investigation of the TS of the composites. The water absorption of composites was studied with the use of standard methodic.

It was shown generally that the mechanical properties of the composites essentially depend on the conditions of obtaining the composites. For example, the strengthening of composites containing sawdust with 5% increases with increasing technological temperature in the range of 90–120°C at constant (17 MPa) pressure. Besides this parameter for the same composite but obtained at a constant temperature (110°C) increases also when the technological pressure increases up to 15 MPa. It is shown that the mechanical properties of the composites are higher than the degree of dispersion of the sawdust particles.

Mechanical strengthening of the composites depends essentially on the recapture factor too. It was observed that the dependence of strengthening on the binder concentration has the maximum at definite concentrations of the choice binder and at increasing its concentration the mechanical properties decrease. Here we see that the mechanical properties of the composite containing the binder higher than the definite amount (for the given composite) are worse in comparison with the same composites but containing less concentration of the same binder. This fact may be explained in terms of the creation of the associates (clusters) of the binder molecules at relatively high concentrations of lasts, which usually leads to the weakening of the composite (Table 1.12).

The technological factor is essentially reflected in the TS of composites. For example, by method Vica it is shown that TS of the composite based on sawdust containing 5% PhES-80 and obtained at temperature 110°C and 15 MPa increases 5 times in comparison with the composite obtained at 10 MPa and the same temperature. Some good results are obtained for composites with PE and colophony.

The value of the water absorption of the composites based on sawdust is in good corresponding with data of mechanical strengthening and TS. Namely the higher the density of composites the lower their water absorption.

KEYWORDS

- **composite materials**
- **energy dispersion**
- **FTIR spectral investigations**
- **micro-X-ray analysis**
- **optical microscope**
- **scanning electron microscope**
- **thermal stability**
- **viscosity**

REFERENCES

1. *Encyclopedic Dictionary of Chemistry*, (1983). M. Soviet Encyclopedia.
2. *Guide for the Production of Particle Boards*, (1990). M. Forestry.
3. Niska, K. O., & Sain, M., (2012). *Wood-Polymer Composites*. 1st Edition–May 29, 2008; eBook ISBN: 9781845694579.
4. Clemons, C. D., (2002). Wood-plastic composites in the USA. *Forest Product Journal, 52*(6).
5. Stark, N., (2005). Influence of moisture on mechanical properties of wood-polypropylene composites. *J. of Thermoplastic Composite Materials, 11*, 321.
6. Marel, J., et al., (2006). The durability of wood-plastic composites. *Wood Design Focus, 16*(3).
7. Klesov, A., (2007). *Wood-Plastic Composites* (p. 698). J. Wiley & Sons.
8. Wang, R., Zheng, S., & Zheng, Y., (2011). *Polymer Matrix Composites and Technology* (p. 568). Elsevier Sci. and Techn.
9. Ellis, W. D., & O'Dell, J. L., (1999). *J. Appl. Pol. Sci., 73*, 2493–2505.
10. Khan, M. A., & Ali, K. M., (1992). *J. Appl. Pol. Sci., 45*, 167.
11. Rozman, H. D., & Kumar, R. N., (1995). *J. Appl. Pol. Sci., 57*, 1291.
12. Uchara, T., Furuno, T., & Jodai, S. J., (1990). *Soc. Mater. Sci. Japan, 39*, 1223.
13. Ellis, W. D., (1994). *Wood Fiber Sci., 26*, 333.
14. Alén, R., (2000). Structure and chemical composition of wood. In: Stenius, P., (ed.), *Forest Products Chemistry* (pp. 11–57). Jyväskylä, Finland: Gummerus Printing.
15. Kilpeläinen, A., Peltola, H., Ryyppö, A., Sauvala, K., Laitinen, K., & Kellomäki, S., (2003). Wood properties of Scots pines (*Pinus sylvestris*) grown at elevated temperature and carbon dioxide concentration. *Tree Physiology, 23*(13), 889–897.
16. Hakkila, P., (1989). *Utilization of Residual Forest Biomass*. Springer, Germany.
17. Korhonen, P., (1997). *Kemiallisten Pääkomponenttien Jakautuminen Puuaineksessa ja Puukuidun Soluseinässä [Master's thesis]*. Department of Chemistry, University of Jyväskylä, Finland.
18. Sjöström, E., (1993). *Wood Chemistry: Fundamentals and Applications*. Academic Press, USA.
19. Viikki, A., (1995). *Puun Polysakkaridien Käyttäytyminen Sulfaattikeitossa [Master's thesis]*. Department of Chemistry, University of Jyväskylä, Finland.
20. Saarela, K. E., Harju, L., Rajander, J., Lill, J. O., Heselius, S. J., Lindroos, A., & Mattson, K., (2005). Elemental analyses of pine bark and wood in an environmental study. *Science of the Total Environment, 343*(1–3), 231–234.
21. Valentín, L., Kluczek-Turpeinen, B., Willför, S., Hemming, J., Hatakka, A., Steffen, K., et al., (2010). Scots pine (*Pinus sylvestris*) bark composition and degradation by fungi: Potential substrate for bioremediation. *Bioresource Technology, 101*(7), 2203–2209.
22. Peltonen, S., (1981). Studies on bark extracts from Scots pine (*Pinus sylvestris*) and Norway spruce (*Picea abies*). Part I. Main chemical composition. *Paperi Ja Puu, 63*, 593–595.
23. Dietrichs, H. H., Garves, K., Behrensdorf, D., & Sinner, M., (1987). Untersuchungen über die kohlenhydrate der rinden einheimischer holzarten. *Holzforschung, 32*, 60–97.
24. Nurmi, J., (1993). Heating values of the above-ground biomass of small-sized trees. *Acta Forestalia Fennica, 236*, 30.
25. 25. Nurmi, J., (1997). Heating values of mature trees. *Acta Forestalia Fennica, 256*, 28.

26. Vávřová, P., Penttilä, T., & Laiho, R., (2009). Decomposition of Scots pine fine woody debris in boreal conditions: Implications for estimating carbon pools and fluxes. *Forest Ecology and Management, 257*(2), 401–412.
27. Lassi, U., & Wikman, B., (2011). *Biomassan Kaasutus Sähköksi, Lämmöksi ja Biopolttoaineiksi*. Highbio–projektijulkaisu. Kokkola: University of Jyväskylä, Kokkola University Consortium Chydenius, Finland.
28. Fengel, D., & Wegener, G., (1989). *Wood: Chemistry, Ultrastructure, Reactions*. Walter de Gruyter, Germany.
29. Bertaud, F., & Holmbom, B., (2004). Chemical composition of earlywood and latewood in Norway spruce heartwood, sapwood, and transition zone wood. *Wood Science and Technology, 38*(4), 245–256.
30. Rhén, C., (2004). Chemical composition and gross calorific value of the above-ground biomass components of young *Picea abies*. *Scandinavian Journal of Forest Research, 19*(1), 72–81.
31. Willför, S., Sundberg, A., Hemming, J., & Holmbom, B., (2005). Polysaccharides in some industrially important softwood species. *Wood Science and Technology, 39*(4), 245–257.
32. Wainhouse, D., Cross, D. J., & Howell, R. S., (1990). The role of lignin as a defense against the spruce bark beetle *Dendroctonus micans*: Effect on larvae and adults. *Oecologia, 85*(2), 257–265.
33. Spiridon, I., Popa, M., & Popa, VI., (1995). On some characteristics of lignin and polyphenolic products separated from spruce bark. *Cellulose Chemistry and Technology, 29*(2), 115–121.
34. Anerud, E., & Jirjis, R., (2011). Fuel quality of Norway spruce stumps–influence of harvesting technique and storage method. *Scandinavian Journal of Forest Research, 26*(3), 257–266.
35. Willför, S., Sundberg, A., Pranovich, A., & Holmbom, B., (2005). Polysaccharides in some industrially important hardwood species. *Wood Science and Technology, 39*(8), 601–617.
36. Ukkonen, K., & Erä, V., (1979). Birch bark extractives. *Kemia-Kemi. 5*, 217–220.
37. Pulkkinen, E., & Nurmesniemi, H., (1980). Studies on chemical composition of the inner bark of *Betula verrucosa*. *Paperi Ja Puu, 4*, 285–288.
38. Pinto, P. C. R. O., Sousa, A. F., Silvestre, A. J. D., Neto, C. P., Gandini, A., Eckerman, C., et al., (2009). *Quercus suber* and *Betula pendula* outer barks as renewable sources of oleochemicals: A comparative study. *Industrial Crops and Products, 29*(1), 126–132.
39. Krutul, D., Zielenkiewicz, T., Antaczak, A., Zawadski, J., Radomski, A., Kupczyk, M., et al., (2011). Influence of the environmental pollution on the chemical composition of bark and wood of trunk, branches, and main roots of birch (*Betula pendula* Roth.). *Annals of Warsaw University of Life Sciences, Forestry, and Wood Technology, 74*, 242–248.
40. Voipio, R., & Laakso, T., (1992). Pienikokoisten puiden maanpäällisen biomassan kemiallinen koostumus. *Folia Forestalia, 789*, 22.
41. Kliosov, A. A., (2007). *Wood-Plastic Composites* (p. 698). John Wiley & Sons, Hoboken, New Jersey.
42. Ellis, W. D., O'Dell, J. L., (1999). Surface morphology of PECVD fluorocarbon thin films from hexafluoropropylene oxide, 1,1,2,2-tetrafluoroethane and difluoromethane. *J. Appl. Polym. Sci., 73*(12), 2493–2505.

43. Patil, Y. P., Gajre, B., Dusane, D., Chavan, S., & Mishra, S., (2000). Effect of maleic anhydride treatment on steam and water absorption of wood polymer composites prepared from wheat straw, cane bagasse, and teak wood sawdust using Novolac as matrix. *J. Appl. Polym. Sci., 77*(13), 2963–2967.
44. brember, K. I., & Shneider, M. H., (1985). The significance of the glass transition of lignin in thermomechanical pulping. The significance of the glass transition of lignin in thermomechanical pulping. *Wood Sci. and Technology, 19*, 75–81.
45. Hausen, B. M., Kuhlwein, A., & Schulz, K. H., (1982). Colophony allergy. A contribution to the origin, chemistry, and uses of colophony and modified colophony products, 1. *Derm Beruf Umwelt., 30*(4), 107–115.
46. Muraganatham, S., Anbalagan, G., & Ramamurthy, N., (2009). FTIR and SEM (EDS) comparative analysis of medical plants. *Eclipta alba* Hass K and *Eclipta prostrate* Linn. *Rom. J. Biophysics, 19*, 285–294.
47. Shirwaikar, A., Malini, S., & Kumari, S. C., (2003). Protective effect of *Pongamia pinnata* flowers against cisplatin and gentamicin-induced nephrotoxicity in rats. *Indian J. Exp. Biol., 1*, 58–62.
48. Aneli, J., Buzaladze, G., Markarashvili, E., Tatrishvili, T., Kemkhadze, L., & Mukbaniani, O., (2015). *Composites Based on Sawdust and Some Organic Binders* (p. 10). Abstracts of communications of 4th International Caucasian symposium on Polymers and advanced materials, Batumi. http://www.icsp4.tsu.ge/data/file_db/icsp/ICSP.pdf (accessed on 06 June 2023).
49. Mukbaniani, O., Aneli, J., Buzaladze, G., Markarashvili, E., & Tatrishvili, T., (2015). *Wood Polymer Composites Based on the Renewable Raw Materials*. Abstracts of presentation of the first SDSU–Georgia stems workshop on nanotechnology and environmental sciences, September 5, Tbilisi, Georgia.
50. https://www.tsu.ge/data/file_db/faculty_zust_sabunebismetk/TOTAL-final-2.pdf (accessed on 06 June 2023).

CHAPTER 2

Composite Materials on the Basis of Straw: New Challenges

OMAR MUKBANIANI,[1,2] JIMSHER ANELI,[2] and TAMARA TATRISHVILI[1,2]

[1]*Department of Macromolecular Chemistry, Ivane Javakhishvili Tbilisi State University, Tbilisi, Georgia*

[2]*Institute of Macromolecular Chemistry and Polymeric Materials, Ivane Javakhishvili Tbilisi State University, Tbilisi, Georgia*

ABSTRACT

In this chapter, the composites are based on dry straw and some organic-inorganic binders. The components were dispersed and mixed in the mixer used for the dispersion of the coffee.

Using optical microscopy and scanned electron microscopy it was enabled to obtain the character of distribution of the ingredients in the composites. The strengthening of materials on bending and impact viscosity were studied in detail.

2.1 INTRODUCTION

The creation of wood polymer composites (WPC) presents a very actual problem today. Obviously, it is due to very large perspectives of the application of these products in life. At this time there are many scientific articles and inventions in the sphere of synthesis of WPC on the basis of different wood materials and binders–thermoplastic and thermosetting polymers [1–7]. For example, different type of wood was combined with different combinations of hexanediol diacrylate, hydroxyethyl methacrylate, and maleic anhydride

Biocomposites: Environmental and Biomedical Applications. Omar Mukbaniani, Tamara Tatrishvili, Neha Kanwar Rawat, & A. K. Haghi (Eds.)

[2]. In this work, treatment slowed the rates of water vapor and liquid water absorption. It gives the hardest and most dimensionally stable WPC. WPC prepared using hydroxyethyl methacrylate was a hardener than specimens made without hydroxyethyl methacrylate and excluded water and moisture more effectively. It was established the effect of maleic anhydride (50%) treatment on steam and water absorption of WPC prepared from wheat straw, cane bagasse, and teak wood sawdust with sizes 425 mcm using novolac as matrix [3]. The values of shock viscosity and water absorption of the composite are 2–3 times better than that for analogous material obtained without treatment. WPC on the basis of cellulose, hemicelluloses, and lignin impregnated with MMA, styrene, acrylonitrile, vinyl chloride, and acrylamide after polymerization initiated with gamma-radiation and electron beams, from one side, and polymerization with different catalysts, from the other one, are characterized with improved physical properties [4]. In the work of authors Sebe & Brook [5], the hydrophobicity of maritime pine wood surfaces was modified by silicone. A generic method for the introduction of a variety of silicones at the surface of pre-treated wood was developed. The initial treatment of wood with maleic anhydride and ally glycidyl ether resulted in oligo-esterified wood-bearing terminal alkenes. The hydrosilylation of these groups, performed with hydride-terminated silicones, led to very hydrophobic surfaces, even after extensive Soxhlet extraction with good solvents for silicones. The silicones appear to be attached to the wood by covalent bonds. The interesting results were obtained in the work [6], where trembling aspen has been impregnated with the coupling agent γ-methacryloxypropyltrimethoxysilane and the hygroscopicity and anti-shrink efficiency before and after water extraction were defined. There was a hygroscopicity depression with treatment, but not enough to account for the observed anti-shrink efficiencies. There was limited reversibility after water extraction. This indicated that the wood was interacting with part of the sorbed silane more strongly than with water and that was a diversity of sorption environments within the wood-silane complex. The wood plastic composite was prepared with simul and MMA in the presence of methanol used as a swelling agent at different proportions [7]. In this WPC Sulfuric acid was incorporated into the impregnating solution to investigate its effect on the polymer loading and tensile strength of the composite. Around 1% acid solution induces the highest polymer loading.

In the work of authors Elvy, Dennis, & Loo-Teck [8], it is established that PMMA does not form bonds with the hydroxyl groups of the cellulose fibers but simply bulks the void spaces within the wood structure Here it was used vinyltriacetoxysilane before impregnation with MMA in order to evaluate the effects of the silane coupling agent on the physical properties of WPC. The composites were prepared using a catalyst accelerator method and the polymerization process was initiated at room temperature. This method involves the impregnation of wood samples using MMA, containing benzoyl peroxide (1%), lauroyl peroxide (0.5%), and accelerator N,N-dimethylaniline (0.5%). This method requires no heating in the initiation process. The material obtained using this method has improved compression strength and hardness. The morphology of the composite was examined with scan electron microscopy.

Various inorganic silicon compounds are used for the treatment of wood [9]. Among these compounds, silicofluorides represent one of the long-known classes of wood preservatives based on silicon. Data on an organic fluorosilicon compound ("silafluofen") are additionally presented. The mode of action of these fluorides is based on their biocidal action. "Water glass," an alkali silicate, was able to enhance the durability of wood but showed some important drawbacks. Because of its high hygroscopicity and its high pH values, increased moisture absorption and strength loss of wood were frequently observed. Wood treated with tetra-alkoxysilanes showed enhanced dimensional stability, especially when the hydrolysis and the condensation of the silanes were controlled to react within the cell wall. Positive results are obtained in the works devoted to the creation of WPC using silicon-organic compounds and sol-gel systems [10], where some compounds such as organofunctional silanes which are mostly applied in combination with tetra-alkoxysilanes (sol-gel process) as well as chlorosilanes and trimethylsilyl derivatives were proposed for a full impregnation treatment of wood. The effects related to the various treatments vary from an increase in dimensional stability, durability, and fire resistance to an enhanced hydrophobization of wood.

Commonly used plants for hay include mixtures of grasses such as ryegrass (Lolium species), timothy, brome, fescue, Bermuda grass, orchard grass, and other species, depending on the region. Hay may also include legumes, such as alfalfa (lucerne) and clovers (red, white, and subterranean). Legumes in hay are ideally cut pre-bloom. Other pasture forbs are also sometimes a part of the mix, though these plants are not necessarily desired as certain forbs are toxic to some animals.

Oat, barley, and wheat plant materials are occasionally cut green and made into hay for animal fodder; however, they are more usually used in the form of straw, a harvest byproduct where the stems and dead leaves are baled after the grain has been harvested and threshed. Straw is used mainly for animal bedding. Although straw is also used as fodder, particularly as a source of dietary fiber, it has lower nutritional value than hay.

It is the leaf and seed material in the hay that determines its quality. Farmers try to harvest hay at the point when the seed heads are not quite ripe, and the leaf is at its maximum when the grass is mowed in the field. The cut material is allowed to dry so that the bulk of the moisture is removed but the leafy material is still robust enough to be picked up from the ground by machinery and processed into storage in bales, stacks, or pits. (Jones, Steven M.; Russell, Mark, Managing Fescue for Horses FSA3042 (PDF), University of Arkansas Division of Agriculture, Cattle grazing fescue often develops a chronic, unthrifty condition, especially apparent during the summer months. In addition, some cattle occasionally develop lameness and lose portions of their feet and tails during fall and winter. Some mares grazing fescue have reproductive problems during the last trimester of pregnancy.

Straw as agricultural waste biomass could be a source of alternative energy to substitute fossil energy for reducing greenhouse gas emissions as well as avoid the local pollution problems from open burning. Rice straw is attractive as a fuel because it is renewable and considered to be carbon dioxide neutral [13–16] but has not yet been commercially used as a feedstock for heat and energy because of insufficient incentives or benefits for farmers to collect rice straw instead of field burning. Direct comparison of straw with coal, still the dominant solid fuel in electricity and heat generation, often reveals inferior properties of straw.

In nature, pentose is found mainly in the form of polysaccharides pentosans $(C_5H_80_4)$ n, as well as part of the pectin and plant gums, such as gum Arabic, and cherry glue. Pentosans are wood (hardwood 22–27% pentosans), hay (10–15%), straw, seed shells, sunflower husk, etc.

Herbaceous plants having a high content of cellulose are widely used for the production of tissue and to produce chemical processing of fibrous semi-finished material for the manufacture of paper. In many countries, such as Pakistan, Thailand, and Peru, non-wood raw materials are the main raw material from which produce cellulose. In China, nearly 87% of cellulose-containing products are made from straw.

Long-fiber raw materials: The sources are long-staple raw cotton and linen and are divided according to the morphological and economic traits in the group: curley and intermediate. It is the most large-capacity kind of fibrous raw material in the world is cotton.

2.1.1 SHORTLY FIBROUS RAW MATERIALS

Perspective views of short fibrous raw material are waste agricultural plants (straw cereals), the annual volume of which is very large. For example, the estimate of the number of fiber waste generated in Russia shows that in 2001 the most common mass of stems of crops suitable for pulp production was approximately 150 million tons. The ratio of individual pieces of straw cereals is shown in Tables 2.1–2.3. The family of cereals applies cane, whose stocks are also quite large.

The peculiarity of the chemical composition of non-wood plant fibers containing a relatively high content of inorganic compounds and elements.

Vegetable materials contain various mineral elements accumulated during the growth of plants. The composition of macro- and micronutrient plants (average) is shown in Table 2.3. Main raw plant macronutrients: potassium, calcium, phosphorus, magnesium, and sulfur. Major minerals are iron, copper, manganese, zinc, molybdenum, and cobalt.

Mineral elements perform important structural and metabolic functions in plant and fungal cells. The content of mineral elements in plant material is usually sufficiently high density and provides requirements for mineral elements cultivated fungus.

TABLE 2.1 The Chemical Composition of Some Plants

Type	Cereal (%)	Rice (%)	Latifoliate (%)
Cellulose	45–55	43–49	57
Hemicellulose	26–32	23–28	23
Lignin	16–21	12–16	25
Ash	2–9	15–20	1
Silica	2–8	9–14	0.5

In the frame of the project, the composites based on straw west materials with different binders such as ecologically clean liquid glass, alkoxy silanes (PhES-50 and PhES-80), wood adhesive, colophon, and vinyl-triethoxysilane were prepared and investigated their structure, as well as physical–mechanical, thermal properties and water absorption.

The composites were prepared in the presence of binders and fillers with different concentrations. Namely, it was prepared composites with two components: straw + binder, tree component systems: straw + binder (1) + binder (2), and four-component systems: straw + binder (1) + binder

TABLE 2.2 The Chemical Composition of Tree Species Not Plant Raw Materials (Mass.%)

Raw Material	Cellulose	Lignin	Pentanones	Extracted Compounds			Ash
				Dichloroethane	Water	Implied 1%NaOH solution	
Wheat straw	34.8–49.2	24.5	19–30	1.5–3.7	–	–	4.6–5.5
Rye straw	40.4–49.3	22.9	27.7–28.8	2.0–2.6	–	–	3.1–4.1
Rice straw	42.9	17.9	27.1	6.2	17.4	54.7	15.3
The stems of cotton	37.5–40.0	21.3–24.7	18.2	2.6–3.1	8.2–12.5	30.2	3.1–5.8
Cane	42.2	20.8	29.3	1.2	–	9.2	4.7–5.6
Bagasse	55.4	19.0	27.5	–	–	1.1	0.9
Esparto	43.8	21.2	26.6	2.5	9.6	–	2.3

(2) + different additives. The composites were prepared by using the method of hot pressing of highly dispersed (50 mcm) powders of ingredients under a pressure of 150 kg/cm^2 and a temperature of 130°C. It was prepared 2 types of samples with two shapes: cylinder (for investigations of water absorption) and rectangle forms (for conducting mechanical investigations) [17, 18].

TABLE 2.3 The Mineral Composition of Plant Substrates

Elements	The Content in the Solid Mass
Macro-Elements	
Potassium (K)	0.5–6%
Calcium (Ca)	0.2–3.5%
Phosphorus (P)	0.1–0.8%
Magnesium (Mg)	0.1–0.8%
Sulfur (S)	0.5–1%
Microelement	
Iron (Fe)	25–300 ppm
Copper (Cu)	4–30 ppm
Manganese (Mn)	15–800 ppm
Zinc (Zn)	15–100 ppm
Molybdenum (Mo)	0.1–5.0 ppm
Cobalt (Co)	< 0.1 ppm

The straw, like the sawdust, contains cellulose structural rings, which, from the other side, contain hydroxyl groups. The used phenylethoxysilane contains ethoxy groups. These groups participate in the etherification reaction with a binder (PhES-50 and PhES-80) during the period of macromolecular reactions:

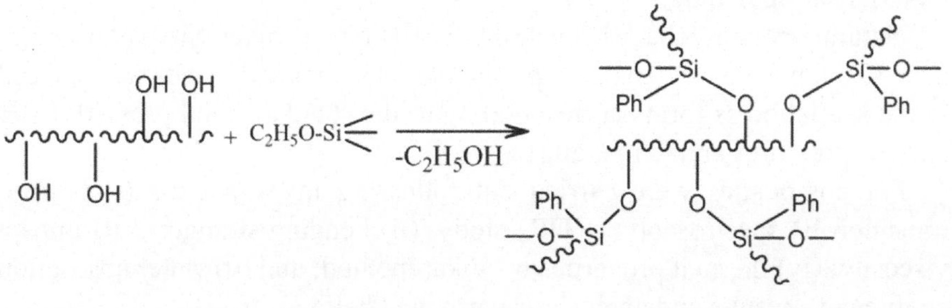

where;

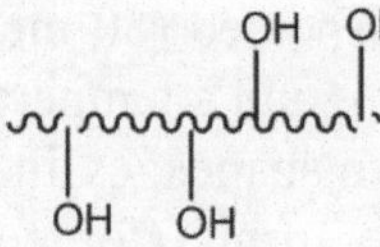

Cellulose chain

Etherification reactions with the use of vinyltrietoxisilane elapse analogically. Processes that occur during the curing are complex and varied.

The processes occurring in the silicate solution, are governed by two reversible reactions:

$$\equiv SiOH + OH- \rightarrow \equiv SiO- + H_2O$$

$$\equiv SiOH + \equiv SiO- \rightarrow \equiv Si–O–Si\equiv + HO-$$

It is possible interaction with dehydration between ≡Si-OH groups contained in the liquid glass molecules and cellulose. The result is three-dimensional structures.

The colophony is in the content of the softwood structure, which contains the isomeric acid rings. One of the ground parts of its structure is:

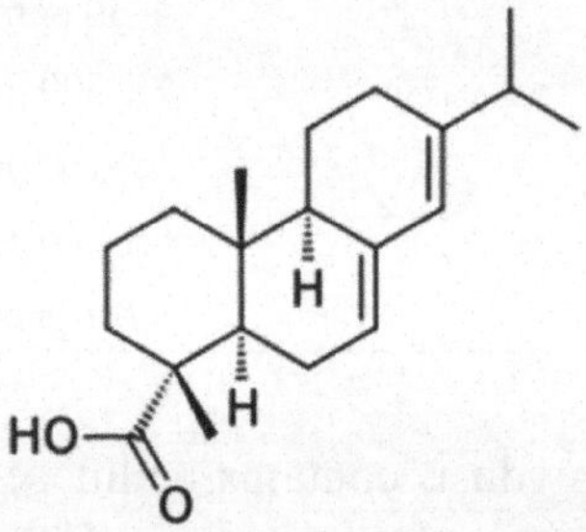

The noted structural ring can introduce to reaction both with PhES-50 or PhES-80 and with liquid glass. In the presence of colophony, it may be realized the donor-acceptor bond with the straw matrix, which is connected with the formation of additive intermolecular forces and leads to increasing material strengthening.

Natural dry straws have been destructed using the mixer. Sizes of particles near 50 mcm were used in our experiments. The obtained blends were placed in the spatial press forms corresponding to the standards and pressed under various pressures and temperatures.

For composites were carried out following investigations (i) Fourier transition IR spectroscopy (FTIR) study; (ii) bending strength; (iii) impact viscosity; (iv) thermal properties by Vikat method; and (v) water absorption coefficient. Results and analysis are provided below.

The microstructure of the samples was studied by an NMM-800RF/TRF type of optical microscope. For composite materials made based on straw and different hardeners and additives scanning electron microscopic (SEM) investigations are provided. Besides energy dispersive, X-ray examinations have been carried out in parallel with the micro-spectral (EDS) examinations.

2.2 FOURIER TRANSFORM INFRARED (FTIR) SPECTROSCOPY INVESTIGATION

Fourier transforms infrared spectroscopy investigation has been carried out in KBr. We have investigated the FTIR spectra for pure straw, binders, two-, three- and four-components composites systems. It is interesting, that in composites based on straw, one can see the absorption pick characteristic for asymmetric valence oscillation of bonds ≡Si-O-Si≡ with a maximum near 1066 cm^{-1}, which possesses siloxane bond in the fragment of cyclo-tetrasiloxane fragment. Besides the sorption strips coincide with one another. In the spectrum, one can see the absorption strips 1267, 1373, 1429, 1515, 1600–1650, 1733, 2800–2950, 3363 typical for methyl groups, C-H bond absorption (-C/C-/CH_3), CH_2 cellulose –lignin, C=C aromatic, C=C alkene, (C=O etheric bond), C-H methyl, methylene and phenyl groups, O-H alcoholic group, respectively.

In the spectrum of PhES-80 one can see two maximums in the region of 1060 and 1132 cm^{-1} corresponding to asymmetrical valence oscillations of the fragment Si-O-Si and absorption stripe in the zone 1349 and 1483 cm^{-1} of Si-OC_2H_5.

Two small bands at 1,610–1,633 are assigned to the absorbed water and β-glycosidic linkages between the sugar units, respectively [19].

The spectrum of straw shows the same basic structure as all wood samples: strong broad OH stretching (3,300–4,000 cm^{-1}), C–H stretching in methyl and methylene groups (2,800–3,000 cm^{-1}), and a strong broad superposition with sharp and discrete absorptions in the region from 1,000 to 1,750 cm^{-1} [20–24]. The absorptions are situated at 1,510 and 1,600 cm^{-1} (aromatic skeletal vibrations). The absorptions are located near 1,700 cm^{-1}.

FTIR spectra of initial straw, PhES-50, PhES-80, and composite materials on their basis are presented in Figures 2.1–2.9.

In the spectrum of straw, one can observe absorption bands with two maxima for asymmetric valence oscillation characteristic for Si-O-Si bonds at 1,060 and 1,132 cm^{-1}, absorption bands for Si-OC_2H_5 at 13,490 and 1,483 cm^{-1}.

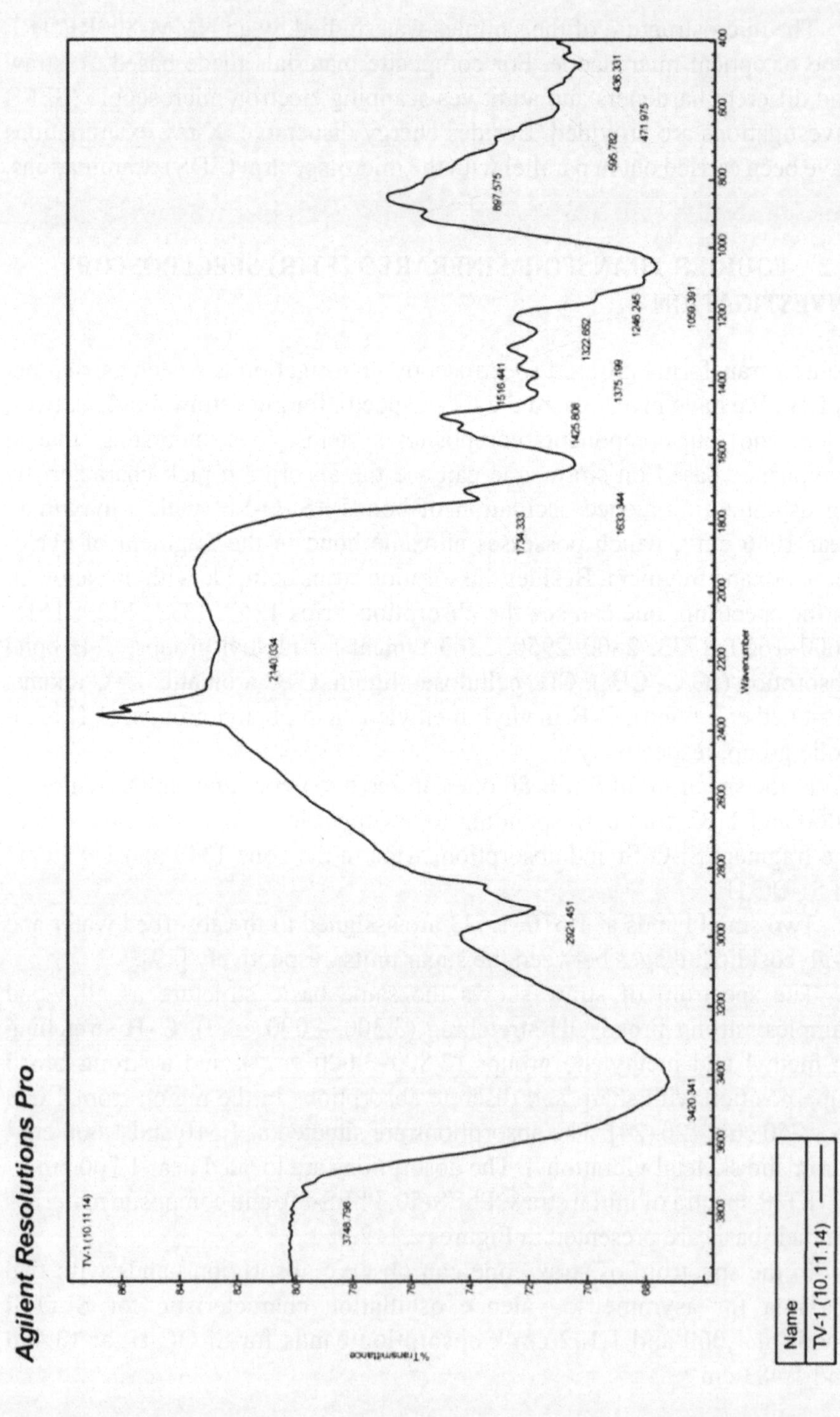

FIGURE 2.1 FTIR spectrum of straw.

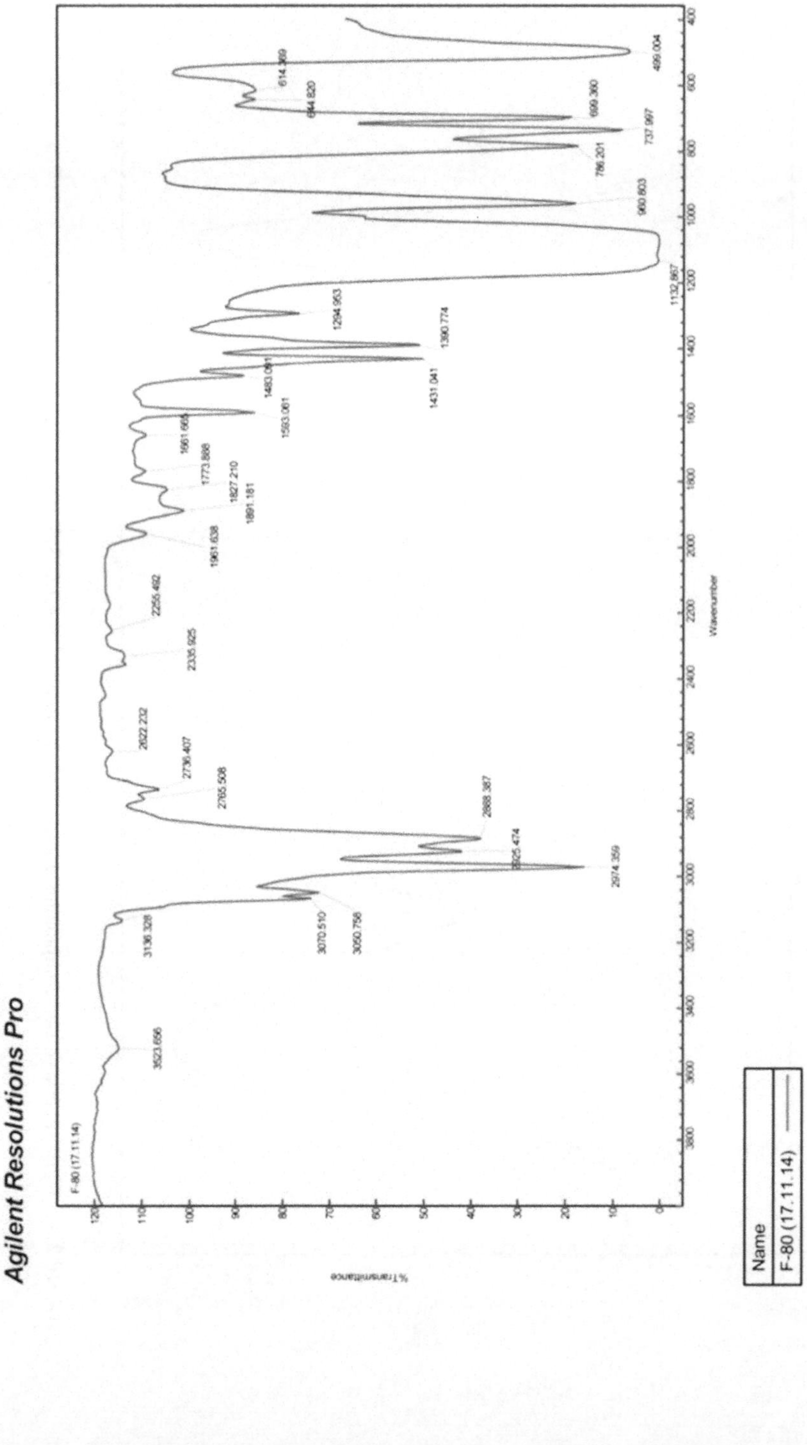

FIGURE 2.2 FTIR spectra of PhES-80.

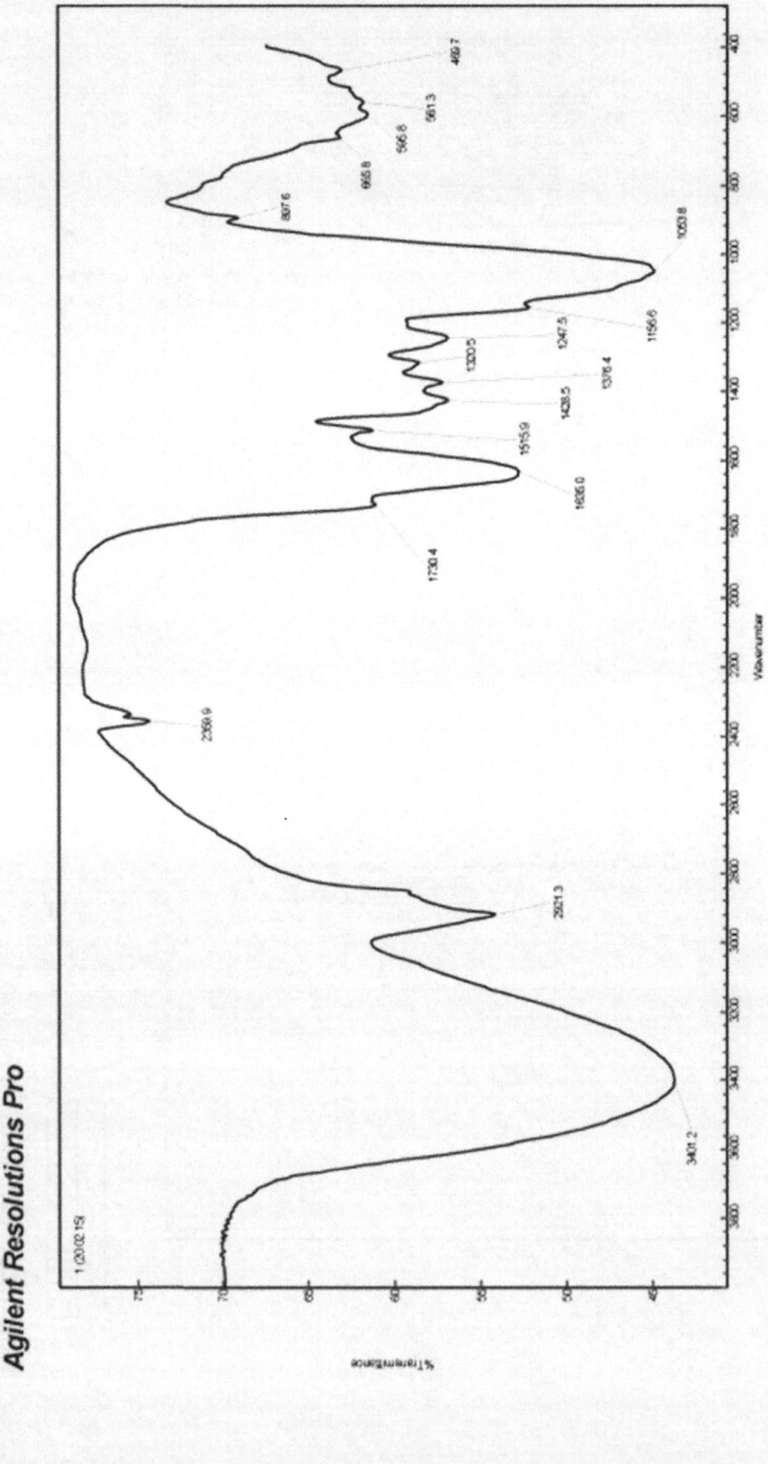

FIGURE 2.3 FTIR spectra of two components composite (1) (Straw 97% + LG 3%).

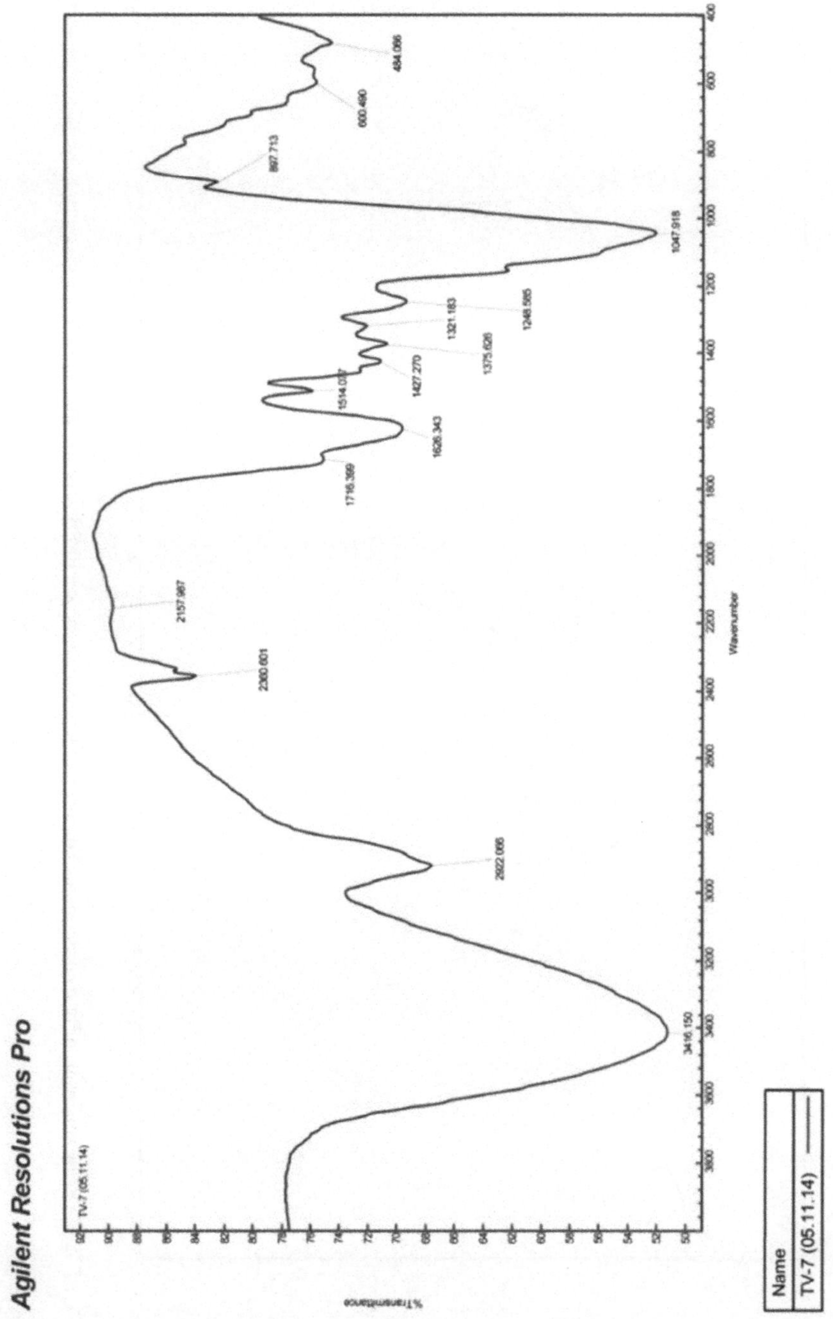

FIGURE 2.4 FTIR spectrum of two components composite (6) (Straw 95% + PhES-50 – 5%).

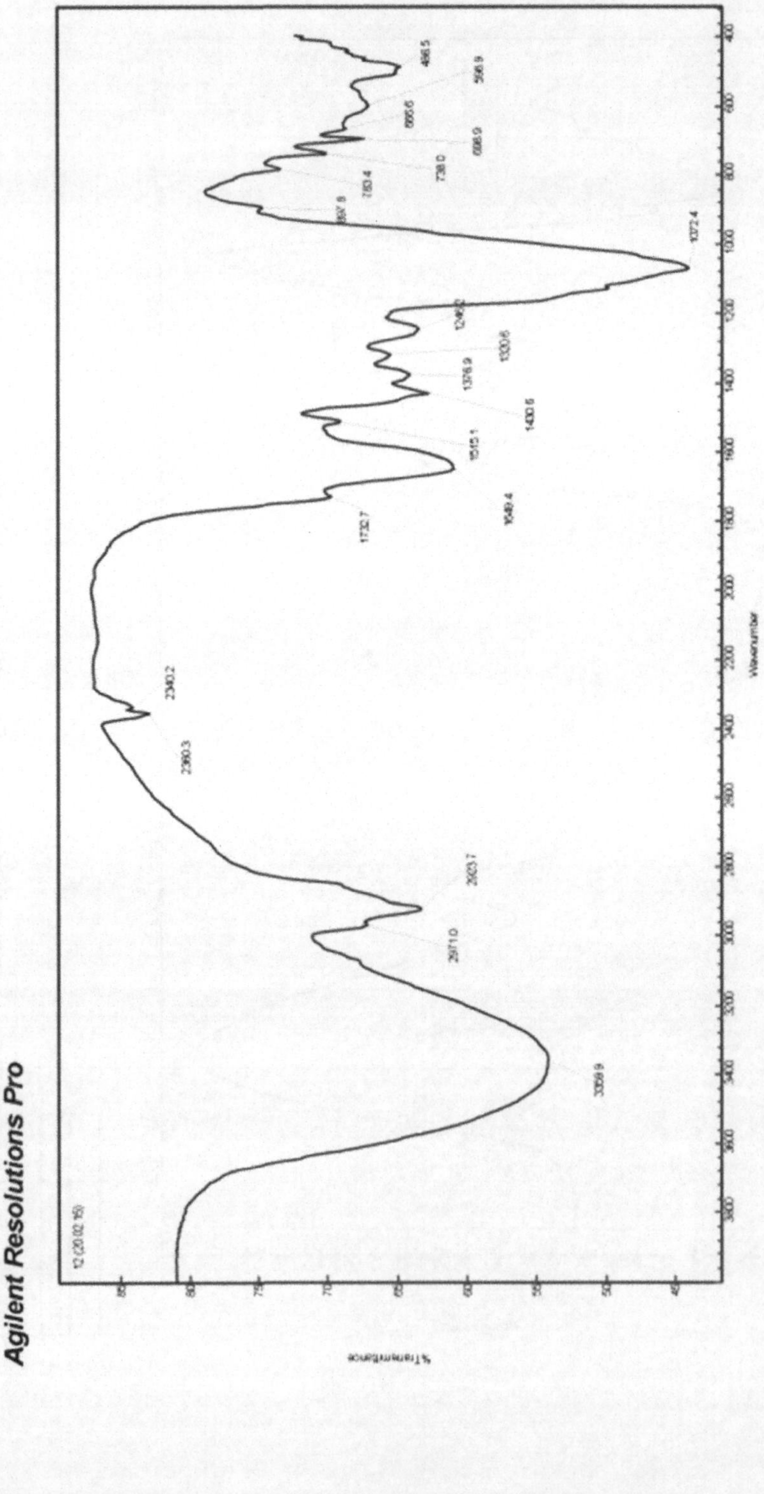

FIGURE 2.5 FTIR spectrum of two components composite (12) (Straw 85% + PhES-80 – 15%).

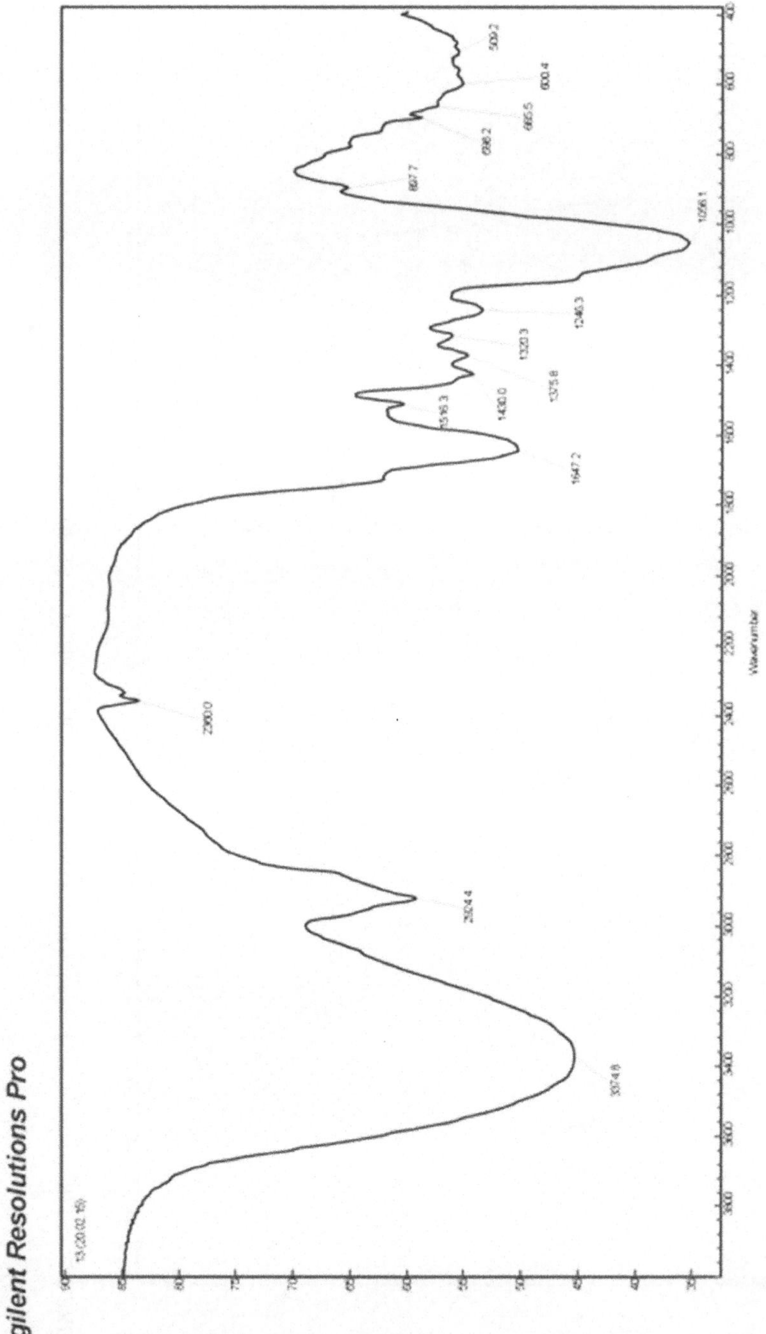

FIGURE 2.6 FTIR spectrum of two components composite (2.13) (Straw 90% + Polyethylene 10%).

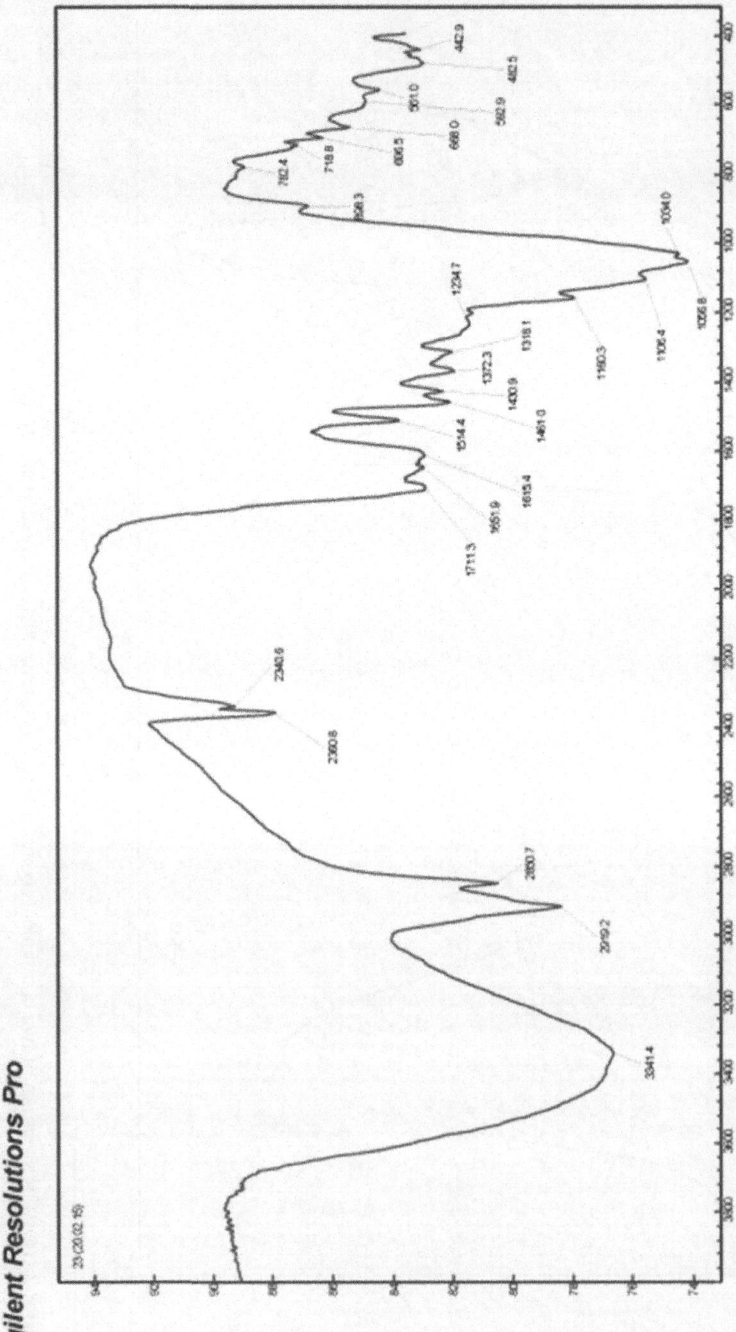

FIGURE 2.7 FTIR spectrum of tri-component composite (2.23) (Straw 90% + PhES-80 – 5% + LG 5%).

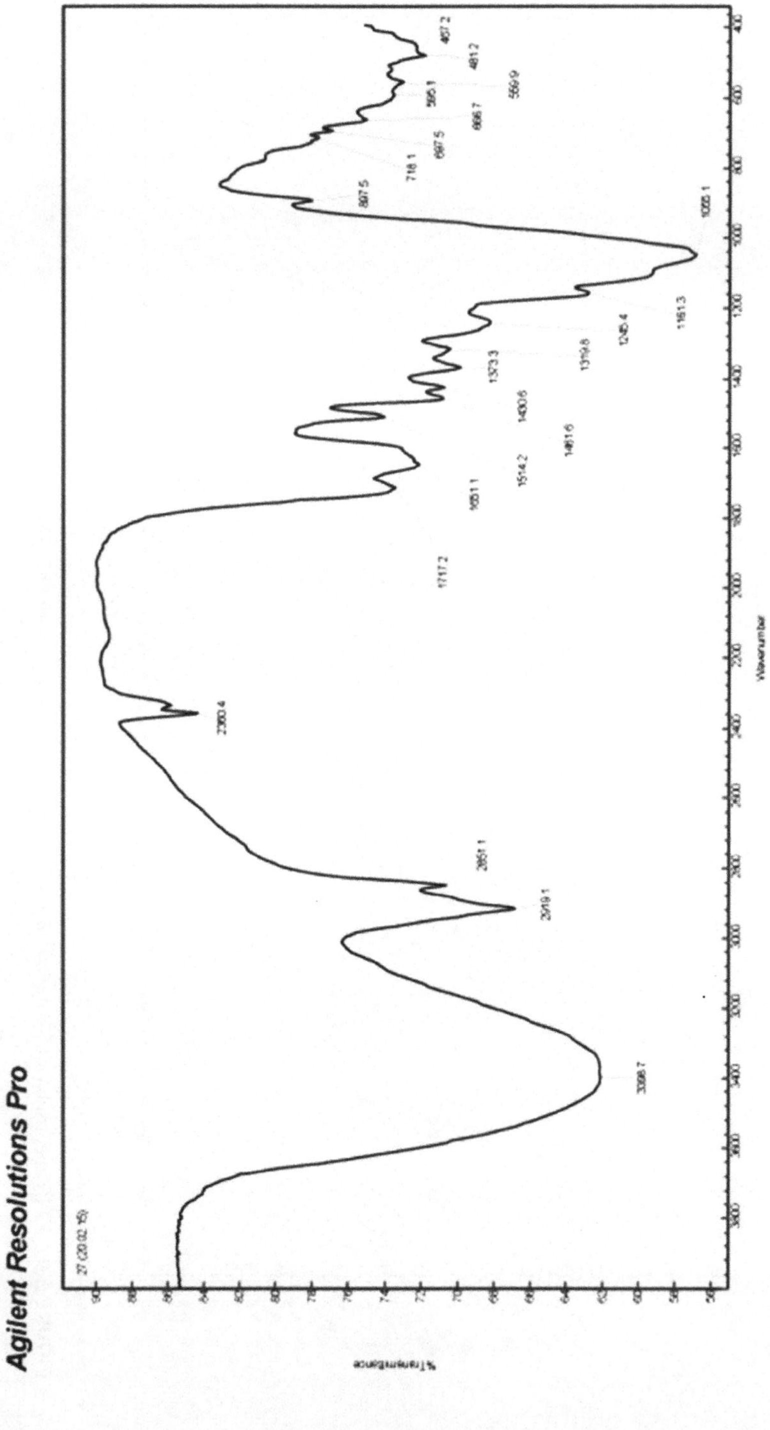

FIGURE 2.8 FTIR spectrum of four-component composite (2.27) (Straw 85% + PhES-80 – 5% + LG 5% + PE 5%).

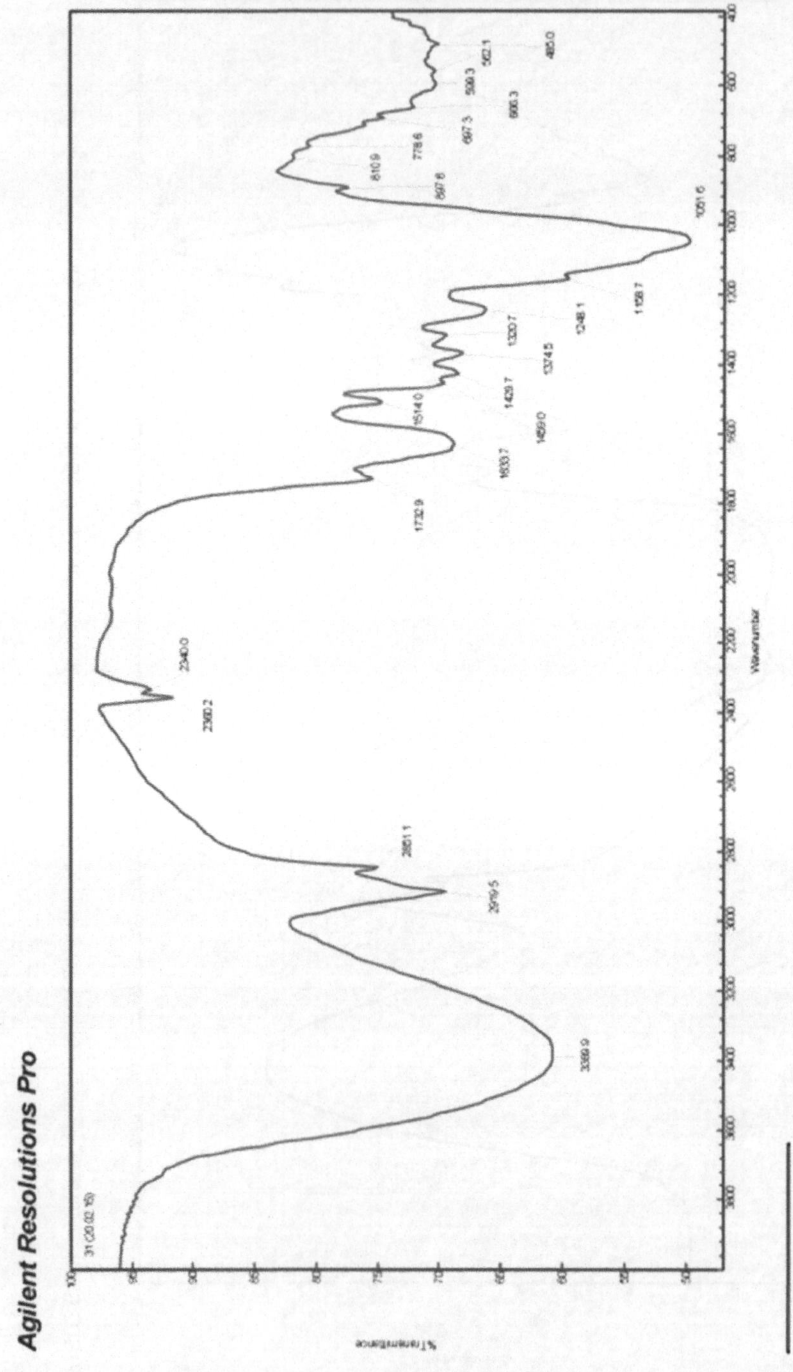

FIGURE 2.9 FTIR spectrum of tri-components composite (2.31) (Straw 94% + PhES-80 – 3% + PE 3%).

2.3 STUDY OF THE MICROSTRUCTURE OF COMPOSITE MATERIALS WITH USE OF OPTICAL MICROSCOPY

The wood composite microstructure of the samples was studied by an NMM-800RF/TRF type of optical microscope. With the aim of investigating the structure of the materials we have prepared a sample by following manipulations: after polishing the surface of the sample with longevity of about 1 cm we transferred it to the emery paper and continued this procedure for 1 h, after which the sample was polished on the coarse calico. After this, the sample was transferred to the object glass of the optical microscope. There were studied the separate parts of the sample by use of the noted optical microscope. The composite obtained under conditions of 150 kg/cm^2 and a temperature near 130°C was studied. Particularly, on the figures there are presented the sifted and not sifted samples of straw and the composites based on hay with different concentrations of the binder. It was prepared The figures of the composites with 2, 3, and 4 components (Figures 2.19–2.28). Here are the figures for polished surfaces. Unfortunately, we could not obtain the polish figures of all samples because some of them were destroyed at polishing. The darkened regions on the figures correspond to the surfaces which did not undergo the polishing (Figures 2.10–2.12).

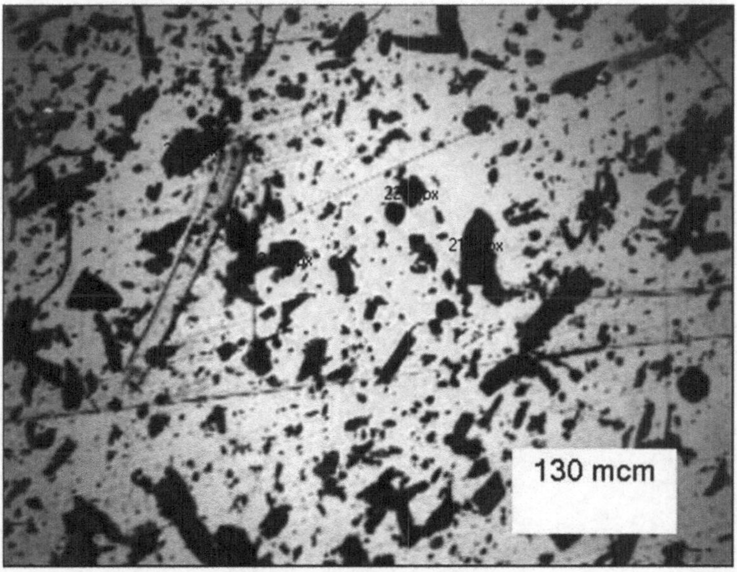

FIGURE 2.10 Micro-structure of sifted straw with different particle sizes. Their average size is 130 mcm.

FIGURE 2.11 Optical microscopic data of non-sifted straw particles with a width of about 360 mcm.

FIGURE 2.12 Optical microscopic data of composite 2.1 (Straw 97%, LG 3%).

The surface is homogeneous. There are no insertions or cracks. The particle's size is 400 mcm. Definition of the sizes of inserts or cracks one can establish with the use of the scale shown in the figures.

In Figure 2.13, one can see the inserts with different sizes of about 500 mcm. In this figure, one can see inserts with average sizes of 500 mcm. Here are not the cracks.

FIGURE 2.13 Optical microscopic data of composite 2.9 (Straw 97% + PhES-80 – 3%. T – 130°C; P = 150 kg/cm^2).

In Figure 2.14, there are inserts with different sizes (average sizes 300 mcm). There are no cracks and one can't see the dividing line.

In Figure 2.15, one can see that the surface is homogeneous and the crack sizes are about 400 mcm. On this surface the differentiation of the particles is very difficult; it does not see the dividing line.

In Figure 2.16, one can see the inserts with average sizes of about 205 mcm, the cracks are not seen.

In Figure 2.17, one can see that the surface is homogeneous. On the surface, there are no inserts or cracks. If you can see the dividing grain by means of the scale, you can establish its size.

FIGURE 2.14 Optical microscopic data of composite 2.10 (Straw 95%; PhES-80 – 5%.).

FIGURE 2.15 Optical microscopic data of composite 2.6 (Straw 95%; PhES-50 – 5%).

FIGURE 2.16 Optical microscopic data of composite 2.29 (Straw 90%; PhES-50 – 5% + PE 5%).

FIGURE 2.17 Optical microscopic data of composite 2.26 (Straw 94%; LG 3%, PE 3%).

2.4 SCANNING ELECTRON MICROSCOPIC (SEM) AND ENERGY DISPERSION MICRO-X-RAY ANALYSIS OF COMPOSITE MATERIALS ON THE BASIS OF STRAW

The composites based on straw were testified by the use of scanned energy-dispersive micro-roentgen analysis (SEM). This work was provided in parallel with investigations of micro-spectral analysis (EDS). In Figure 2.18, there are the present results obtained with the use of these methods.

The composites on the basis of straw and a variety of binders and additives have been tested on a scanning electron microscope (SEM). In parallel for samples energy dispersive X-ray spectral analysis was carried out. In Figures 2.18, 2.22, 2.27, 2.33, 2.36, and 2.41, the micrograms for composites are given.

In Figures 2.18, 2.22, 2.27, 2.33, 2.36, and 2.41, one can see outside the epidermis of the composites on the basis of straw, which is well organized partial corrugated structure. These figures show that the cell has a linear profile epidermal Lemma. On the surface, one can observe the surface grooves and heterogeneous structure.

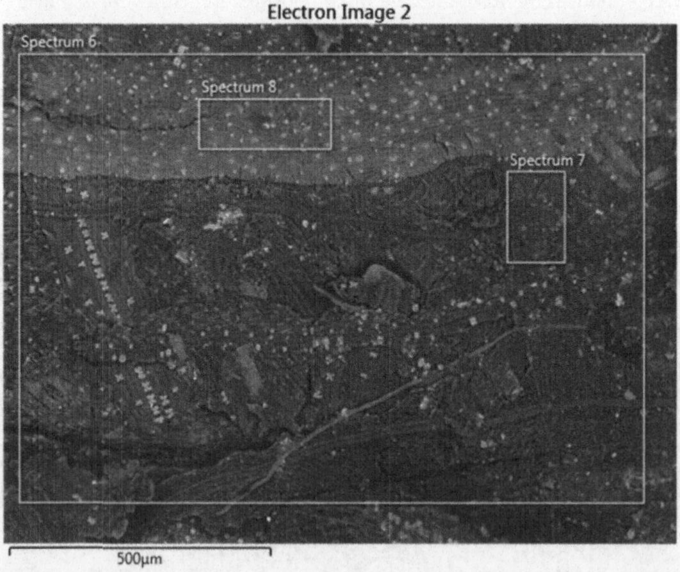

FIGURE 2.18 Scanning electron microscopic micrograms of composite 2.2 (Straw 95% + Liquid glass 5%).

The outer surface of the lemma contains a variety of sizes and papillae conversation threads. Along the length of both internal and external epidermal surfaces are located the silicon atoms, which create biomass in a relatively stable block, which is formed around the straw structure [1] (Figures 2.19–2.21 and Table 2.4).

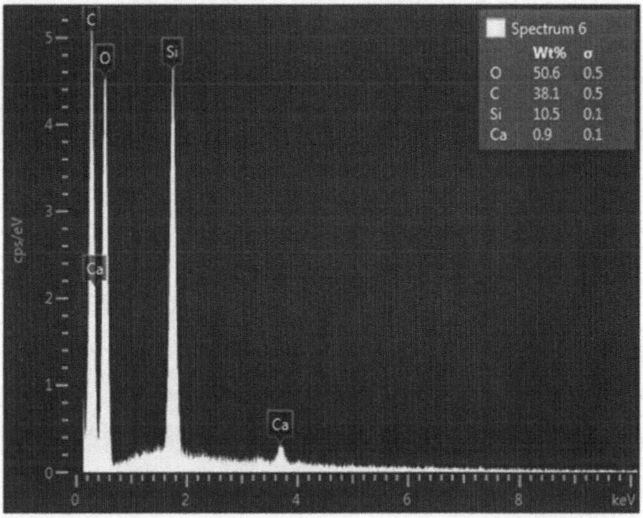

FIGURE 2.19 Energy dispersion micro-X-ray spectral analysis of composite 2.2 (Straw 95% + LG 5%) (Spectrum 6).

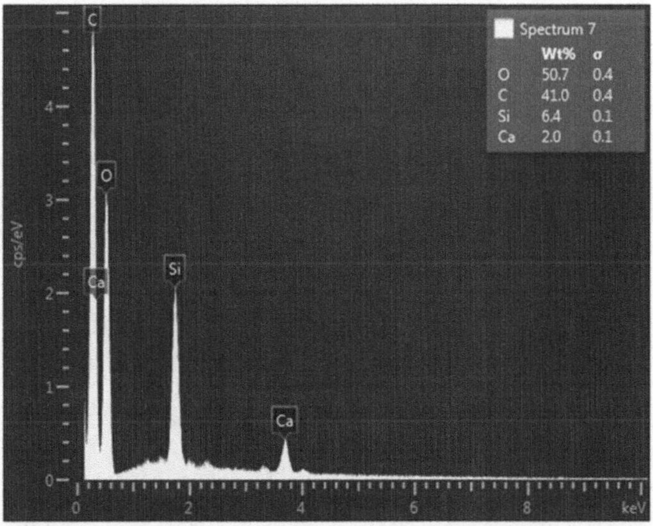

FIGURE 2.20 Energy dispersion micro-X-ray spectral analysis of composite 2.2 (Straw 95% + LG 5%) (Spectrum 7).

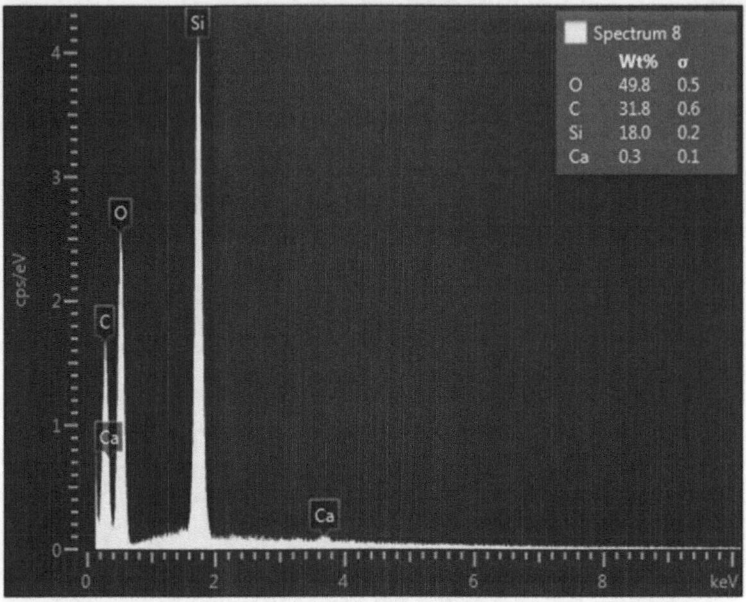

FIGURE 2.21 Energy dispersion micro-X-ray spectral analysis of composite 2.2 (Straw 95% + LG 5%) (Spectrum 8).

TABLE 2.4 Energy Dispersion Micro-X-Ray Spectral Analysis of Composite 2.2 (Straw 95% + LG 5%)

Result Type	**Weight (%)**		
Spectrum Label	**Spectrum 6**	**Spectrum 7**	**Spectrum 8**
C	38.08	41.01	31.84
O	50.57	50.66	49.81
Si	10.49	6.36	18.05
Ca	0.86	1.98	0.30
Total	**100.00**	**100.00**	**100.00**

Statistics	**C**	**O**	**Si**	**Ca**
Max.	41.01	50.66	18.05	1.98
Min.	31.84	49.81	6.36	0.30
Average	36.98	50.34	11.63	1.04
Standard deviation	4.68	0.46	5.93	0.86

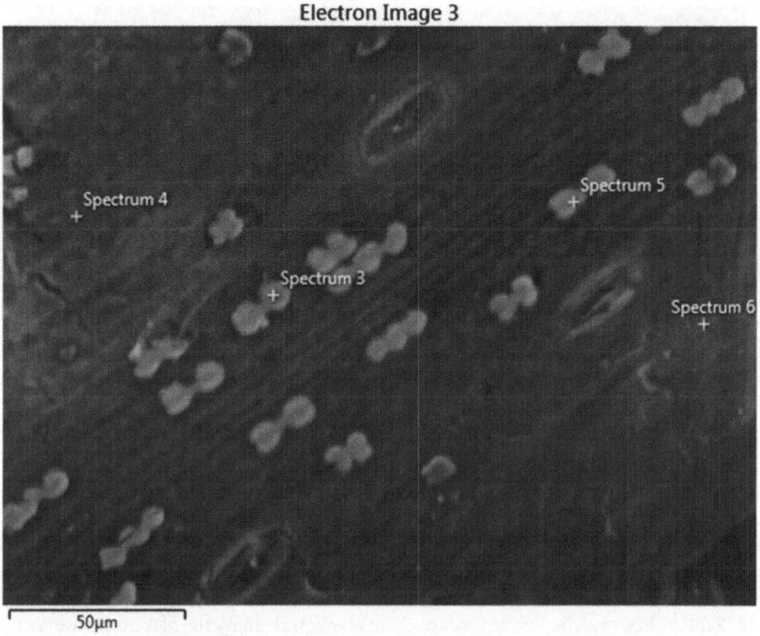

FIGURE 2.22 Scanning electron microscopic micrograms of composite 2.6 (Straw 95% + PhES-50 – 5%).

In Figures 2.19–2.45 and Tables 2.4–2.11, the SEM micrograms and energy-dispersive X-ray micro-chemical spectral analysis results are presented. In the figures, the X-ray reflexes for each element and the quantitative composition of some elements are presented. It is shown that composites contain primarily C, O, Si, Ca, and aluminum. It should be noted that the spectral content of these elements in different districts of the composite is slightly different, which indicates that the composite spectrum is different in different districts (Figures 2.23–2.45).

2.5 COMPOSITE'S MECHANICAL STRENGTHENING

This investigation was provided on the apparatus for strengthening and pressing solid composite materials. To testing were underwent the composites obtained at different temperatures and pressures. Testing on impact bending of the same samples was conducted on the apparatus of Sharpy type.

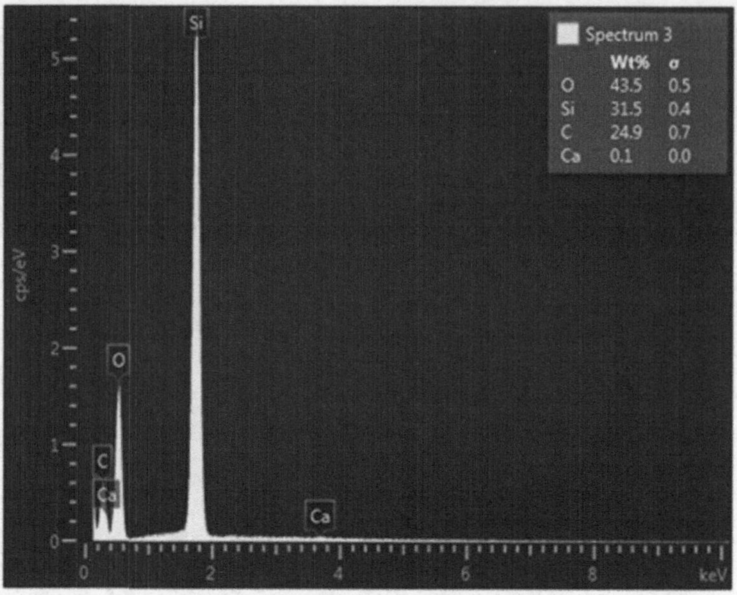

FIGURE 2.23 Energy dispersion micro-X-ray spectral analysis of composite 2.2: Straw 95% + PhES-50 – 5% (Spectrum 3).

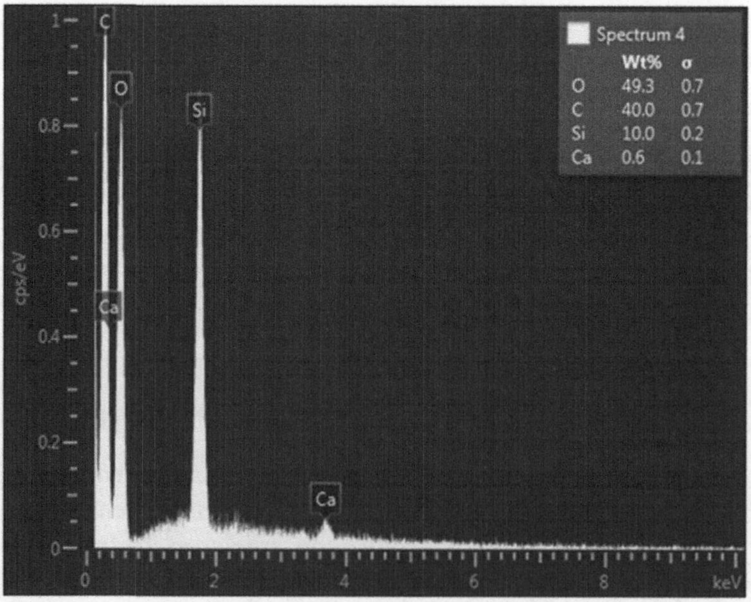

FIGURE 2.24 Energy dispersion micro-X-ray spectral analysis of composite 2.6: Straw 95% + PhES-50 – 5% (Spectrum 4).

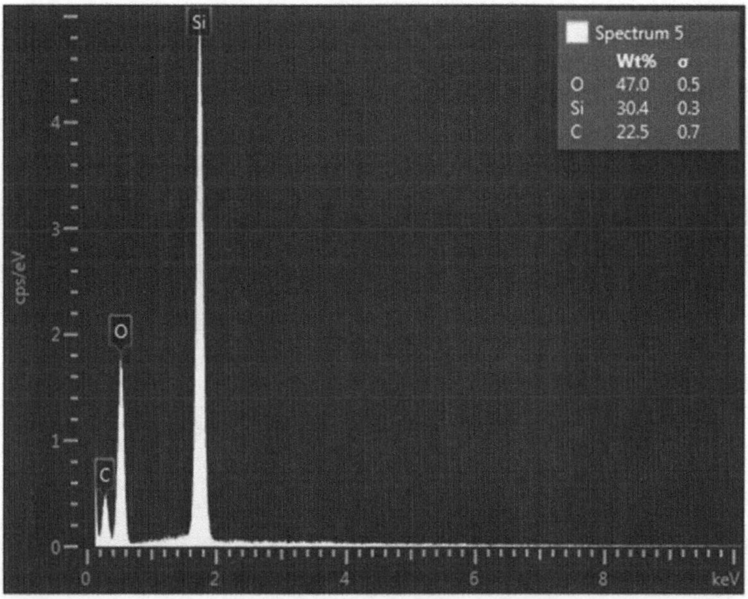

FIGURE 2.25 EDMR of composite 2.2: Straw 95% + PhES-50 – 5% (Spectrum 5).

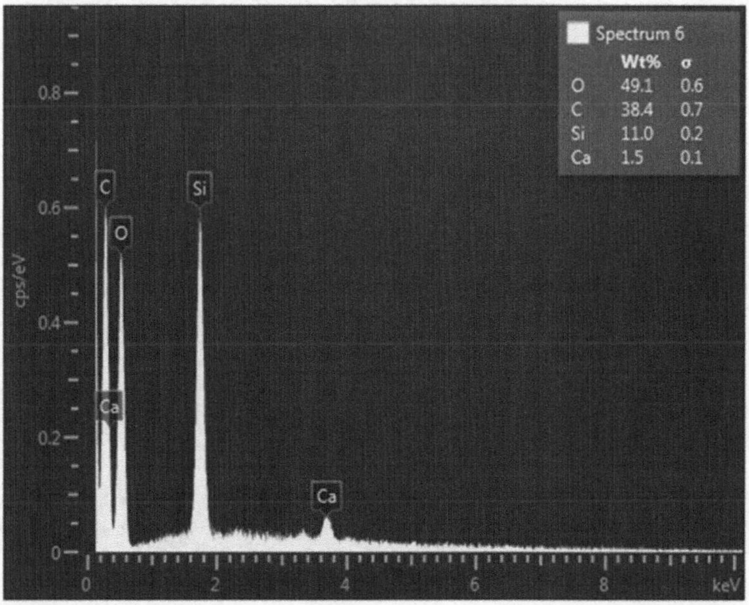

FIGURE 2.26 Energy dispersion micro-X-ray spectral analysis of composite 2.6: Straw 95% + PhES-50 – 5% (Spectrum 6).

TABLE 2.5 Energy Dispersion Micro-X-Ray Spectral Analysis of Composite 2.6 (Straw 95% + PhES-50 – 5%)

Result Type	**Weight (%)**			
Spectrum Label	**Spectrum 6**	**Spectrum 3**	**Spectrum 5**	**Spectrum 4**
C	38.41	24.88	22.53	40.03
O	49.12	43.45	47.04	49.33
Si	10.99	31.53	30.44	9.99
Ca	1.48	0.15	–	0.65
Total	**100.00**	**100.00**	**100.00**	**100.00**

Statistics	**C**	**O**	**Si**	**Ca**
Max.	40.03	49.33	31.53	1.48
Min.	22.53	43.45	9.99	0.15
Average	31.46	47.23	20.74	–
Standard deviation	9.04	2.73	11.85	–

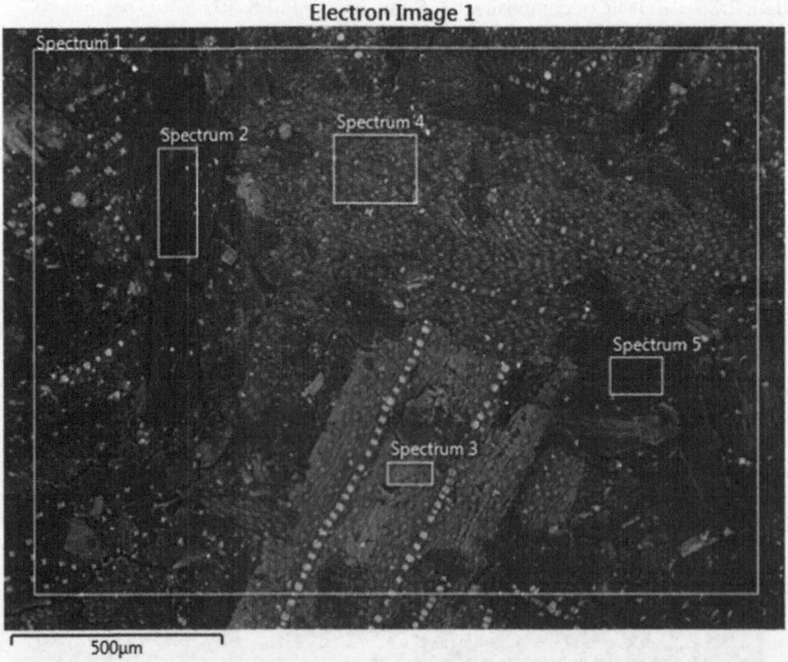

FIGURE 2.27 Scanning electron microscopic micrograms of composite 2.14 (Straw 95% + PE 10%).

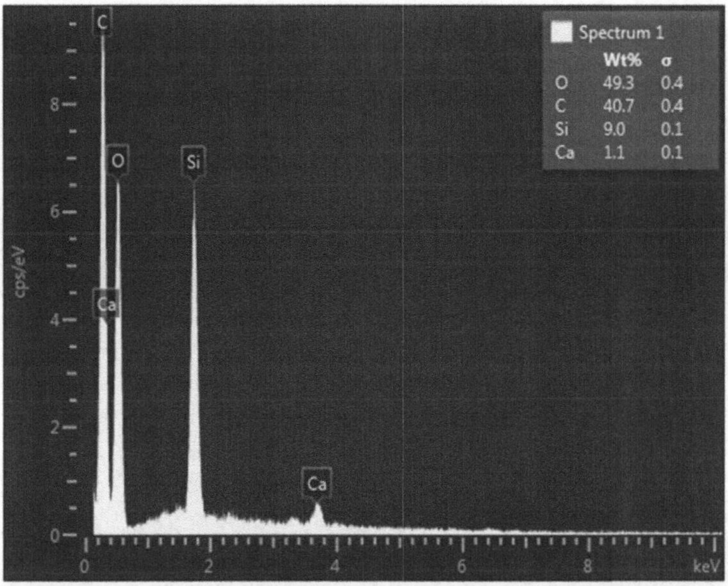

FIGURE 2.28 Energy dispersion micro-X-ray spectral analysis of composite 2.14: Straw 95% + PE 10% (Spectrum 1).

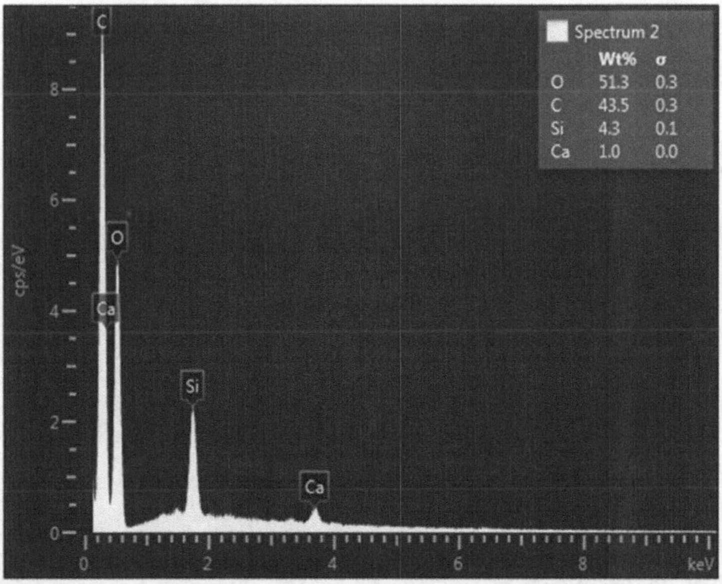

FIGURE 2.29 Energy dispersion micro-X-ray spectral analysis of composite 2.14: Straw 90% + PE 10% (Spectrum 2).

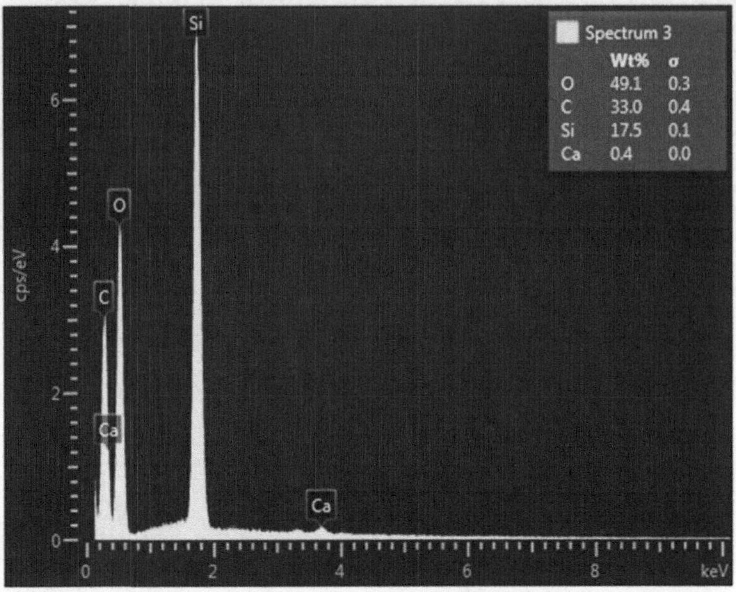

FIGURE 2.30 Energy dispersion micro-X-ray spectral analysis of composite 2.14: Straw 90% + PE 10% (Spectrum 3).

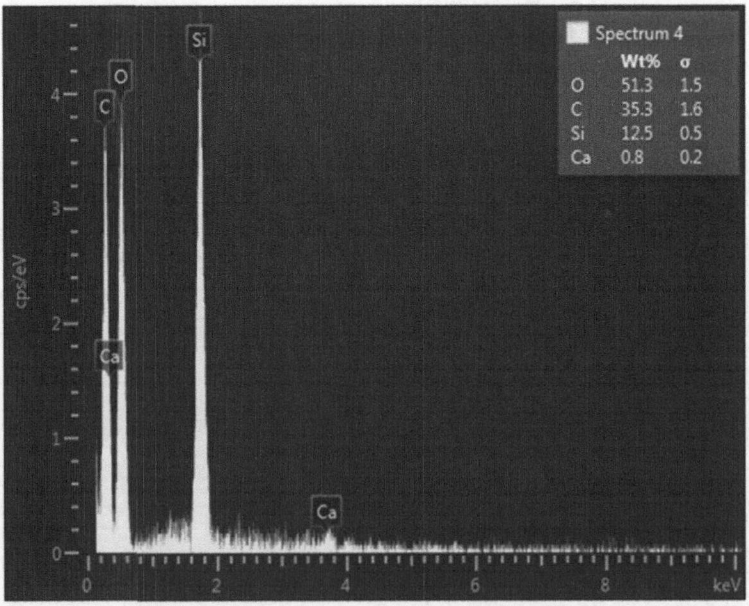

FIGURE 2.31 Energy dispersion micro-X-ray spectral analysis of composite 2.14: Straw 90% + PE-50 – 5% (Spectrum 4).

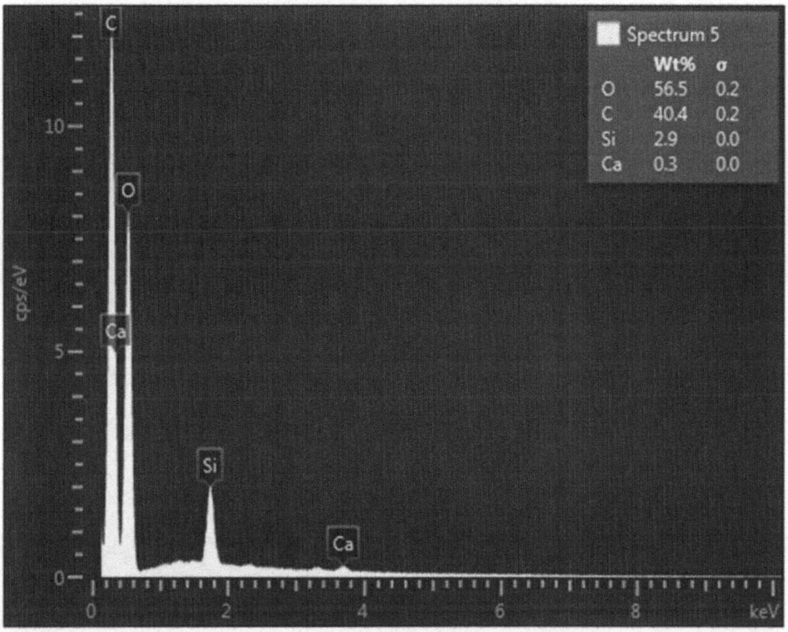

FIGURE 2.32 Energy dispersion micro-X-ray spectral analysis of composite 2.14: Straw 90% + PE 10% (Spectrum 5).

TABLE 2.6 Energy Dispersion Micro-X-Ray Spectral Analysis of Composite 2.14 (Straw 90% + PE 10%)

Result Type	**Weight (%)**				
Spectrum Label	**Spectrum 1**	**Spectrum 2**	**Spectrum 3**	**Spectrum 4**	**Spectrum 5**
C	40.66	43.45	33.02	35.30	40.38
O	49.29	51.28	49.14	51.33	56.45
Si	8.97	4.27	17.46	12.53	2.87
Ca	1.08	1.00	0.38	0.84	0.30
Total	**100.00**	**100.00**	**100.00**	**100.00**	**100.00**

Statistics	**C**	**O**	**Si**	**Ca**
Max.	43.45	56.45	17.46	1.08
Min.	33.02	49.14	2.87	0.30
Average	38.56	51.50	9.22	0.72
Standard deviation	4.27	2.96	5.99	0.36

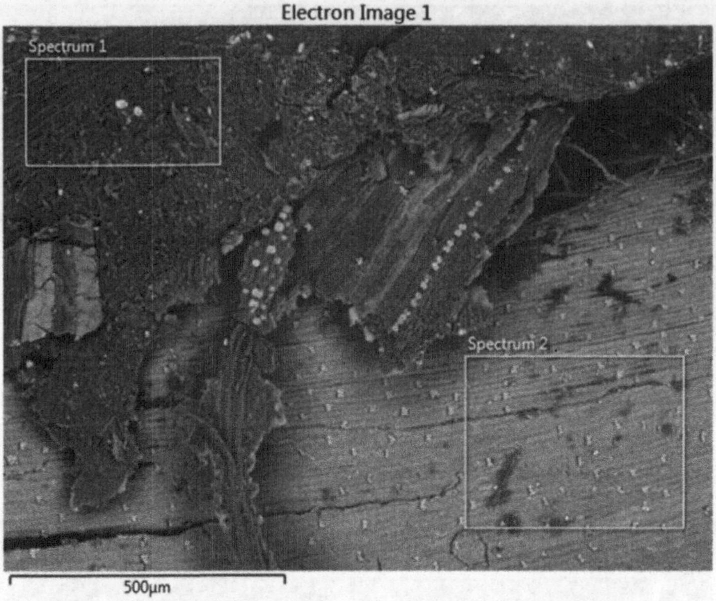

FIGURE 2.33 Scanning electron microscopic micrograms of composite 2.18 (Straw 90% + Colophony 10%).

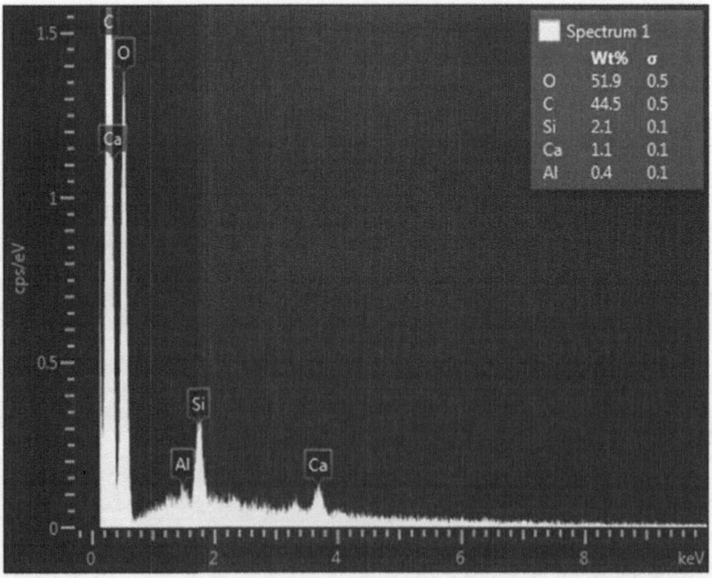

FIGURE 2.34 Energy dispersion micro-X-ray spectral analysis of composite 2.14: Straw 90% + Colophony 10% (Spectrum 1).

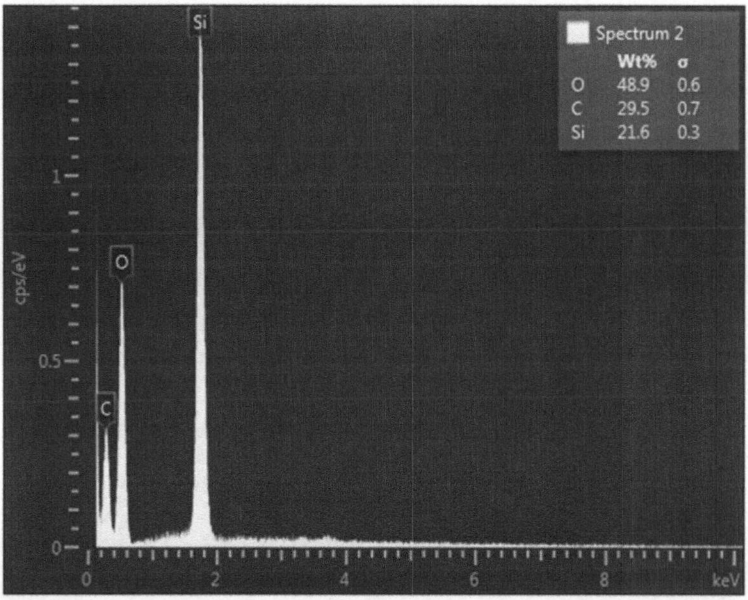

FIGURE 2.35 Energy dispersion micro-X-ray spectral analysis of composite 2.14: Straw 90% + Colophony 10% (Spectrum 2).

TABLE 2.7 Energy Dispersion Micro-X-Ray Spectral Analysis of Composite 2.14 (Straw 90% + Colophony 10%)

Result Type	**Weight (%)**	
Spectrum Label	**Spectrum 2**	**Spectrum 1**
C	29.46	44.54
O	48.90	51.88
Al	–	0.39
Si	21.64	2.11
Ca	–	1.08
Total	**100.00**	**100.00**

Statistics	**C**	**O**	**Al**	**Si**	**Ca**
Max.	44.54	51.88	0.39	21.64	1.08
Min.	29.46	48.90	0.39	2.11	1.08
Average	37.00	50.39	–	11.88	–
Standard deviation	10.66	2.11	–	13.81	–

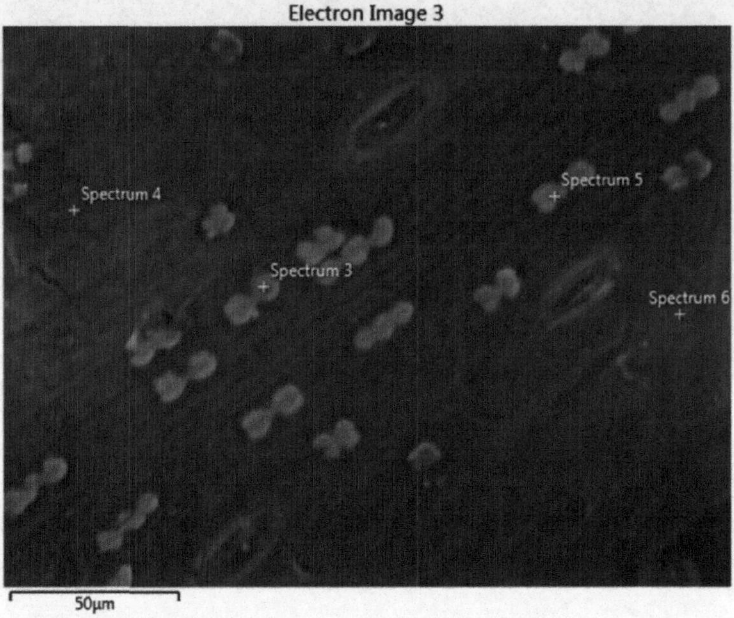

FIGURE 2.36 Scanning electron microscopic micrograms of composite 2.27 (Straw 80% + PhES-80 – 10% + LG 10%).

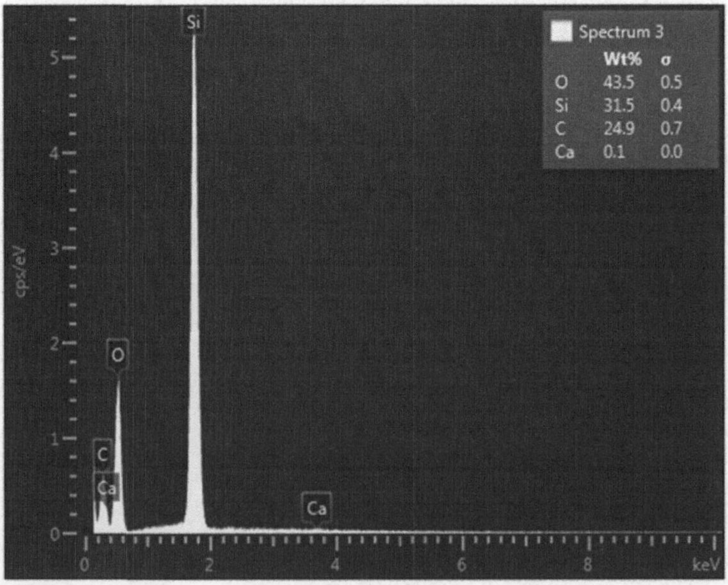

FIGURE 2.37 Energy dispersion micro-X-ray spectral analysis of composite 2.27: Straw 80% + PhES-80 – 10% + LG 10% (Spectrum 3).

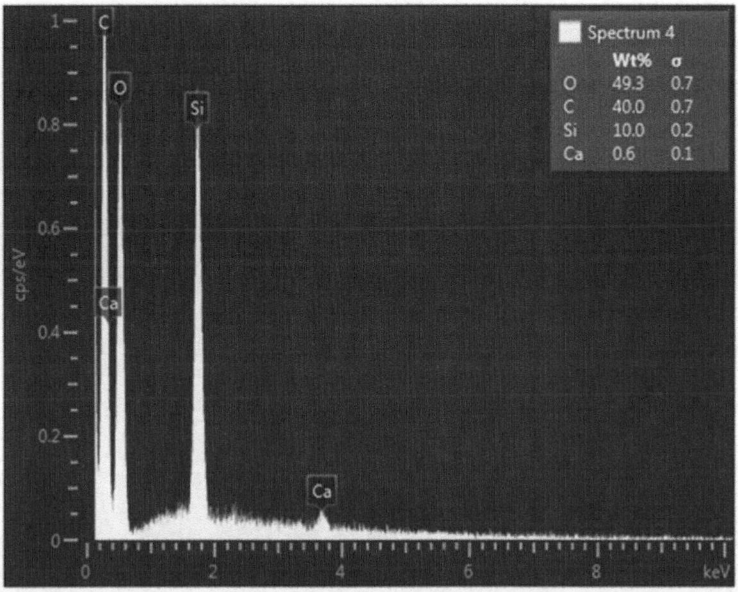

FIGURE 2.38 Energy dispersion micro-X-ray spectral analysis of composite 2.27: Straw 80% + PhES-80 – 10% + LG 10% (Spectrum 4).

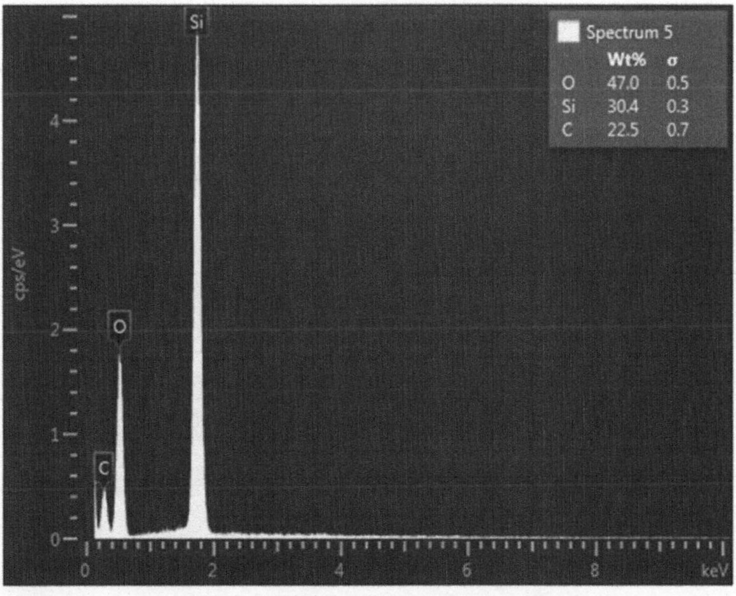

FIGURE 2.39 Energy dispersion micro-X-ray spectral analysis of composite 2.27: Straw 80% + PhES-80 – 10% + LG 10% (Spectrum 5).

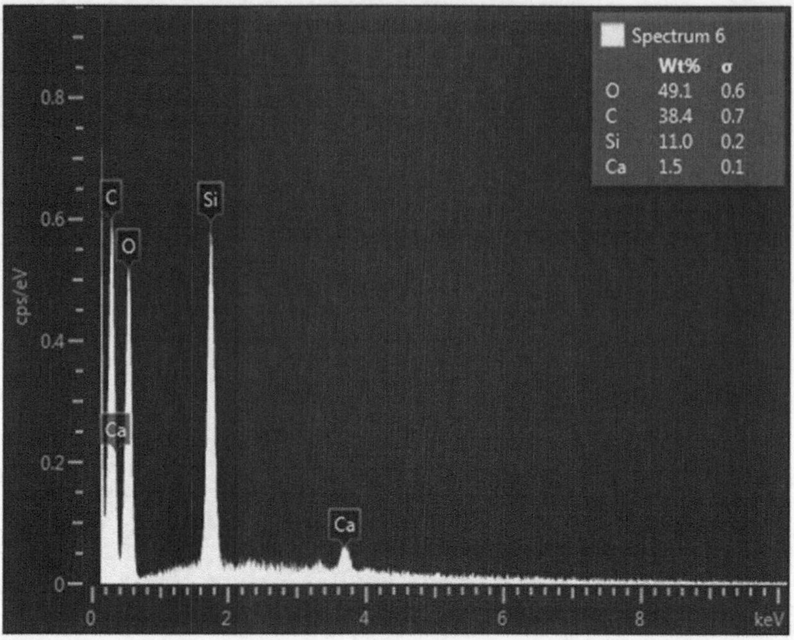

FIGURE 2.40 Energy dispersion micro-X-ray spectral analysis of composite 2.27: Straw 80% + PhES-80 – 10% + LG 10% (Spectrum 6).

TABLE 2.8 Energy Dispersion Micro-X-Ray Spectral Analysis of Composite 2.14: Straw 90% + Colophony 10% (Composite 2.14: Straw 80% + PhES-80 – 10% + LG 10%)

Result Type	**Weight (%)**			
Spectrum Label	**Spectrum 6**	**Spectrum 3**	**Spectrum 5**	**Spectrum 4**
C	37.41	22.88	22.53	40.03
O	49.12	45.45	47.04	49.33
Si	11.99	31.53	30.44	9.99
Ca	1.48	0.15	–	0.65
Total	**100.00**	**100.00**	**100.00**	**100.00**

Statistics	**C**	**O**	**Si**	**Ca**
Max.	40.03	49.33	31.53	1.48
Min.	22.53	43.45	9.99	0.15
Average	31.46	47.23	20.74	–
Standard deviation	9.04	2.73	11.85	–

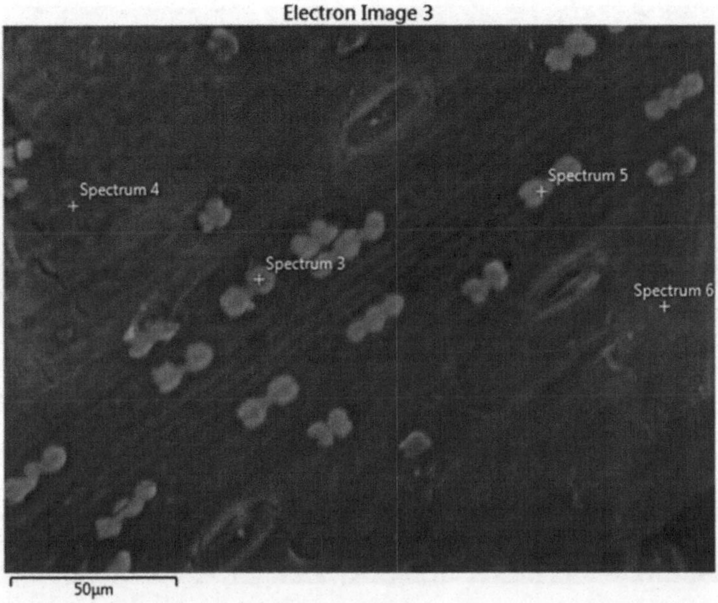

FIGURE 2.41 Scanning electron microscopic micrograms of composite 2.38 (Straw 85% + PhES-80 – 5% + PE 10%).

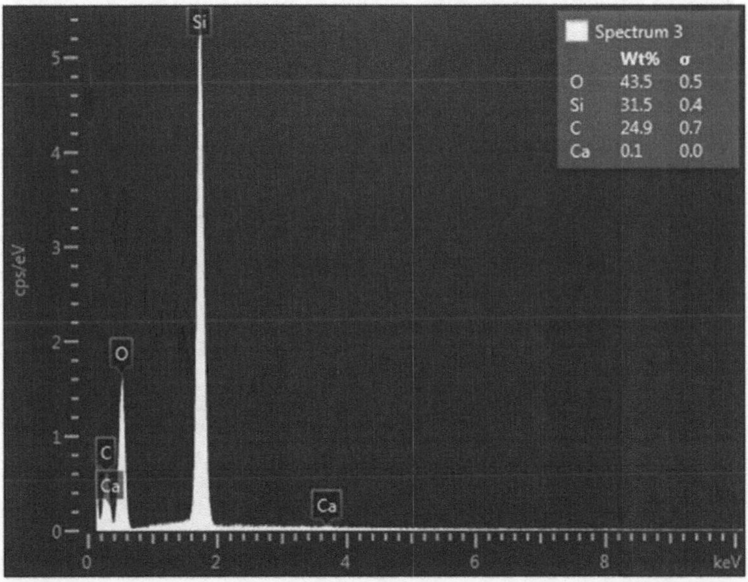

FIGURE 2.42 Energy dispersion micro-X-ray spectral analysis of composite 2.38: Straw 85% + PhES-80 – 5% + PE 10% (Spectrum 3).

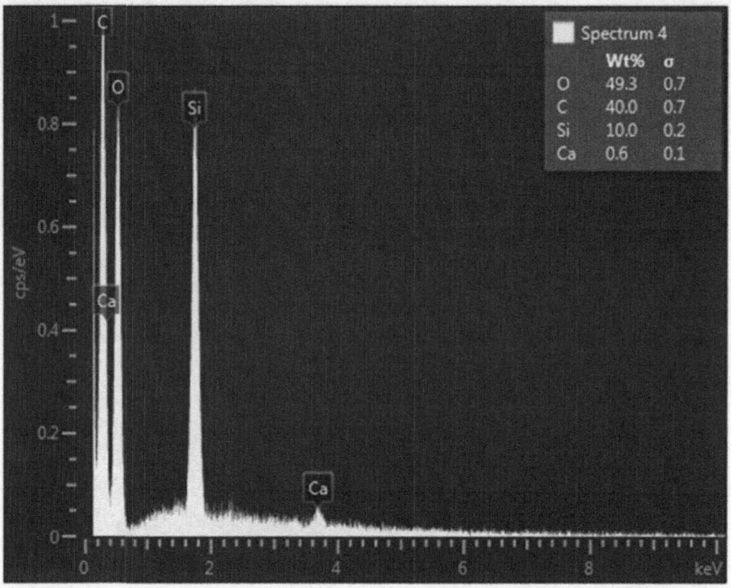

FIGURE 2.43 Energy dispersion micro-X-ray spectral analysis of composite 2.38: Straw 85% + PhES-80 – 5% + PE 10% (Spectrum 4).

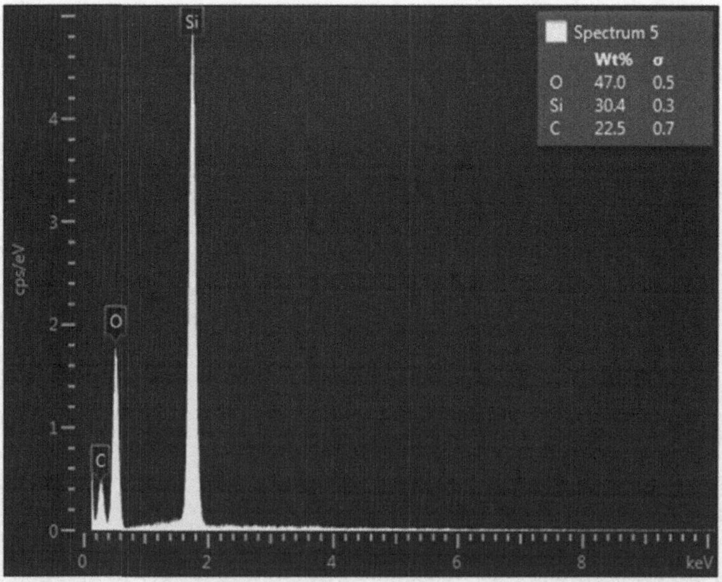

FIGURE 2.44 Energy dispersion micro-X-ray spectral analysis of composite 2.38: Straw 85% + PhES-80 – 5% + PE 10% (Spectrum 5).

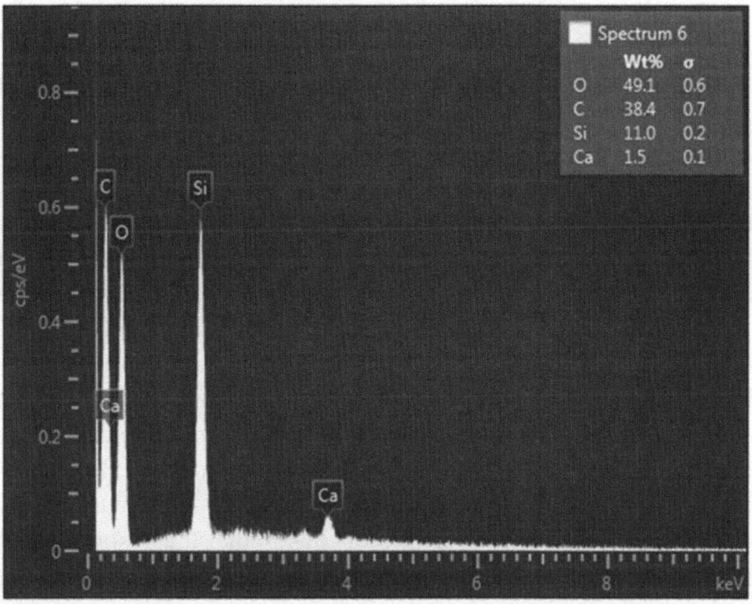

FIGURE 2.45 Energy dispersion micro-X-ray spectral analysis of composite 2.38: Straw 85% + PhES-80 – 5% + PE 10% (Spectrum 6).

TABLE 2.9 Energy Dispersion Micro-X-Ray Spectral Analysis of Composite 2.38 (Straw 85% + PhES-80 – 5% + PE 10%)

Result Type	**Weight (%)**			
Spectrum Label	**Spectrum 6**	**Spectrum 3**	**Spectrum 5**	**Spectrum 4**
C	37.41	22.88	22.53	40.03
O	49.12	45.45	47.04	49.33
Si	11.99	31.53	30.44	9.99
Ca	1.48	0.15	–	0.65
Total	**100.00**	**100.00**	**100.00**	**100.00**

Statistics	**C**	**O**	**Si**	**Ca**
Max.	40.03	49.33	31.53	1.48
Min.	22.53	43.45	9.99	0.15
Average	31.46	47.23	20.74	–
Standard deviation	9.04	2.73	11.85	–

Table 2.10 includes the numerical values of the mechanical strengthening on bending and impact viscosity.

2.5.1 COMPOSITE'S MECHANICAL STRENGTHENING ON BENDING

Mechanical strengthening of the composites at bending deformation was investigated by means of equipment used during the first stage of the project. The testing was provided for the composites, which were obtained at different pressures and temperatures for 10 min. In Table 2.10, the numerical data of the bending strength obtained by us composites are presented.

The data of Table 2.10 shows that the parameter of strengthening at bending the composite's ingredients type and their concentration and also the prehistory of obtaining (temperature, pressure) have an essential influence. The formation of composites proceeded for 10 min.

First of all, in accordance with the data of Table 2.10 shows that the dependence of mechanical characteristics for composites containing one binder (Liquid glass, PhES-50 and PhES-80 – composites 2.1–2.4, 2.5–2.8, 2.9–2.12) on the concentration has extreme character – maximums at definite concentrations. This fact leads to the conclusion that the best characteristics have the composites, in which the compatibility of ingredients reaches their maximal values. The comparison of the composite's characteristics shows also, that at relatively low concentrations the composites based on straw and liquid glass at any compositions are characterized with lower strengthening than that for PhES-50 and PhES-80 (compare the composite 1 with composites 2.5 and 2.9, composite 2 with composites 2.6 and 2.10. This result indicates the different interactions of binders with the straw's particles. Particularly the molecules of PhES create a new bond with the participation of active groups in accordance with the above-considered reactions (it must be noted that PhES-80 is more active to some extent in comparison with PhES-50). The composites containing both liquid glass and PhES are characterized by higher parameters of strengthening than that with one type of binder (compare the composites 2.1 vs 2.19, and composites 2 vs 20). The additive increase of the strengthening is revealed in composites containing polyethylene (PE) (compare the composites 2.1 vs 2.24, composites 2.5 vs 2.24, and composites 2.6 vs 2.29). The main factor in increasing the strengthening of materials, probably, is the value of the interaction forces between both binder and filler and between two binders (for example, between PhES and liquid glass). At two concentrations of PE the composites strength increases, which indicates the higher compatibility of PE macromolecules with other ingredients, although it takes place only at a definite proportion between lasts.

Testify of the obtained composites on the bending strength shows some regulations, which are characterized for considered here composites. So, for composites containing PhES-80 (composites 2.9–2.12), it is appeared an

TABLE 2.10 Mechanical Parameters of Composites Based on the Straw

SL. No.	Composite (Mass%)	Pressure	Temperature (T°C)	Strength on Bending (Kg/cm^2)	Impact Viscosity (kJ/m^2)
1.	Straw 97% + LG 3%	150	130	11.2	53.6
2.	Straw 95% + LG 5%	150	130	11.2	39.2
3.	Straw 90% + LG 10%	150	130	18.6	74.5
4.	Straw 85% + LG 15%	150	130	18.6	69.5
5.	Straw 97% + PhES-50 – 3%	250	130	22.3	62.7
6.	Straw 95% + PhES-50 – 5%	250	130	19.1	64.8
7.	Straw 90% + PhES-50 – 10%	150	130	18.6	68.2
8.	Straw 85% + PhES-50 – 15%	150	130	18.9	55.7
9.	Straw 97% + PhES-80 – 3%	150	130	20.1	65.2
10.	Straw 95% + PhES-80 – 5%	150	130	23.4	70.6
11.	Straw 85% + PhES-80 – 10%	150	130	18.6	61.6
12.	Straw 85% + PhES-80 – 15%	150	130	15.8	66.0
13.	Straw 95% + PE 5%	150	130	20.4	34.2
14.	Straw 90% + PE 10%	150	130	19.2	65.5
15.	Straw 85% + PE 15%	150	130	16.0	62.0
16.	Straw 80% + PE 20%	150	130	19.9	39.7
17.	Straw 95% + Colophony 5%	150	130	20.7	67.8
18.	Straw 90% + Colophony 10%	150	130	24.2	73.9
19.	Straw 85% + Colophony 15%	150	130	22.9	76.1
20.	Straw 80% + Colophony 20%	150	130	17.9	56.9

TABLE 2.10 *(Continued)*

SL. No.	Composite (Mass%)	Pressure	Temperature (T°C)	Strength on Bending (Kg/cm^2)	Impact Viscosity (kJ/m^2)
21.	Straw 95% + Wood glue 5%	150	130	20.7	64.2
22.	Straw 90% + Wood glue 10%	150	130	18.7	59.3
23.	Straw 85% + Wood glue 15%	150	130	20.4	65.0
24.	Straw 80% + Wood glue 20%	150	130	20.4	65.0
25.	Straw 94% + PhES-50 – 3% + LG 3% + Hardener	150	130	24.8	69.35
26.	Straw 90% + PhES-50 – 5% + LG 5%	150	130	17.3	70.39
27.	Straw 80% + PhES-50 – 10% + LG 10%	150	130	16.7	52.26
28.	Straw 80% + PhES-50 – 15% + LG 5%	150	130	18.6	66.7
29.	Straw 80% + PhES-50 – 5% + LG 15%	150	130	16.9	17.3
30.	Straw 94% + PhES-80 – 3% + LG 3% + Hardener	150	130	23.4	68.6
31.	Straw 90% + PhES-80 – 5% + LG 5%	150	130	23.4	62.7
32.	Straw 91% + PhES-50 – 3% + LG 3% + PE 3%	150	130	24.4	72.16
33.	Straw 85% + PhES-50 – 5% + LG 5% + PE 5%	150	130	16.7	59.5
34.	Straw 91% + PhES-80 – 3% + LG 3% + PE 3%	150	130	22.0	65.3
35.	Straw 85% + PhES-80 – 5% + LG 5% + PE 5%	150	130	18.3	63.7
36.	Straw 94% + PhES-50 – 3% + PE 3%	150	130	18.1	63.4
37.	Straw 90% + PhES-50 – 5% + PE 5%	150	130	23.0	66.0
38.	Straw 85% + PhES-50 – 5% + PE 10%	150	130	20.5	74.4
39.	Straw 94% + PhES-80 – 3% + PE 3%	150	130	17.6	61.7
40.	Straw 90% + PhES-80 – 5% + PE 5%	150	130	16.0	58.0

TABLE 2.10 *(Continued)*

SL. No.	Composite (Mass%)	Pressure	Temperature (T°C)	Strength on Bending (Kg/cm^2)	Impact Viscosity (kJ/m^2)
41.	Straw 85% + PhES-80 – 5% + PE 10%	150	130	19.2	67.2
42.	Straw 85% + PhES-50 – 5% + PVA 10%	150	130	19.0	65.5
43.	Straw 88% + PhES-50 – 3% + LG 3% + PE 3% + VinSi(OEt)$_3$ – 3%	150	130	23.5	72.5
44.	Straw 80% + PhES-50 – 5% + LG 5% + PE 5% + VinSi(OEt)$_3$ – 5%	150	130	15.2	68.5
45.	Straw 88% + PhES-80 – 3% + LG 3% + PE 3% + VinSi(OEt)$_3$ – 3%	150	130	18.3	72.1
46.	Straw 80% + PhES-80 – 5% + LG 5% + PE 5% + VinSi(OEt)$_3$ – 5%	150	130	15.2	63.4
47.	Straw 90% + PhES-80 – 5% + PE 5%	150	130	36.8	65.7
48.	Straw 94% + LG 3% + PE 3%	150	130	26.6	60.1
49.	Straw 90% + LG 5% + PE 5%	150	130	27.4	61.7
50.	Straw 94% + LG 4% + PhES-80 – 2%	150	130	18.1	68.5
51.	Straw 94% + LG 2% + PhES-80 – 4%	150	130	18.6	73.8
52.	Straw 85% + LG 5% + PE-80 – 10%	150	130	16.9	64.5
53.	Straw 85% + LG 10% + PE 5%	150	130	16.3	75.3
54.	Straw 85% + LG 7.5% + PE 7.5%	150	130	19.5	75.3
55.	Straw 85% + PhES-50 – 7.5% + PE 7.5%	150	130	15.7	65.6
56.	Straw 85% + PhES-50 – 10% + PE 5%	150	130	11.1	55.7

extreme dependence of numerical values of bending strength on the binder's concentration (we have respectively 20.09–23.44 – 18.6 – 15.82). In this series, the maximal value corresponds to a concentration of 5% of PhES-80. For the liquid glass (composites 2.1–2.4) the strength grows with increasing concentration. These data without a doubt show that the maximum strength must be at a concentration higher than presented in the table data, while for composites with PhES-50 (composites 2.5–2.8), this maximum corresponds to composite with binder lower than 10%. The same information may be made for composites with colophony.

For composites containing three components with PhES-50 and liquid glass (LG) (composites 2.19–2.21), the strength decreases at decreasing of the binder's concentration, while for composites with PhES-80 with concentrations 3–5% (composites 2.22–2.23) the strength practically doesn't change. In this case, maximum strength must be in the range of both low and high concentrations.

For two component composites with PE at 10 and 15% (composites 2.13, 2.14) and colophony (composites 2.15, 2.16) we have rather high characteristics, but in these cases, for finding the strength's maximums, we must know the strength values for composites, the concentrations of binders of which will be lower or higher than ones presented in Table 2.10 (10 and 15%).

High values of strength have characterized the composites containing three and four components, namely the systems containing PhES-50 and PhES-80, LG, and PE with an amount of 3 or 5% (composites 2.24, 2.26, 2.29, 2.30). Good data has a composite 34 with 5 ingredients, which contains additively vinylthrietoxisilane with definite proportion. With more high data is characterized the composite 2.38 with 5% binder (PhES-80, PE, and colophony). The strength of the last composite is equal to 36.84 kgF/cm^2. With good data are characterized the composites with 5% LG and 5% PE.

The testing of the bending strength shows that the determining factors in the definition of strength in bending are the composition and the type of ingredients. The proportion of them and compatibility in the composites play also an important role.

2.5.2 IMPACT ELASTICITY OF COMPOSITES

This characteristic is a very important exploitation parameter. It draws attention, especially in the processes of instantaneous loading. Table 2.10 there are presented the data of the noted parameter.

Table 2.10 data shows once more the efficiency of composite components in the formation of their properties. The concentration variation of the binder changes the composite's properties by a definite level. So, for composites, 2.1–2.4 and 2.5–2.8 the comparison of this parameter shows that it is definite dependence on the binder's concentration. Particularly from the composites contained LG the maximal impact elasticity has the composite 3 contained 10% of LG. For systems with PhES-50 maximum of this parameter appears at 10% binder and for composites with PhES-80 the maximum of the noted parameter corresponds to composites containing the binder at lower than 10%. For composites with colophony maximum of this parameter must be corresponded to one, containing this binder with a concentration not lower than 15%. Especially we must pay attention to the high strength of the composites with colophony. This fact may be explained in the following manner: colophony as the plant product diffused to the straw structure pores in the process of composite formation (pressure, temperature) and promotes the homogenization of composites even at high concentrations. This fact shows the good compatibility of colophony with straw, which is one of the important factors defining the high level of impact elasticity. These composites show rather a high impact elasticity at relatively high concentrations of colophony than the analogs containing the same concentrations of other binders.

High strength is characterized by the system with PhES-50 and LG (composites 4.19 and 4.20), although for the system, in which the same components with 10% are introduced (composite 4.21) the impact elasticity decreases essentially. This parameter as in the case of bending strength increases when the composite with LG and PhES is introduced to PE with definite concentration (composite 4.24). The more high result is obtained in the case of composite 30 containing three components with PhES-50 (5%) and PE (10%). In this case, PE shows once more the good binder properties, the compatibility of which we have noted above.

The same high characteristics have composites with 5 ingredients (34, 36), which contain vynylthriethoxysilane.

There were conducted the works for the establishment of the level of impact elasticity in the case of composites containing more high concentrations of binders. We have prepared the composites with concentrations of binders 10 and 15%. However, the impact elasticity for these composites was lower than in the case of analogous binders at lower concentrations. The composites with LG and PhES exhibit lower compatibility in comparison with colophony and consequently are characterized by lower strength. Besides it is known that if the binder at relatively high concentrations doesn't introduce to the filler matrix or doesn't participate in the reactions it stays in its phase, or forms the so-called "soft phase." The last play the role of plasticizer, which is the reason for defining of decrease in the composite's strength. So, the experimental results allow us to conclude that one of the main factors for obtaining the composites we must select the binder's concentration carefully. The same conclusion may be made for a selection of binder type.

2.6 THERMAL STABILITY (TS) BY VICA

There were conducted the investigations on study of TS of composites by method Vica. Data from these measurements are presented in Figures 2.46–2.51.

From Figure 2.46, one can see that the concentration of LG affects the TS of composite material. Particularly, the maximum of the TS of composites with 3% LG appears at 80°C, but at more high temperatures (140–150°C), the opposite process starts – the deepening of the indenter decreases as a result of thermal widening of the sample. The result for the composite with 5% of binder shows another behavior – softening is raised at more high temperatures. This fact may be ascribed as increasing the softening as a result of increasing the soft phase.

In Figure 2.47, we see that PhES-80 creates the soft phase and promotes to material's softening. The existence of three binders in the composites leads to an increase in the composite's softening because of existing of a low number of the chemical bounds, and the sum of the binder phase increases of the proportion binder/filler and consequently increases the concentration of the soft phase, which promotes to increasing of material's softening. It is clear that elimination of this phenomenon may be removed by a selection of the technological conditions of the formation of the composites.

The composites in which the role of binders play PhES-50 and LG are characterized by high TS (curve 3). The reason for this fact may be a good interaction between the binder's molecules and fillers particles. Curve 2 seems paradoxical in comparison with curve 1. The reason for this fact may be connected with the introduction of PE to the composite, which increases the sum of the binders and promotes to increase of the composite's softening at relatively low temperatures. Composite with PhES-80 gives composite higher TS in comparison with the analogous with PhES-50 (compare the curves 3 and 4).

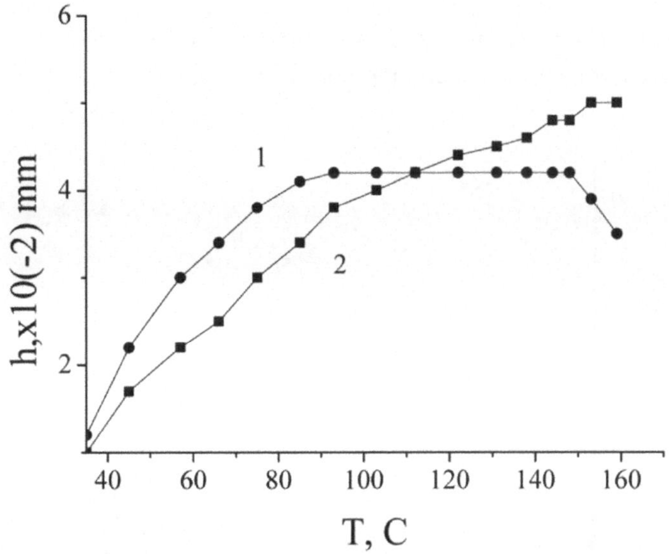

FIGURE 2.46 Dependence of the softening on temperature for composites containing a straw. Curve 1 – LG 3% (composite 2.1), and 2 – LG 5% (composite 2.2).

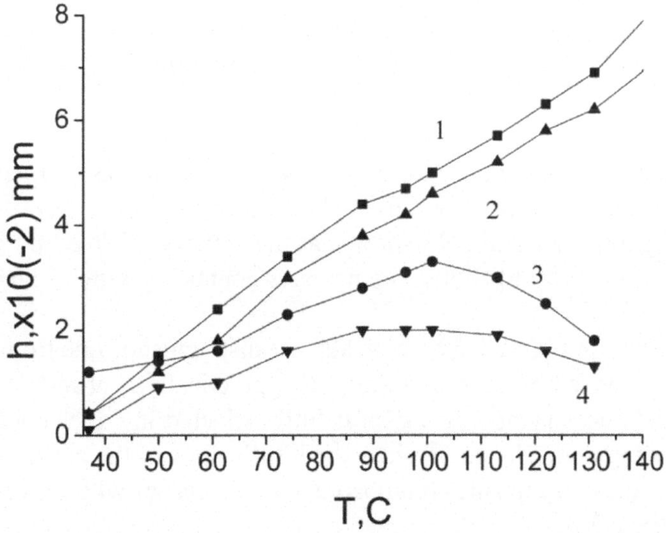

FIGURE 2.47 Dependence of the softening on temperature for composites containing a straw. Curve 1 – PhES-80 – 5% (composite 2.10); 2 – PhES-50 – 3% + LG 3% + PE 3% (composite 2.32); 3 – PhES-50 – 3% + LG 3% (composite 2.1); and 4 – PhES-80 – 5% + LG 5% (composite 2.31).

It seems that $VinSi(OEt)_3$ promotes to increase in TS, which shows a good interaction of this substance with other binders (Figure 2.48). PhES-80 shows more high activity in comparison with PhES-50.

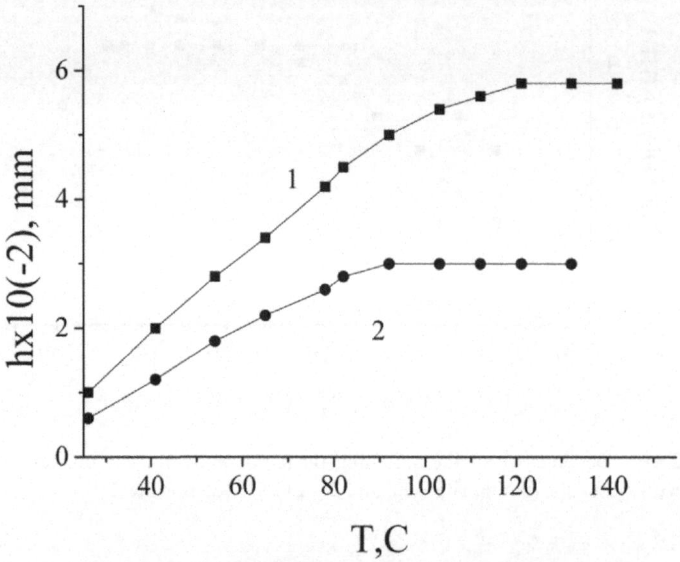

FIGURE 2.48 Dependence of the softening on temperature for composites containing a straw. Curve 1 – PhES-50 – 5% + LG 5% + PE 5% + $VinSi(OEt)_3$ – 5% (composite 2.44); and 2 – PhES-80 – 3% + LG 3% + PE 3% + $VinSi(OEt)_3$ – 3% (composite 2.45).

The data presented in Figure 2.49 shows that in this case, PhES-80 does not show such distinguishes relatively PhES-50 as in Figure 2.21. A comparison of Figures 2.21 and 2.23 show that the activity of PhES-80 increases in comparison with PhES-50 in composites containing vinyltriethoxysilane (Figure 2.50).

The curves in Figure 2.50 practically are distinguished one from another, which is happened because of $VinSi(OEt)_3$ practically is absorbed by other binders and consequently, its effect is little reflected in the character of the corresponding dependence. Besides the limit values of the curves show the high TS of these composites (compare the noted curves with curves 1 and 2 from Figure 2.51).

In accordance with curves in Figure 2.51 the gradation of the TS starts from 50°C and increases in the row: (comp. 2.18) < (comp. 2.23) < (comp. 2.22) So, the composites containing the wood glue are more stable than the analogous based on colophony with the same concentration. This result

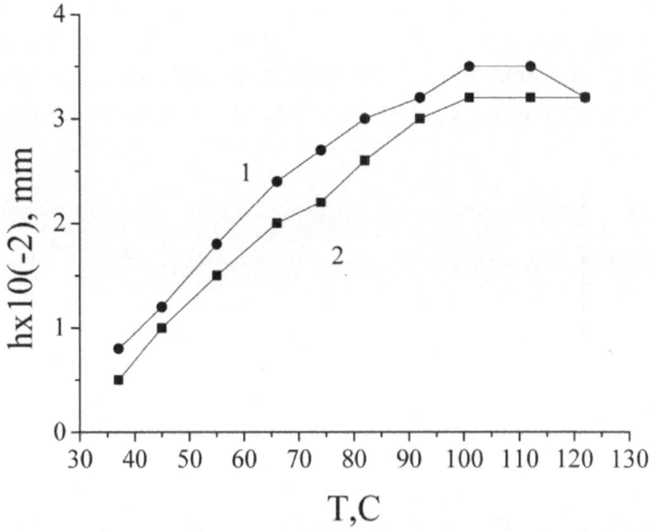

FIGURE 2.49 Dependence of the softening on temperature for composites containing a straw. Curve 1 – PhES-80 – 5% + LG 5% + PE 5% (composite 2.35); and 2 – PhES-50 – 5% + LG 5% + PE 5% (composite 2.33).

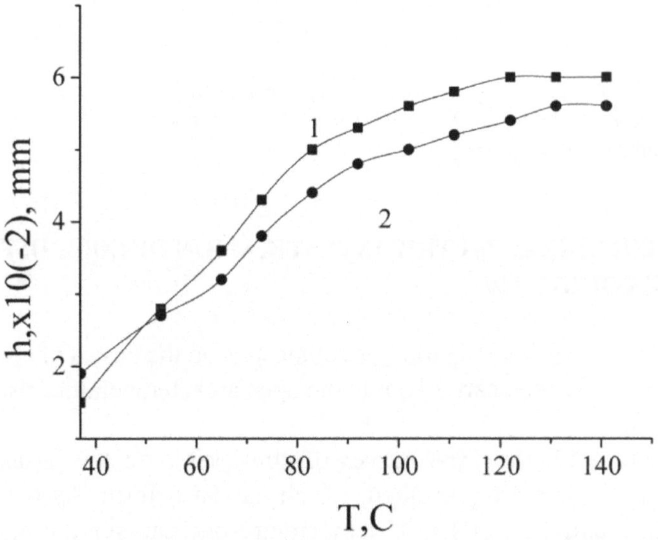

FIGURE 2.50 Dependence of the softening for composites containing a straw. Curve 1 – PhES-50 – 5% + LG 5% + PE 5% + $VinSi(OEt)_3$ – 5% (composite 2.44); and 2 – PhES-80 – 3% + LG 3% + PE 3% (composite 2.34).

shows that the softening of colophony takes place at less temperatures in comparison with wood glue. Besides the composite with wood glue at 10% is more stable than that at 15%. This result shows once more the negative role of the binder with relatively high concentrations in the composites.

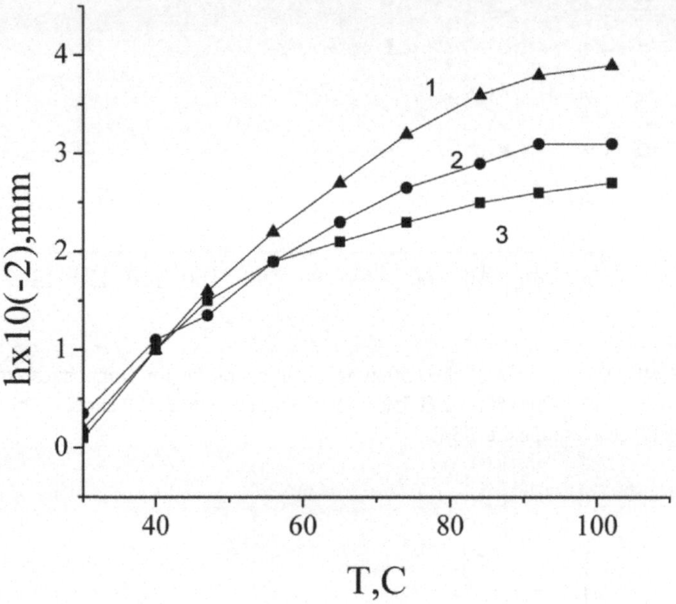

FIGURE 2.51 Dependence of the softening on the temperature for composites containing a straw. Curve 1 – Colophony 10% (composite 2.18); 2 – Wood glue 15% (composite 2.23); and 3 – Wood glue 10% (composite 2.22).

2.7 THERMOGRAVIMETRIC INVESTIGATION OF COMPOSITES ON THE BASIS OF STRAW

Thermogravimetric investigation of composites on the basics of straw on the basis of straw has been carried out in the open area, temperature rise rate was v=10°C/min.

In Figure 2.52, the TGA curves of composites on the basis of straw and various binders are presented, which has an S form. As is seen from Figures 2.52 and 2.53 at 100°C temperature one can see the mass losses which are connected with chemical reactions of functional groups of straw with functional groups of binders.

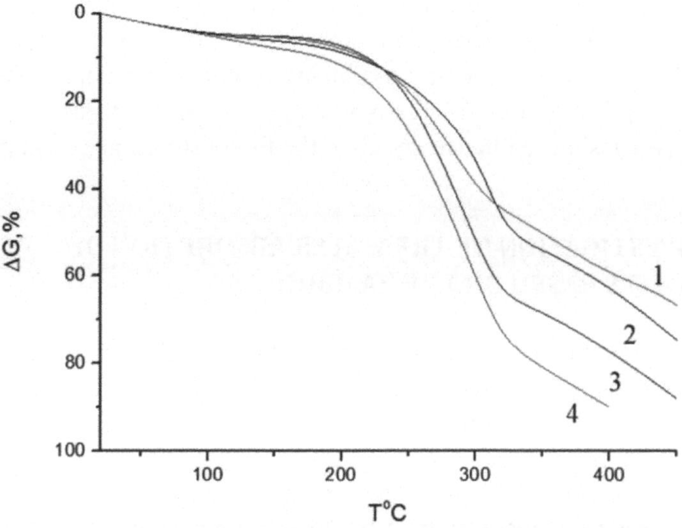

FIGURE 2.52 Thermogravimetric curves of composites based on the straw: Curve 1 – Straw 95% + PhES-50 – 5% (composite 2.6); 2 – Straw 80% + PhES-80 – 5% + LG 5% + PE 5% + VinSi(OEt)$_3$ – 5% (composite 2.46); 3 – Straw 85% + PhES-50 – 5% + LG 5% + PE 5% (composite 2.33); and 4 – Straw 97% + LG 3% (composite 2.1).

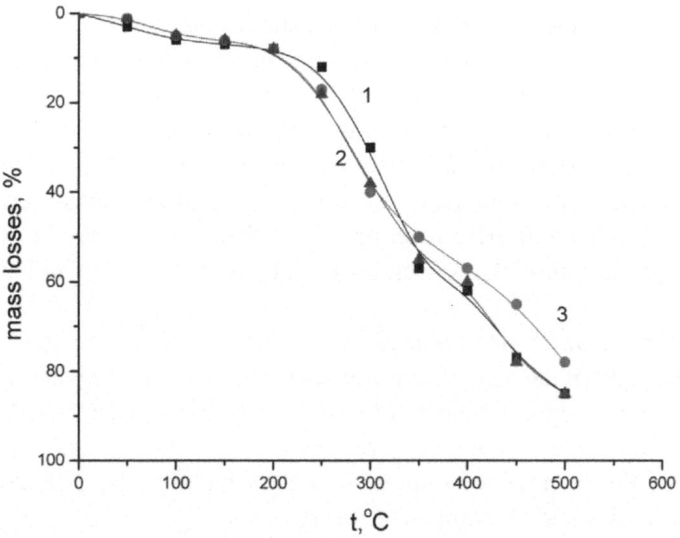

FIGURE 2.53 Thermogravimetric curves of composites based on the straw: where Curve 1 corresponds to composite: Straw 90% + PhES-80 – 5% + PE 5% (composite 2.40); 2 – Straw 90% + PhES-80 – 5% + PE 10% (composite 2.41); and 3 – Straw 90% + PhES-50 – 5% + PE 5% (composite 2.37).

At 250°C temperature for composites (2.6 and 2.47) containing PhES and vinyltriethoxysilane the destruction process runs more slowly, which may be explained by the high value of crosslinking reactions in composites.

The main destruction process is in composite on the basis of straw and liquid glass runs more intensively up to the full destruction of composites.

2.8 INVESTIGATION OF THE WATER ABSORPTION OF COMPOSITES BASED ON THE BAMBOO

Water absorption is one of the important exploitation characteristics for wood composites because this material is used often in wet areas. The experimental results provided on our composites are presented in Table 2.11.

The analysis of Table 2.11 data shows that–for the straw composites containing PhES of both types the general phenomenon is the increase of water absorption at increasing of the binder's concentration (compare the composites 1–4 and composites 5–8). Increasing the binder concentration is directly connected with the material's morphology. It may be supposed that composites it is created an environment, which promotes the diffusion of water molecules to the composite's matrix, in which there are the micro-empties, which are connected one to another by channels with definite sizes. These channels promote from one side to diffusion of low molecular substances and the creation of the hydrogen bonds between binder molecules and water ones. In the preliminary experiments, we obtained an unexpected result – decreasing in the composite weight during 24 four stays in the water (instead of increasing) because of a reaction between water molecules and some products included in the wood structure as a result of evaporation of the secondary products of these reactions. This phenomenon may be connected with the formation of the associates on the basis of binder molecules and water ones.

Another situation was realized in the case of the composites with LG. Here the weight of the composites after 24 h exposition in the water increases. Here it may be supposed that the LG at relatively low concentrations adsorbs the water molecules easily. At increasing concentrations of LG, the micro-empties of the composites saturate and consequently, the diffusion of the water molecules and all composites decreases.

With low water, absorption is differed the composites containing colophony. Here it is exhibited the ability of this binder to fill well of this empty.

The composites containing binary binders are characterized by different water absorption, the amount of which depends on the character of interactions

TABLE 2.11 Water Absorption of Composites on the Basis of Straw and Various Binders

SL. No.	Composite (Mass %)	Mass (g)	Volume (cm^3)	Density (g/cm^3)	Mass Composites After 3 h Exposition in Water (g)	Mass of Composites After 24 h Exposition in Water (g)	Water Absorption After 24 h Exposition in Water (%)
1.	Straw 97% + LG 3%	4.36	3.00	1.45	4.90	6.34	45.38
2.	Straw 95% + LG 5%	3.88	3.17	1.23	4.26	4.46	14.77
3.	Straw 90% + LG 10%	4.57	3.00	1.52	4.85	6.13	34.34
4.	Straw 85% + LG 15%	4.29	2.99	1.43	4.55	5.42	86.36
5.	Straw 97% + PhES-50 – 3%	4.03	2.99	1.35	4.21	4.47	10.93
6.	Straw 95% + PhES-50 – 5%	4.58	2.99	1.53	4.78	5.09	11.26
7.	Straw 90% + PhES-50 – 10%	4.53	3.18	1.42	4.79	5.55	22.72
8.	Straw 85% + PhES-50 – 15%	3.87	2.83	1.37	4.36	5.26	36.19
9.	Straw 97% + PhES-80 – 3%	3.57	2.64	1.35	3.70	3.86	8.24
10.	Straw 95% + PhES-80 – 5%	4.43	3.34	1.33	4.62	4.95	11.71
11.	Straw 90% + PhES-80 – 10%	4.62	3.36	1.38	4.80	5.28	14.40
12.	Straw 85% + PhES-80 – 15%	3.90	2.64	1.48	4.39	4.90	25.64
13.	Straw 95% + PE-80 – 5%	4.58	3.34	1.37	4.76	6.11	33.41
14.	Straw 90% + PE 10%	4.26	3.17	1.34	4.40	4.84	13.78
15.	Straw 85% + PE 15%	4.56	3.56	1.36	4.75	5.10	11.95
16.	Straw 80% + PE 20%	3.57	2.82	1.27	3.63	4.01	12.32
17.	Straw 95% + Colophony 5%	3.80	2.64	1.44	3.95	5.54	45.79
18.	Straw 90% + Colophony 10%	3.75	2.83	1.33	3.91	4.04	7.88
19.	Straw 85% + Colophony 15%	4.24	3.00	1.41	4.42	4.96	17.23
20.	Straw 90% + Colophony 20%	3.36	2.46	1.37	3.50	4.52	34.52

TABLE 2.11 *(Continued)*

SL. No.	Composite (Mass %)	Mass (g)	Volume (cm^3)	Density (g/cm^3)	Mass Composites After 3 h Exposition in Water (g)	Mass of Composites After 24 h Exposition in Water (g)	Water Absorption After 24 h Exposition in Water (%)
21.	Straw 95% + Wood glue 5%	3.95	2.82	1.40	4.12	5.60	41.77
22.	Straw 90% + Wood glue 10%	4.39	3.18	1.38	4.84	5.37	28.55
23.	Straw 85% + Wood glue 15%	4.26	3.18	1.34	4.70	5.62	31.92
24.	Straw 80% + LG 20%	3.52	2.46	1.43	3.69	4.73	34.38
25.	Straw 94% + PhES-50 – 3% + LG 3%	3.61	2.64	1.37	3.77	4.62	27.98
26.	Straw 90% + PhES-50 – 5% + LG 5%	3.46	2.46	1.41	3.60	3.92	13.29
27.	Straw 80% + PhES-50 – 10% + LG 10%	3.77	2.82	1.34	4.20	4.64	22.97
28.	Straw 80% + PhES-50 – 15% + LG 5%	3.53	2.46	1.43	3.77	4.97	40.79
29.	Straw 80% + PhES-50 – 5% + LG 15%	3.40	2.64	1.29	3.62	4.01	17.94
30.	Straw 94% + PhES-80 – 3% + LG 3% + Hardener	4.74	3.34	1.42	4.91	5.28	11.50
31.	Straw 90% + PhES-80 – 5% + LG 5%	4.52	3.17	1.43	4.72	5.08	12.52
32.	Straw 91% + PhES-50 – 3% + LG 3% + PE 3%	2.80	2.11	1.32	2.90	3.14	12.14
33.	Straw 85% + PhES-50 – 5% + LG 5% + PE 5%	3.56	2.64	1.35	3.69	4.01	12.64
34.	Straw 91% + PhES-80 – 3% + LG 3% + PE 3%	2.79	2.11	1.32	2.91	3.19	14.34
35.	Straw 85% + PhES-80 – 5% + LG 5% + PE 5%	3.59	2.64	1.36	3.76	4.89	36.21
36.	Straw 94% + PhES-50 – 3% + PE 3%	3.73	2.64	1.41	3.88	4.98	33.51

TABLE 2.11 *(Continued)*

SL. No.	Composite (Mass %)	Mass (g)	Volume (cm^3)	Density (g/cm^3)	Mass Composites After 3 h Exposition in Water (g)	Mass of Composites After 24 h Exposition in Water (g)	Water Absorption After 24 h Exposition in Water (%)
37.	Straw 90% + PhES-50 – 5% + PE 5%	3.79	2.82	1.34	3.91	4.98	31.40
38.	Straw 85% + PhES-50 – 5% + PE 10%	4.25	3.18	1.34	4.52	5.02	18.03
39.	Straw 94% + PhES-80 – 3% + PE 3%	3.69	2.64	1.40	3.82	4.60	24.66
40.	Straw 85% + PhES-50 – 10% + PE 5%	3.83	2.82	1.36	4.02	5.97	55.87
41.	Straw 90% + PhES-80 – 5% + PE 5%	4.25	3.36	1.27	4.47	4.90	15.39
42.	Straw 85% + PhES-50 – 5% PVA 10%	1.89	1.58	1.20	2.11	2.60	37.57
43.	Straw 88% + PhES-50 – 3% + LG 3% + PE 3% + $VinSi(OEt)_3$ – 3%	3.66	2.64	1.39	3.80	4.29	17.21
44.	Straw 80% + PhES-50 – 5% + LG 5% + PE 5% + $VinSi(OEt)_3$ – 5%	3.34	2.46	1.36	3.45	3.95	18.26
45.	Straw 88% + PhES-80 – 3% + LG 3% + PE 3% + $VinSi(OEt)_3$ – 3%	3.69	2.64	1.40	3.90	5.01	35.77
46.	Straw 80% + PhES-80 – 5% + LG 5% + PE 5% + $VinSi(OEt)_3$ – 5%	3.55	2.64	1.34	3.67	4.31	21.41
47.	Straw 90% + LG 5% + PE 5%	3.63	2.64	1.38	3.79	4.42	21.76
48.	Straw 94% + LG 3% + PE 3%	3.51	2.64	1.33	3.68	4.66	32.76
49.	Straw 85% + LG 5% + PE 5% + Colophony 5%	3.68	2.64	1.39	3.84	4.73	28.53
50.	Straw 94% + PhES-80 – 2% + LG 4% + PE 3%	2.77	1.94	1.42	3.03	4.21	51.98
51.	Straw 94% + PhES-80 – 4% + LG 2%	2.93	2.28	1.28	3.14	4.26	45.39

TABLE 2.11 *(Continued)*

SL. No.	Composite (Mass %)	Mass (g)	Volume (cm^3)	Density (g/cm^3)	Mass Composites After 3 h Exposition in Water (g)	Mass of Composites After 24 h Exposition in Water (g)	Water Absorption After 24 h Exposition in Water (%)
52.	Straw 85% + LG 5% + PE 10%	3.66	2.64	1.38	3.76	4.34	18.57
53.	Straw 85% + LG 10% + PE 5%	3.51	2.64	1.32	3.66	4.30	22.50
54.	Straw 85% + LG 7.5% + PE 7.5%	3.51	2.46	1.42	3.64	4.48	27.63
55.	Straw 85% + PhES-50 – 7.5% + PE 7.5%	3.78	2.82	1.34	3.90	5.02	32.80
56.	Straw 85% + PhES-50 – 10% + PE 5%	3.73	2.64	1.43	3.86	4.99	33.78

of them with water molecules. The experiment shows that the water of the wood composites may be changed by the selection of type and concentration of the binders.

2.9 CONCLUSIONS OF COMPOSITES ON THE BASIS OF STRAW

There are obtained the composites based on dry straw and some organic-inorganic binders. The technology of production of these composites is rather simple. The components were dispersed and mixed in the mixer used for the dispersion of the coffee. The concentration of the binders varied in the range of 3–20 wt.%. Obtained by this method the blends were placed to the spatial forms corresponding to the standards for a measure of physical-mechanical properties, TS, and water absorption under pressure up to 170 kg/cm^2 and temperatures in the range 100–150°C for 10 min.

By use of spectroscopic methods (FTIR, scanned electron microscope, optical microscope, and energy-dispersive micro-X-ray spectral analysis) it was provided the investigations of chemical and morphological structures of the composites based on straw and some binders (liquid glass, PhES-50, PhES-80, PE, colophony, and wood glue). FTIR spectroscopy was used to establish the creation of some new chemical bonds between phases of composites. The micro-X-ray analysis helped us to define the content and concentration of the separate atoms included in composites. Using optical microscopy and scanned electron microscopy it was enabled to obtain the character of distribution of the ingredients in the composites.

On the basis of physical-mechanical investigations, it was studied the strengthening of materials on bending and impact viscosity. It was established that the dependence of these parameters on the binder concentration has extreme character – at definite concentrations of binders (differ from one another) the strengthening of composites is characterized by maximums. This phenomenon is described with the creation of structure, in which the ingredient particles in the composite are distributed more homogeneously, than in the case of structures containing the ingredients with other concentrations.

These maximums are distinctly for systems containing one ingredient: liquid glass, PhES-50 and PhES-80, colophony, PE, and wood glue.

TS of the composites was investigated with the use of the method Vica. Basically, the results obtained from the investigation of TS are in accordance with data obtained from the study of mechanical properties – the best characteristics of these properties correspond to the composites with one and the same concentrations of binders, since these structures are both mechanically and thermally stable. The analogical conclusion may be made from data from thermal gravimetry investigations.

Tendency ascribed relatively mechanical strengthening and TS saves after investigations of water absorption, although numerical data of this parameter have a larger scatter.

KEYWORDS

- **composite materials**
- **energy dispersion**
- **Fourier transform infrared**
- **mechanical strengthening**
- **microstructure**
- **optical microscopy**
- **raw materials**
- **scanning electron microscope**

REFERENCES

1. Kliosov, A. A., (2007). *Wood-Plastic Composites* (p. 698). John Wiley & Sons, Hoboken, New Jersey.
2. Ellis, W. D., & O'Dell, J. L., (1999). Surface morphology of PECVD fluorocarbon thin films from hexafluoropropylene oxide, 1,1,2,2-tetrafluoroethane and difluoromethane. *J. Appl. Polym. Sci., 73*(12), 2493–2505.
3. Patil, Y. P., Gajre, B., Dusane, D., Chavan, S., & Mishra, S., (2000). Effect of maleic anhydride treatment on steam and water absorption of wood polymer composites prepared from wheat straw, cane bagasse, and teak wood sawdust using Novolac as a matrix. *J. Appl. Polym. Sci., 77*(13), 2963–2967.
4. Li, Y., (2002). *Wood Polymer Composites*. Northeast Forestry University, Harbin, China.
5. Sebe, G., & Brook, M. A., (2001). Hydrophobization of wood surfaces: Covalent grafting of silicone polymers. *Wood Sci. and Technology, 35*, 269–282.
6. Brember, K. I., & Shneider, M. H., (1985). The significance of the glass transition of lignin in thermomechanical pulping. The significance of the glass transition of lignin in thermomechanical pulping. *Wood Sci. and Technology, 19*, 75–81.
7. Husain, M. M., Khan, M. A., Idriss, K. M., & Mustafa, A. I., (1996). Effect of sulfuric acid in the wood plastic composite. *Radiat. Phys. Chem., 47*(1), 149–153.

8. Elvy, S. B., Dennis, G. R., & Loo-Teck, Ng., (1995). Effects of coupling agent on the physical properties of wood-polymer composites. *J. Material PROCESSING and Technology, 48*(1–4), 365–371.
9. Mai, C., & Militz, H., (2004). Modification of wood with silicon compounds. Inorganic silicon compounds and sol-gel systems: A review. *Wood Science and Technology, 37*(5), 339–348.
10. Mai, C., & Militz, H., (2004). Modification of wood with silicon compounds. Treatment systems based on organosilicon compounds a review. *Wood Science and Technology, 37*(6), 453–461.
11. Park, B. D., Wi, S. G., Lee, K. H., Singh, A. P., Yoon, T. H., & Kim, Y. S., (2003). Characterization of anatomical features and silica distribution in rice husk using microscopic and micro-analytical techniques. *Biomass and Bioenergy, 25*, 319–327.
12. Ch. Turmanova, S., Dimitrova, A. S., & Vlaev, L. T., (2008). Study of polypropene composites filled with rice husks ash. *Oxidation Communications, 31*(2), 465–481.
13. Atchison, J. E., (1976). *Agricultural Residue and Other Non-Wood Plant Fiber Science. 191*(4228), Feb. 20, Materials Issue.
14. Jiang, G., (2007). *Conference of Global Warming in China.* http://en.wikipedia.org/wiki/Jiang_Gaoming (accessed on 06 June 2023).
15. Doweidar, H., El-Damrawi, G., El-Egili, K., Moustafa, Y. M., Ramadan, R. M., Abdelghany, M., & Kamal, M., (2009). Rice straw ash as a raw material for glass production. *Glass Technology–European Journal of Glass Science and Technology Part A, 50*(4), 196–202(7).
16. Vail, J. G., (1952). *Soluble Silicates* (Vol. 1, 2). Reinhold, New York.
17. Frederik Kotz, Klaus Plewa, Werner Bauer, Norbert Schneider, Nico Keller, Tobias Nargang, et al., (2016). "Liquid Glass: A Facile Soft Replication Method for Structuring Glass," *Advanced Materials,* https://doi.org/10.1002/adma.201506089.
18. Isler, R., (1982). *Silica Chemistry. Part 2.* MIR, Moscow, 1982, 712 pp.
19. Saikia, C. N., Ali, F., Goswami, T., & Grosh, A. C., (1995). Esterification of high α-cellulose extracted from *Hibiscus cannabinus* L. *Ind. Crops Prod., 4*, 233–239.
20. Mukbaniani, O., Aneli, J., Buzaladze, G., Markarashvili, E., & Tatrishvili, T., (2015). Composites on the basis of straw with some organic and inorganic binders. *Oxid. Commun.* (In press).
21. Aneli, J., Mukbaniani, O., Buzaladze, G., Tatrishvili, T., & Markarashvili, E., (2015). *New Biocomposites Based on Straw and Different Binders Jimsheri Aneli* (p. 39). Abstracts of communications of the 6th Nordic wood biorefinery conference, NWBC 2015, Helsinki, Finland.
22. https://eventegg.com/enefm-2015/ (accessed on 06 June 2023).
23. Markarashvili, E., Buzaladze, G., Kalatozishvili, L., Otiashvili, D., Tatrishvili, T., Aneli, J., & Mukbaniani, O., (2015). *New Composites on the Basis of Liquid Glass and Straw* (p. 87). Abstracts of communications of 4th International Caucasian symposium on polymers and advanced materials, Batumi. http://www.icsp4.tsu.ge/data/file_db/icsp/ICSP.pdf (accessed on 06 June 2023).
24. Aneli, J., Buzaladze, G., Markarashvili, E., Tatrishvili, T., & Mukbaniani, O., (2015). New biocomposites based on straw and different binders. In: *3rd International Congress on Energy Efficiency and Energy Related Materials ENEFM 2015, Oludeniz Mugla/Turkey, 105.* https://eventegg.com/enefm-2015 (accessed on 06 June 2023).
25. Owen, N. L., & Thomas, D. W., (1989). Infrared studies of "hard" and "soft" woods. *Appl. Spectroscopy, 43*, 451–455.

CHAPTER 3

Composite Materials on the Basis of Leaves: Innovations and Applications

OMAR MUKBANIANI,[1,2] JIMSHER ANELI,[2] and TAMARA TATRISHVILI[1,2]

[1]*Department of Macromolecular Chemistry, Ivane Javakhishvili Tbilisi State University, Tbilisi, Georgia*

[2]*Institute of Macromolecular Chemistry and Polymeric Materials, Ivane Javakhishvili Tbilisi State University, Tbilisi, Georgia*

ABSTRACT

In this chapter, the new composites on the basis of leaves and in-situ obtained of polymeric matrix. The bonds enhanced the mechanical and other exploitation properties of composites with existence of intensive chemical bonds between phases. The chemical interphases bonds are also reflected on mechanical properties and on level of water absorption of the new composites and have the same order than that for analogs given for the composites based on leaves with other binders.

3.1 INTRODUCTION

There were fulfilled the works on creation and investigation of new composites on the basis of dry leaves of different woods and binders as liquid glass (LG), phenyletoxysilan of two types – PhES-50 and PhES-80, polyethylene (PE) of low density, colophony, and wood glue. All binders besides LG had a powder like view (sizes of the particles – 10–100 mcm).

Biocomposites: Environmental and Biomedical Applications. Omar Mukbaniani, Tamara Tatrishvili, Neha Kanwar Rawat, & A. K. Haghi (Eds.)

A structurally complete leaf of an angiosperm consists of a petiole (leaf stalk), a lamina (leaf blade), and stipules (small structures located to either side of the base of the petiole). Not every species produces leaves with all these structural components. In certain species, paired stipules are not obvious or are absent altogether. A petiole may be absent, or the blade may not be laminar (flattened). The tremendous variety shown in leaf structure (anatomy) from species to species is presented in detail below under morphology. The petiole mechanically links the leaf to the plant and provides the route for transfer of water and sugars to and from the leaf. The lamina is typically the location of the majority of photosynthesis. The upper (adaxial) angle between a leaf and a stem is known as the axil of the leaf. It is often the location of a bud. Structures located there are called "axillary."

Leaves are normally extensively vascularized and typically have networks of vascular bundles containing xylem, which supplies water for photosynthesis, and phloem, which transports the sugars produced by photosynthesis. Many leaves are covered in trichomes (small hairs) which have diverse structures and functions [1–6].

The major tissue systems present are:

- The epidermis, which covers the upper and lower surfaces;
- The mesophyll tissue inside the leaf, which is rich in chloroplasts (also called chlorenchyma);
- The arrangement of veins (the vascular tissue).

These three tissue systems typically form a regular organization at the cellular scale. Specialized cells that differ markedly from surrounding cells, and which often synthesize specialized products such as crystals, are termed idioblasts [7].

The epidermis is the outer layer of cells covering the leaf. It is covered with a waxy cuticle which is impermeable to liquid water and water vapor and forms the boundary separating the plant's inner cells from the external world.

Parallel-veined, parallel-ribbed, parallel-nerved, Penni parallel – veins run parallel for the length of the leaf, from the base to the apex. Commissural veins (small veins) connect the major parallel veins. Typical for most monocotyledons, such as grasses.

As the leaves we used mostly leaves from sycamore tree, which we dried and then milled. All ingredients of the composites (except dry leaves) were presented in amount 3, 5, 10, 15, and 20 wt.%. The obtained composites may be divided conventionally on the several groups. Two-, three-, and four-component systems:

1. Leaves + LG (liquid glass 3–20%);
2. Leaves + PhES (phenetoxysilsesquioxane 50 and 80; 3–20%);
3. Leaves + PE (3–20%);
4. Leave + Colophony (3–20%);
5. Leaves + wood glue (3–20%);

From literature it is known that the leaves mainly contain sufficient amounts of starch, cellulose, hemicellulose, and pectic substances, lignin, polyphenols. All these above-mentioned compounds may react with proposed binders. For example, phenylethoxysilane (PhES-50 and PhES-80) contains ethoxy groups. The noted groups participate to the etherification reaction with a leaf through the macromolecular and intra-molecular reactions. Etherification reactions proceed analogically at use of vinyltrietoxisilane. Processes that occur during the curing are complex and varied.

The processes occurring in the silicate solution, governed by two reversible reactions:

$$\equiv SiOH + HO- \rightarrow \equiv SiO- + H_2O$$

$$\equiv SiOH + \equiv SiO- \rightarrow \equiv Si-O-Si\equiv + HO-$$

Colophony (rosin) derives from pine resin, tall oil and stump extractives. It is used as is or in chemically modified forms: hydrogenated, disproportionated, esterified, polymerized [12].

Composites were obtained with use of method described in the early periods. Natural dry leaf was destructed in the mixer like of coffee mill. The fraction of particles with middle sizes near 50 mcm was used in our experiments. The ingredients LG and dispersive powders of PhES, PE, and $VinSi(OEt)_3$ with amount of 3–20 wt.% were added to the powder of leaf and carefully mixed in the mixer. The obtained blends were placed to the spatial press-forms corresponding to the standards and pressed under pressures 100–150 kg/cm^2 and temperatures 80–170°C.

The samples had two types of forms – cylindrical (with sizes d=1.5 cm, h=2 cm) for investigation of water absorption and parallelepipeds (with sizes 12.0×1.50×0.70) for investigation of mechanical properties strengthening on bending and impact viscosity.

3.2 FOURIER TRANSFORM INFRARED (FTIR) SPECTROSCOPY INVESTIGATION

We have investigated the FTIR spectra for pure leaves, two-, three-, and four-components composites systems. In Figure 3.1, the peak at 1,263, 1,055 cm^{-1} assigned to the C–O stretching vibration and indicates the esters. The weak absorption band of 796, 700, 617 cm^{-1} indicate the presence chloride, bromide in our plant systems [13]. It is interesting, that in composites based on leaves one can see the absorption bands characteristic for asymmetric valence oscillation of bonds ≡Si-O–Si≡ with maximum near 1,055–1,070 cm^{-1}, which possess to siloxane bond in the fragment of cyclotetrasiloxane fragment. Besides of the sorption strips are coincided one to other. In the spectrum one can see the absorption bands 1243, 1373, 1449, 1515, 1613–1620, 1731, 2800–2950, 3363 typical for methyl groups, C-H bond absorption (–C/C–/CH_3), CH_2 cellulose – lignin, C=C aromatic, C=C alkene, (C=O etheric bond), C–H methyl, methylene, and phenyl groups, O–H alcoholic group, respectively (Figures 3.2–3.4).

This O-H stretching also indicate the phenolic compound, which have excellent antioxidant properties [14]. It must be noted that the peak at 3,375–3,400 cm^{-1} covers the entire region with a very broad peak.

3.3 STUDY OF THE MICROSTRUCTURE OF COMPOSITE MATERIALS

Leaf composite microstructure of the samples was studied by NMM-800RF/TRF type of optical microscope.

Particularly, in these figures there are presented the samples of leaves and the composites based on the leaves with different concentrations of the binder [15–18].

In Figure 3.5, optical microscopic data of the leaf is presented (magnification x100), leaf particle is quite large and all different sizes, so the scale of the figure, which is about 150–500 microns, is within the limits.

Magnification of the sample: leaf 97% + 3% LG presented in Figure 3.6 is x100. Materials show different colors for different inserts, inserts do not have a clearly defined border, so the scale is indicated in the figure. At first glance it does not have open pores, and well pressed.

For the same sample (Figure 3.7), the magnification is about x100, and the materials show different colors for different inserts. The inserts do not have pronounced borders, so the scale is indicated in the figure. At first

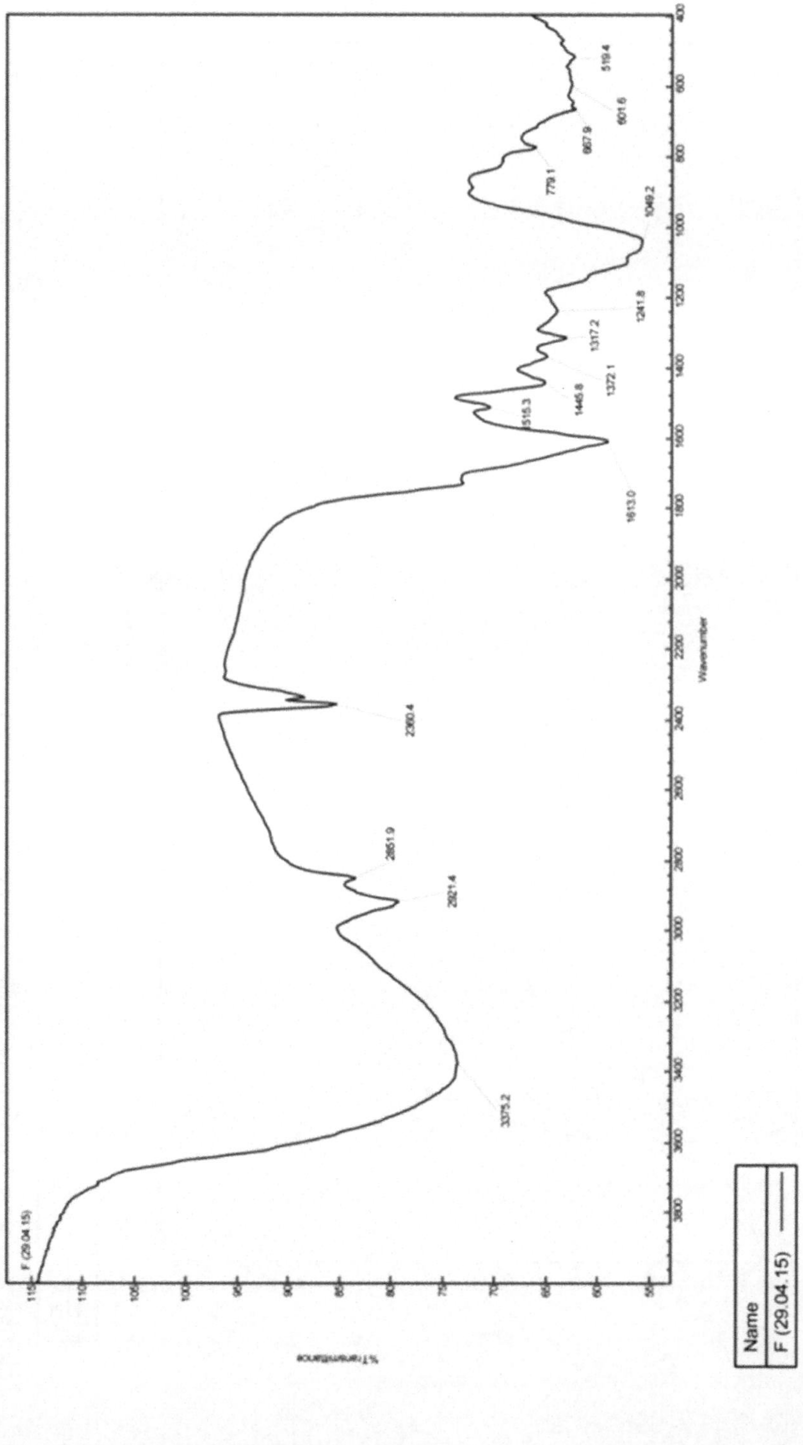

FIGURE 3.1 FTIR spectra of leaves.

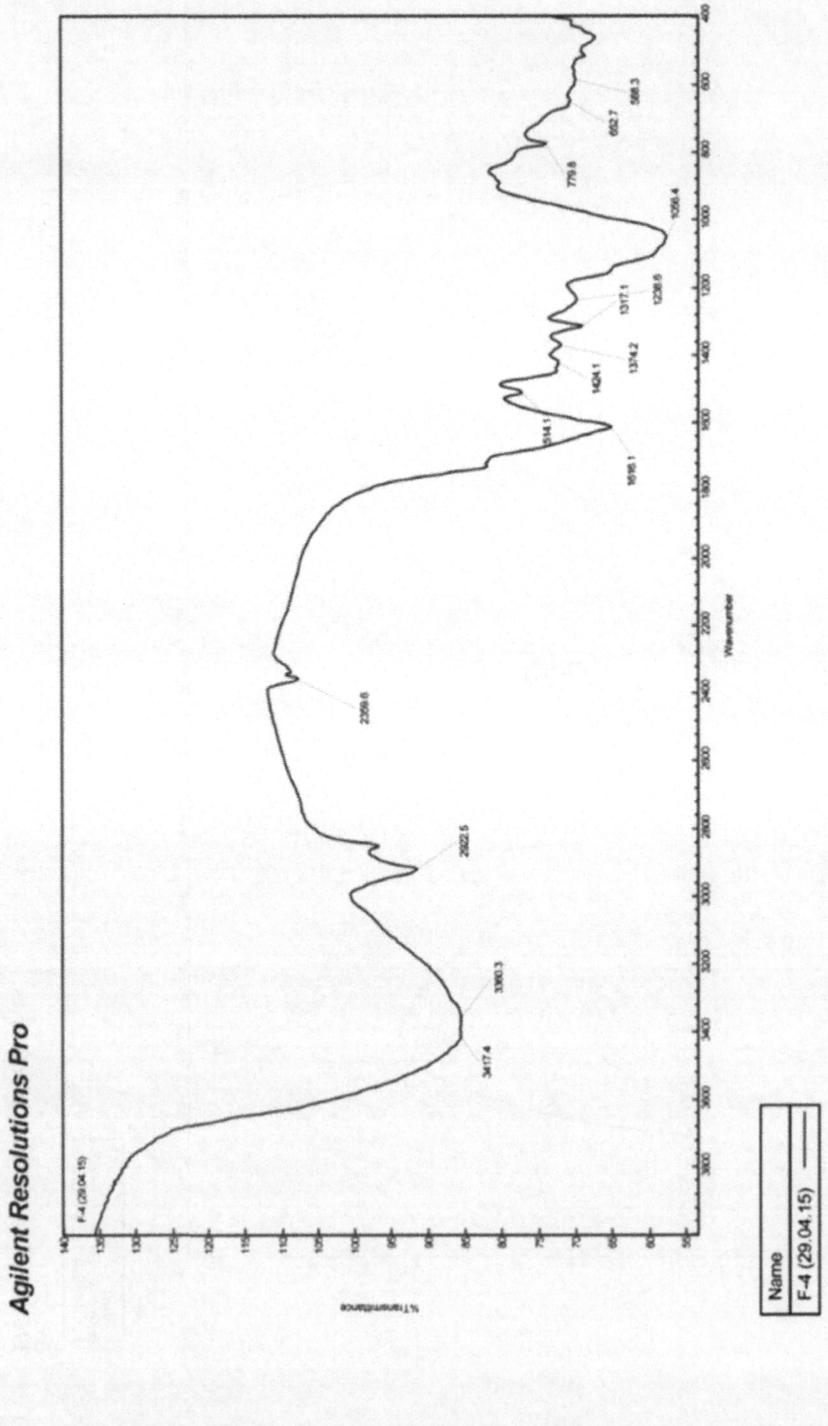

FIGURE 3.2 FTIR spectra of composites leaves (85%) + LG (15%).

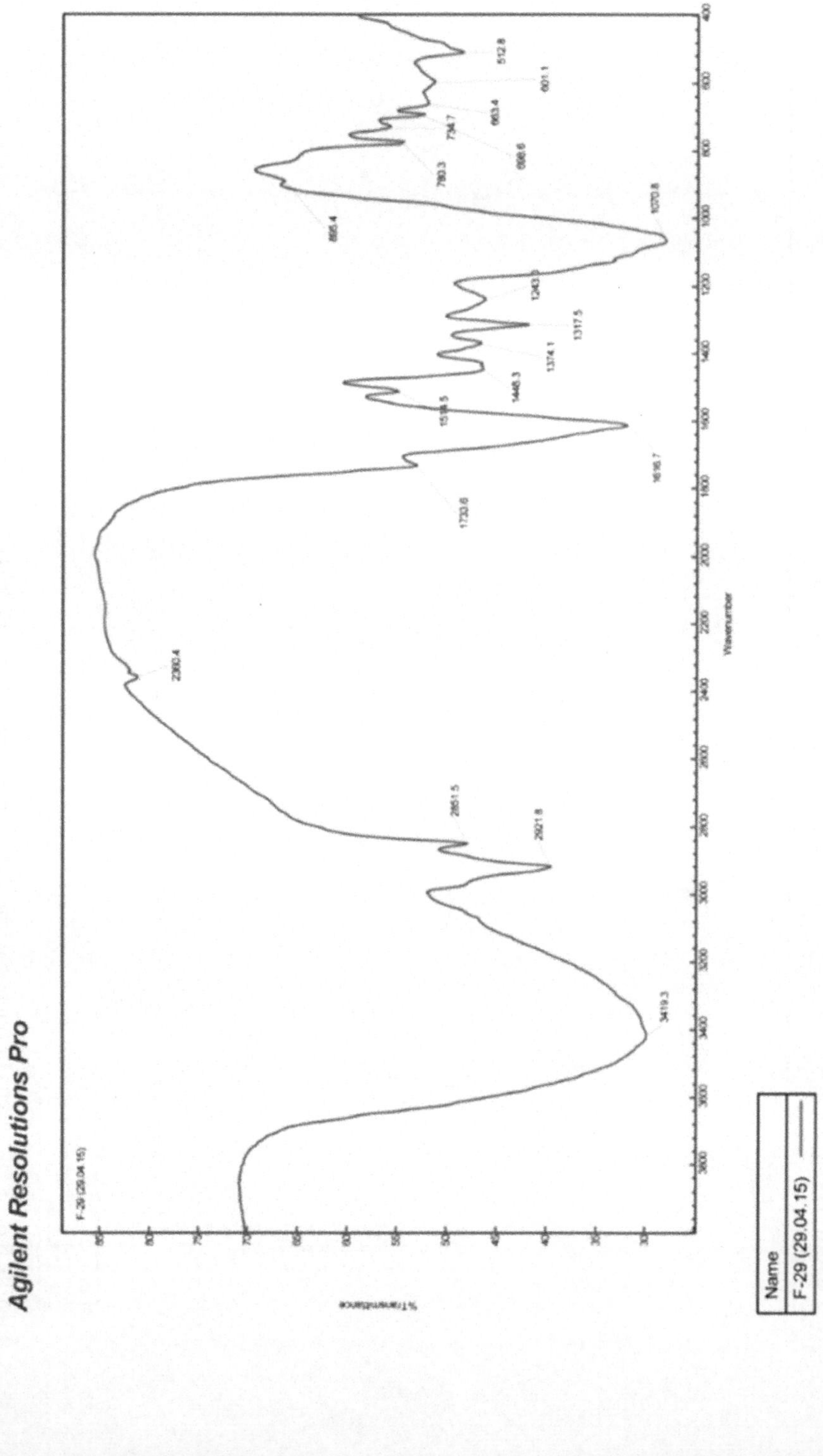

FIGURE 3.3 FTIR spectra of composite leaf (90%) + PhES-50 (5%) + PE (5%).

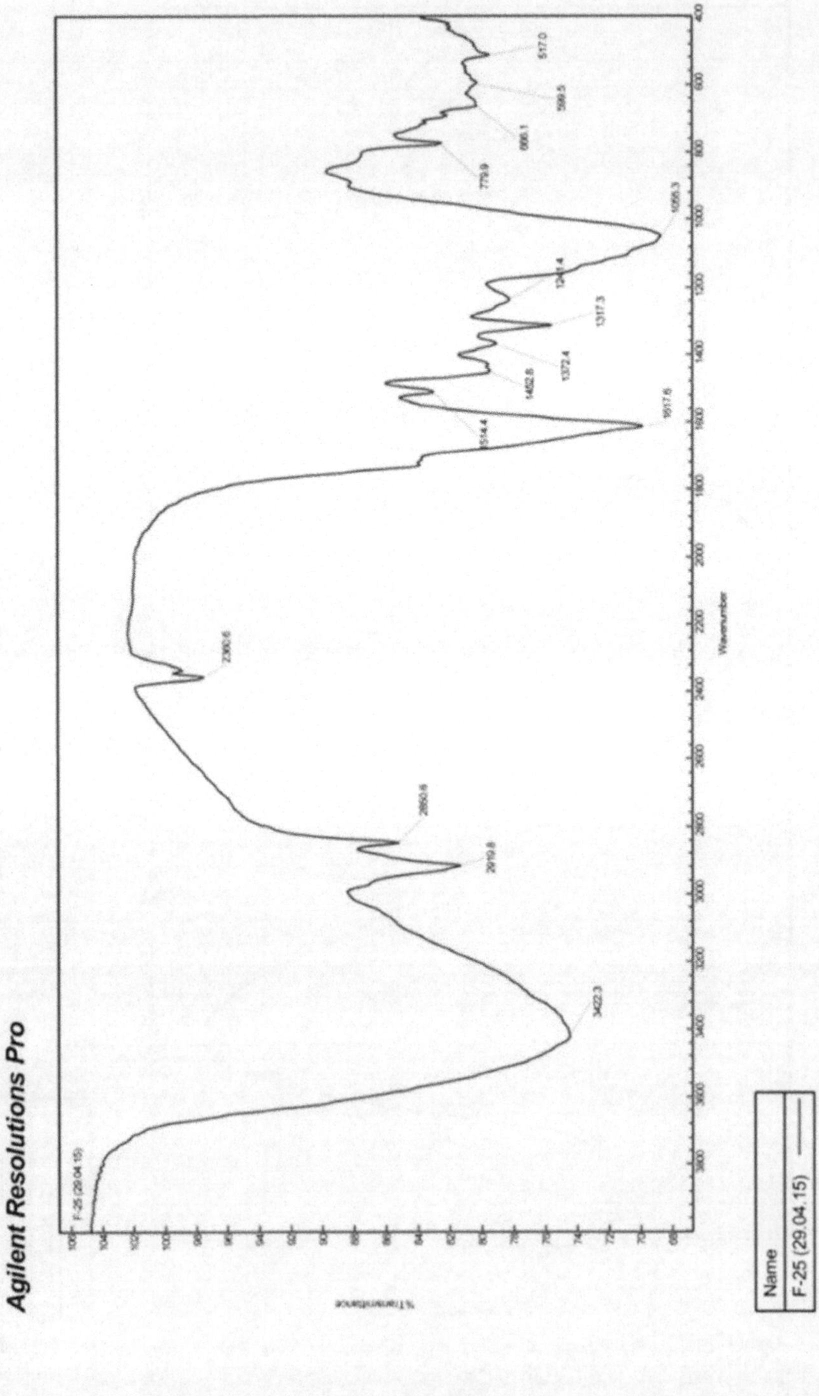

FIGURE 3.4 FTIR spectra of composite leaf (85%) + PhES-50 (5%) +LG 5% +PE (5%).

glance it does not have open pores, white inserts are about 10–20 microns in size and material of a likely long.

FIGURE 3.5 Optical microscopic data of leaves.

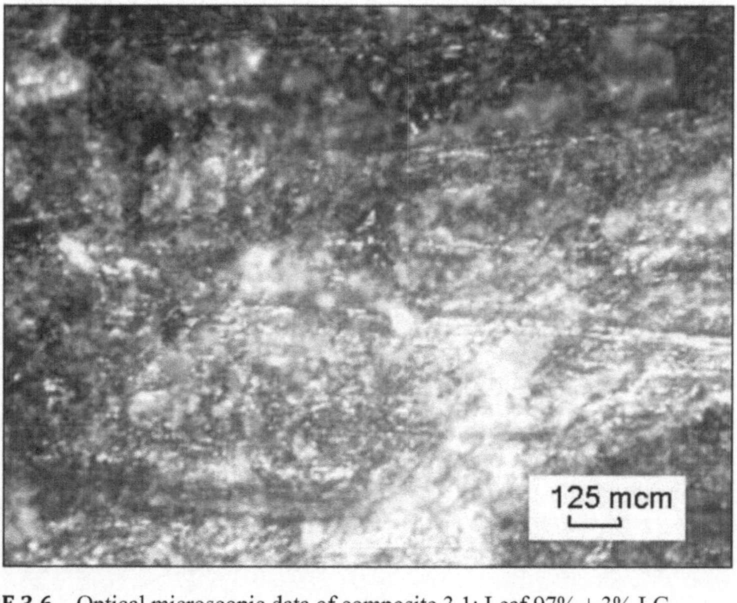

FIGURE 3.6 Optical microscopic data of composite 3.1: Leaf 97% + 3% LG.

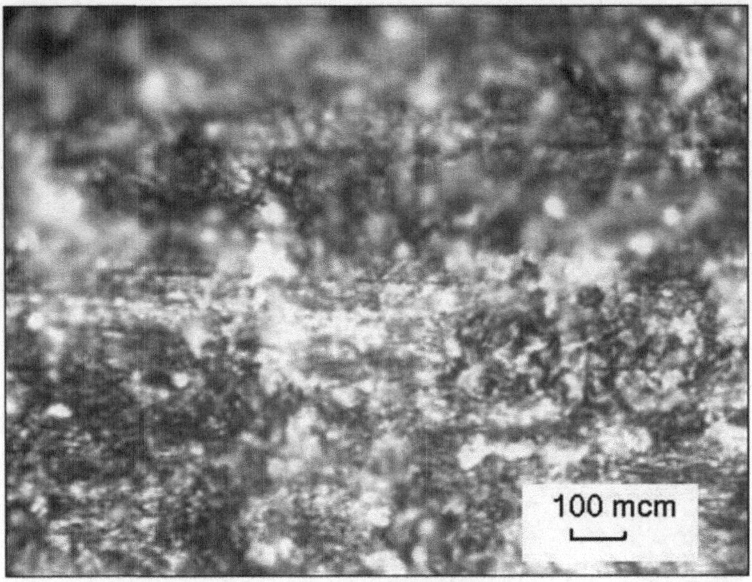

FIGURE 3.7 Optical microscopic data of composite 3.1: Leaf 97% + 3% LG.

In Figure 3.8, the magnification is about x100, for sample 10, materials show different colors for different inserts, inserts do not have a pronounced border, so the scale is indicated in the figure. Material has open pores. The insertion sizes are about 125 mcm.

Presented in Figures 3.9–3.12, microstructures of composites based on leaves allow us to have some imagination about cracks, micro-inclusions, grains in the leaves, on the basis of which it may be made an estimation about their existence a geometrical size.

As it seen from in Figure 3.9, the material is well pressed (it is not possible an establishing of the grains sizes, the binder is well homogeneous, because on the surface the inclusions are not seemed). The different color of this figure shows that this composite contained several ingredients.

Analysis of Figure 3.10 shows that the material is well pressed, and distribution of the binder is homogeneous, but contains many components.

In accordance with Figure 3.11, the composite is homogeneous and may be contained the fibers. The color corresponds to many components system.

Opinion about Figure 3.12 is coinciding to one made for Figure 3.11.

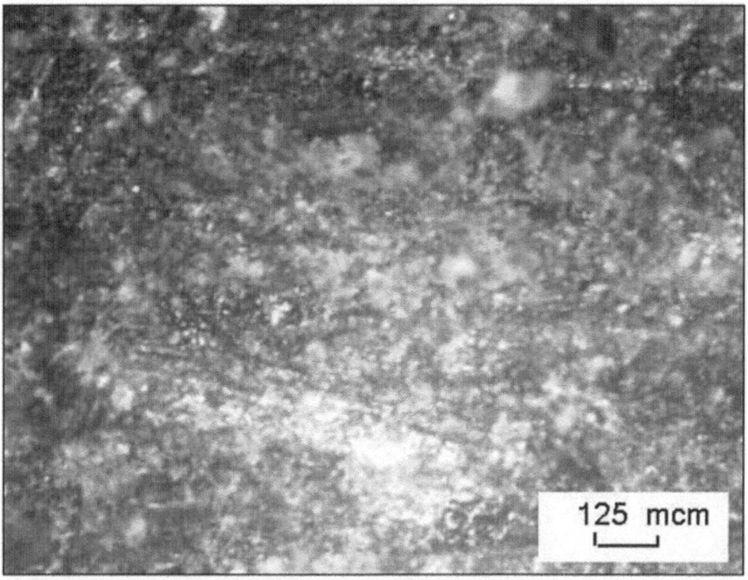

FIGURE 3.8 Optical microscopic data of composite 3.10: Leaf 95% + PhES-80 – 5%.

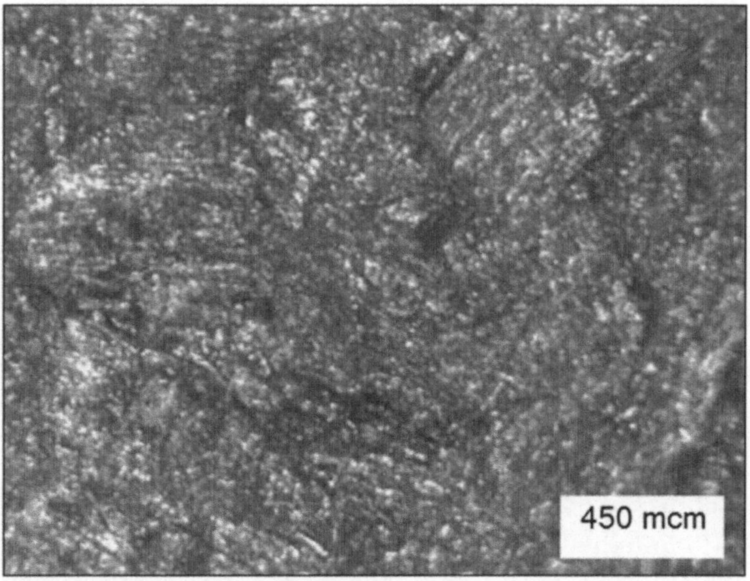

FIGURE 3.9 Optical microscopy figure of composite 3.35: Leaf (85%) + LG (5%) + PhES-80 (5%) + PE (5%).

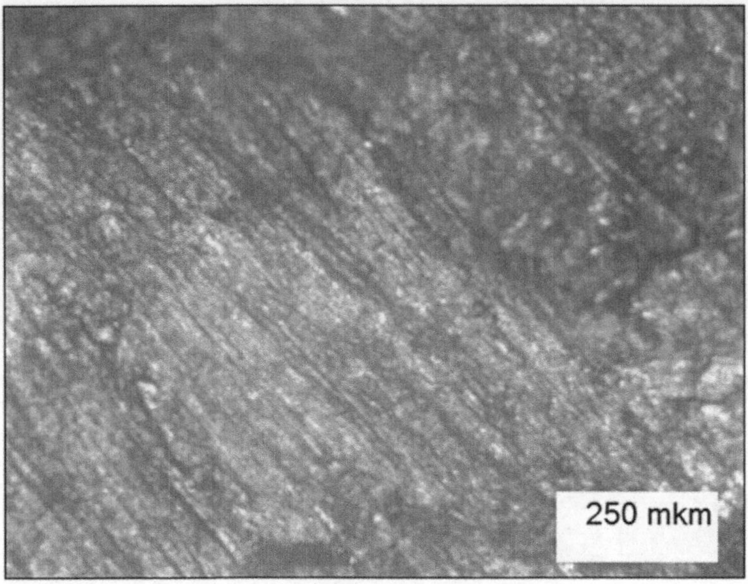

FIGURE 3.10 Optical microscopy figure of composite 3.35: Leaf (85%) + LG (5%) + PhES-80 (5%) + PE (5%) x50.

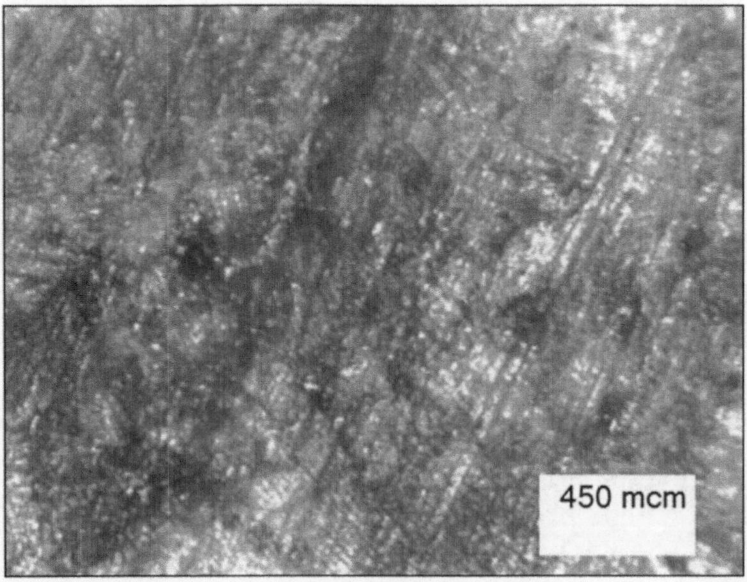

FIGURE 3.11 Sample 3.35 (leaf (90%) + PhES-80 (5%) +PE (5%) x50.

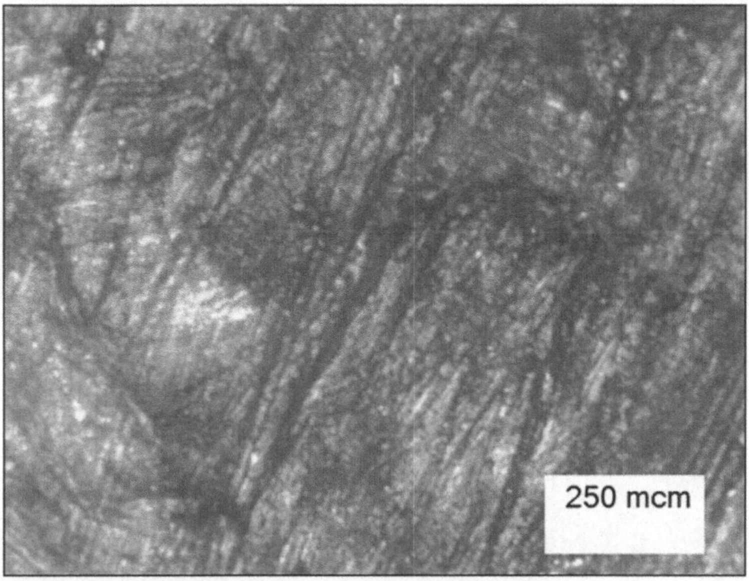

FIGURE 3.12 Sample 3.35 (leaf (90%) + PhES-80 (5%) + PE (5%).

3.4 SEM AND ENERGETIC DISPERSIVE X-RAY MICRO-SPECTRAL ANALYSIS OF THE COMPOSITES BASED ON LEAVES

Relatively stable Si-O carcass and biomass formed around it creates leaves shell. The silicon atoms are presented elsewhere but they are mainly distributed on the protrusions and roughness inner and outer of the epidermis and are connected with basis part of the leaf (Figure 3.13).

There were provided the investigations of composites with use of SEM and EDS methods at application of the microscope Nikon Eclipse LV 150. The SEM micrograms were obtained at different magnifications (x100–x1000). The surface analysis, chemical analysis and visualization of the composites were provided with use of SEM.

Analysis of investigations of the outer part of leaves epidermis (3.9) shows that they have well organized corrugated structure. The outer structure of the lemma includes the spinal inclusions, which have a linear profile. In the lemma epidermis cells placed corrugated sharpens and channels are connected by fiber like inclusions. The outer surface of the lemma includes the hills and roughness with different sizes, which often are torn in the composites and therefore are not considered (Figures 3.14–3.39 and Tables 3.1–3.5).

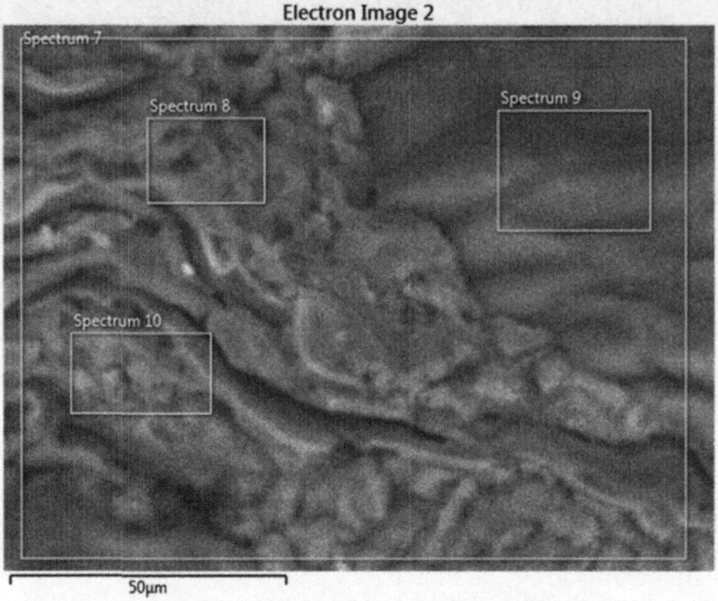

FIGURE 3.13 Scanning electronic microscopic photo for composite 3.10: Leaves 95% + PhES-80 – 5%.

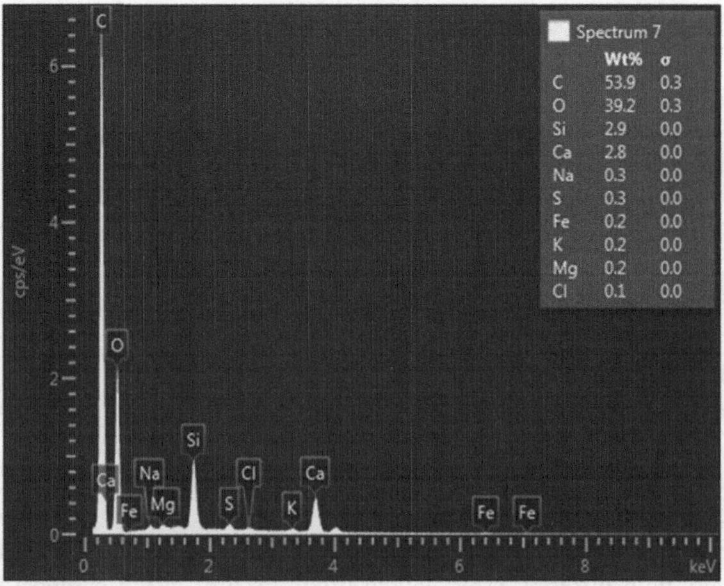

FIGURE 3.14 Energy dispersion micro-X-ray spectral analysis of composite 3.10: Leaf 95% + PhES-80 – 5% (Spectrum 7).

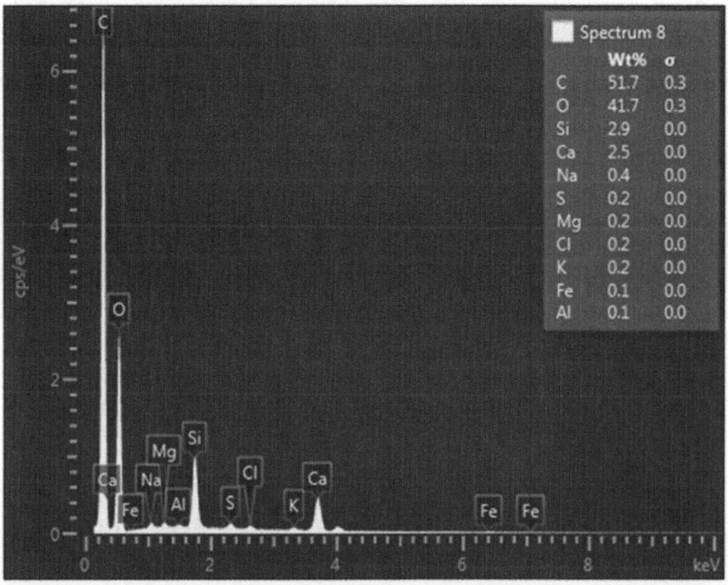

FIGURE 3.15 Energy dispersion micro-X-ray spectral analysis of composite 3.10: Leaf 95% + PhES-80 – 5% (Spectrum 8).

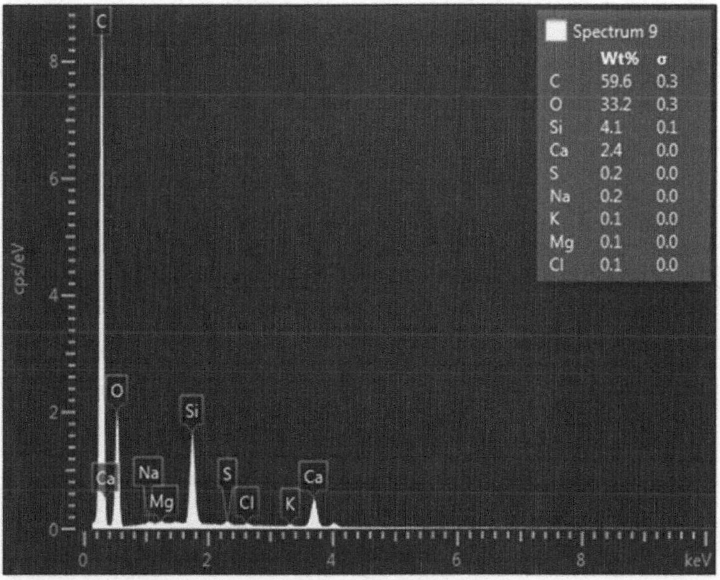

FIGURE 3.16 Energy dispersion micro-X-ray spectral analysis of composite 3.10: Leaf 95% + PhES-80 – 5% (Spectrum 9).

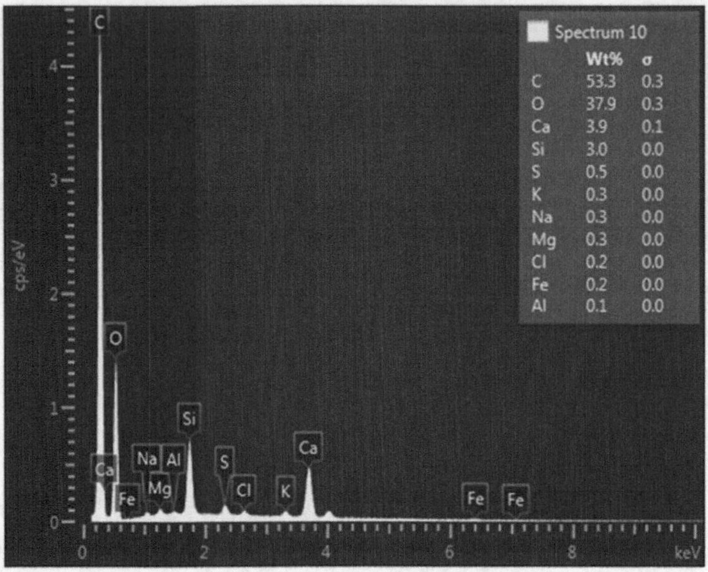

FIGURE 3.17 Energy dispersion micro-X-ray spectral analysis of composite 3.10: Leaf 95% + PhES-80 – 5% (Spectrum 10).

TABLE 3.1 Results of Energy Dispersion Micro-X-Ray Analysis for Composites: Leaves 95% + Liquid Glass 5%

Result Type	**Weight (%)**			
Spectrum Label	**Spectrum 10**	**Spectrum 7**	**Spectrum 8**	**Spectrum 9**
C	53.31	53.86	51.67	59.60
O	37.91	39.17	41.67	33.16
Na	0.28	0.29	0.35	0.17
Mg	0.27	0.18	0.20	0.12
Al	0.11	–	0.10	–
Si	3.03	2.93	2.85	4.08
S	0.47	0.29	0.23	0.20
Cl	0.24	0.14	0.17	0.12
K	0.32	0.18	0.16	0.14
Ca	3.87	2.75	2.49	2.41
Fe	0.19	0.20	0.11	–
Total	**100.00**	**100.00**	**100.00**	**100.00**

Statistics	**C**	**O**	**Na**	**Mg**	**Al**	**Si**	**S**	**Cl**	**K**	**Ca**	**Fe**
Max.	59.60	41.67	0.35	0.27	0.11	4.08	0.47	0.24	0.32	3.87	0.20
Min.	51.67	33.16	0.17	0.12	0.10	2.85	0.20	0.12	0.14	2.41	0.11
Average	54.61	37.98	0.27	0.19	–	3.22	0.30	0.17	0.20	2.88	–
Standard deviation	3.46	3.58	0.07	0.06	–	0.58	0.12	0.05	0.08	0.68	–

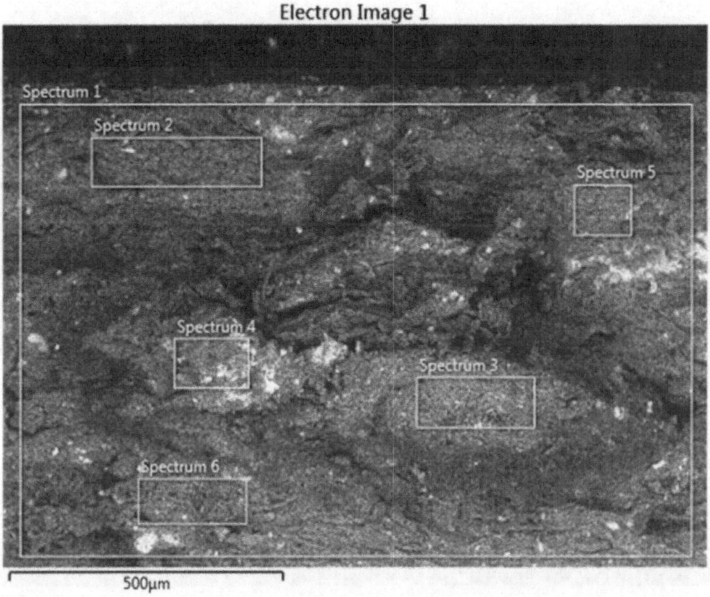

FIGURE 3.18 Scanning electronic microscopic photo for composite 3.2: Leaves 95% + Liquid glass 5%.

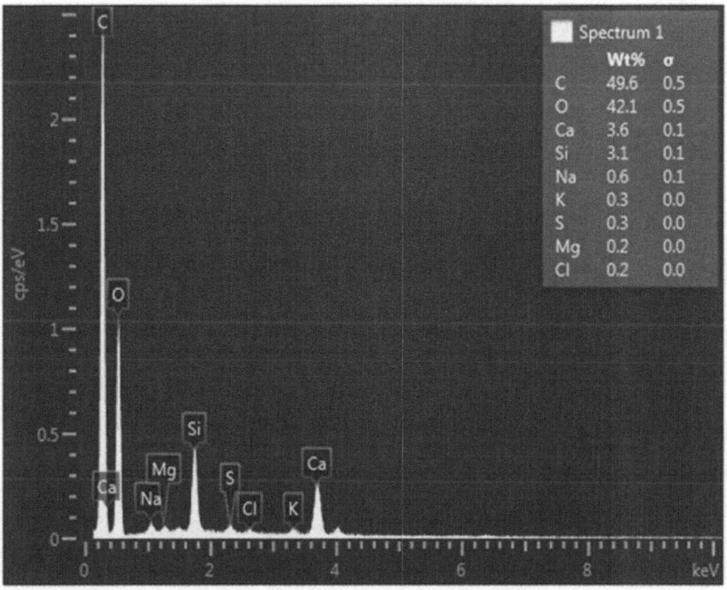

FIGURE 3.19 Energy dispersion micro-X-ray spectral analysis of composite 3.2: Leaves 95% + Liquid glass 5% (Spectrum 1).

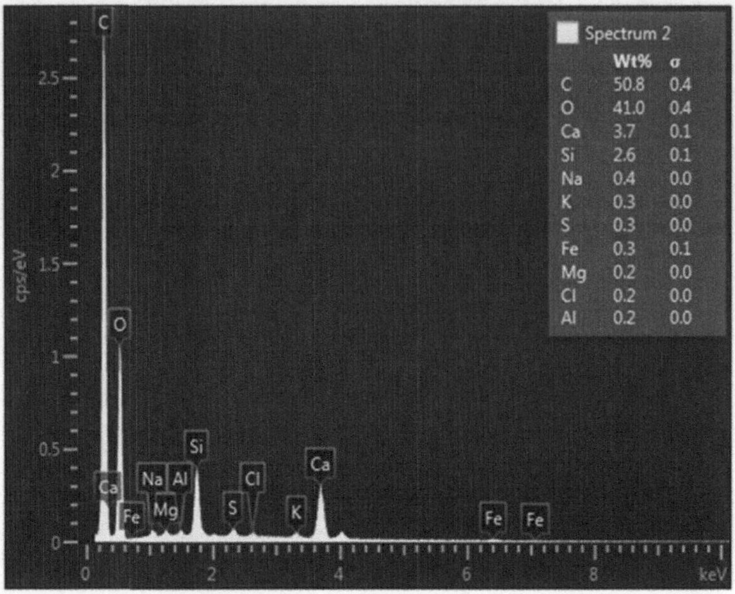

FIGURE 3.20 Energy dispersion micro-X-ray spectral analysis of composite 3.2: Leaves 95% + Liquid glass 5% (Spectrum 2).

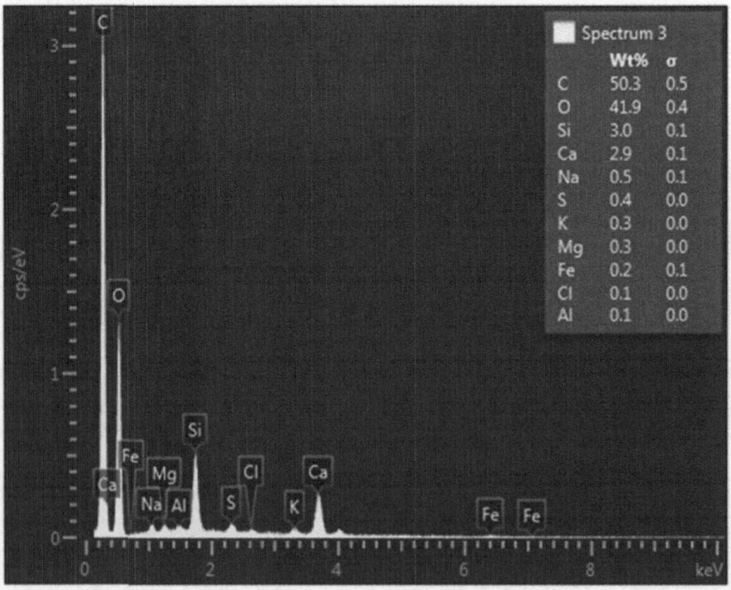

FIGURE 3.21 Energy dispersion micro-X-ray spectral analysis of composite 3.2: Leaves 95% + Liquid glass 5% (Spectrum 3).

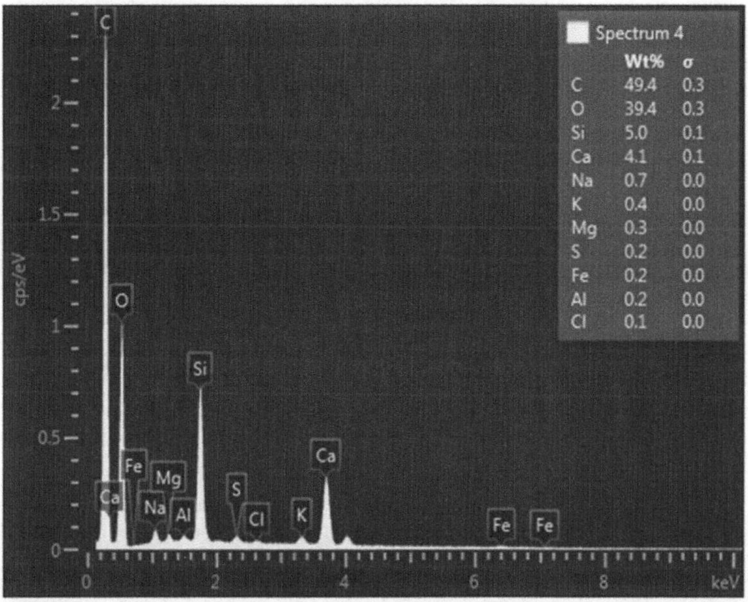

FIGURE 3.22 Energy dispersion micro-X-ray spectral analysis of composite 3.2: Leaves 95% + Liquid glass 5% (Spectrum 4).

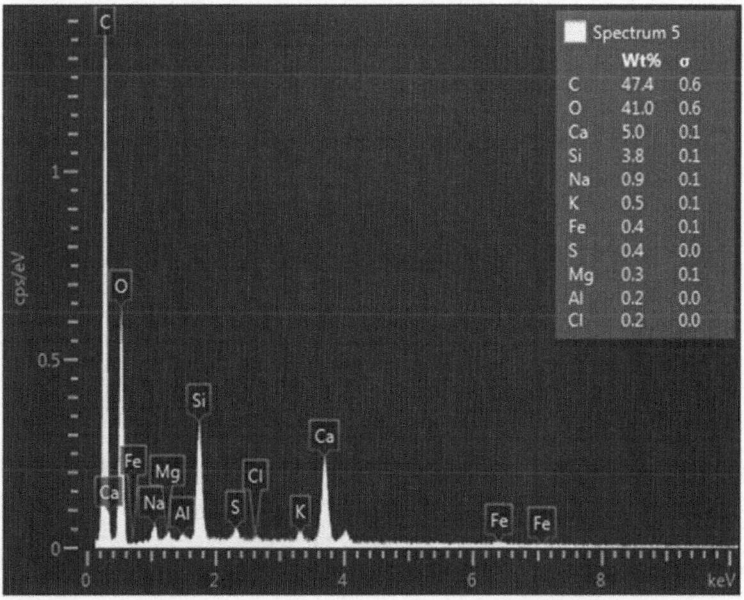

FIGURE 3.23 EDRMS for composite 3.2: Leaves 95% + Liquid glass 5% (Spectrum 5).

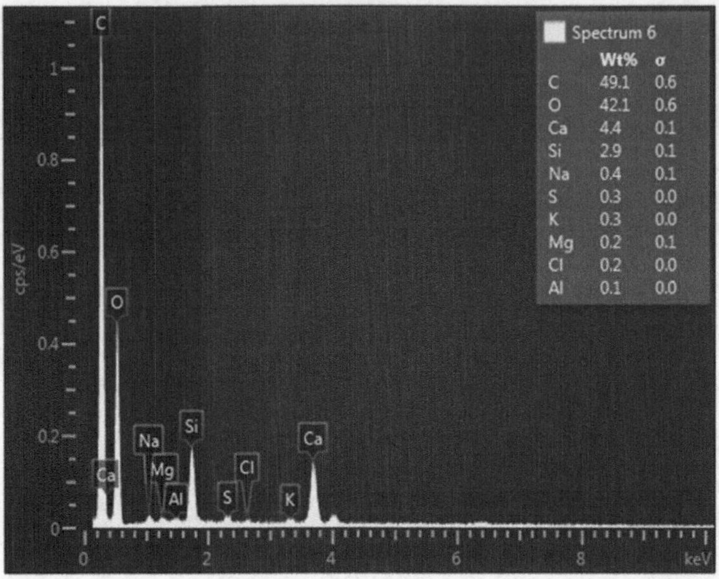

FIGURE 3.24 EDRMS for composite 3.2: Leaves 95% + Liquid glass 5% (Spectrum 6).

TABLE 3.2 Results of Energy Dispersion Micro-X-Ray Analysis for Composites: Leaves 95% + Liquid Glass 5%

Result Type	**Weight (%)**					
Spectrum Label	**Spectrum 6**	**Spectrum 1**	**Spectrum 2**	**Spectrum 3**	**Spectrum 4**	**Spectrum 5**
C	49.06	49.61	50.77	50.35	49.42	47.39
O	42.10	42.14	41.01	41.89	39.38	40.98
Na	0.44	0.56	0.41	0.48	0.69	0.93
Mg	0.21	0.25	0.25	0.28	0.27	0.26
Al	0.14	–	0.17	0.14	0.16	0.20
Si	2.88	3.13	2.62	2.96	5.01	3.79
S	0.32	0.29	0.31	0.36	0.24	0.38
Cl	0.15	0.16	0.19	0.14	0.11	0.17
K	0.30	0.30	0.31	0.32	0.39	0.53
Ca	4.39	3.56	3.68	2.87	4.12	4.96
Fe	–	–	0.29	0.21	0.21	0.40
Total	**100.00**	**100.00**	**100.00**	**100.00**	**100.00**	**100.00**

Statistics	**C**	**O**	**Na**	**Mg**	**Al**	**Si**	**S**	**Cl**	**K**	**Ca**	**Fe**
Max.	50.77	42.14	0.93	0.28	0.20	5.01	0.38	0.19	0.53	4.96	0.40
Min.	47.39	39.38	0.41	0.21	0.14	2.62	0.24	0.11	0.30	2.87	0.21
Average	49.43	41.25	0.58	0.25	–	3.40	0.32	0.15	0.36	3.93	–
Standard deviation	1.18	1.05	0.20	0.03	–	0.88	0.05	0.03	0.09	0.73	–

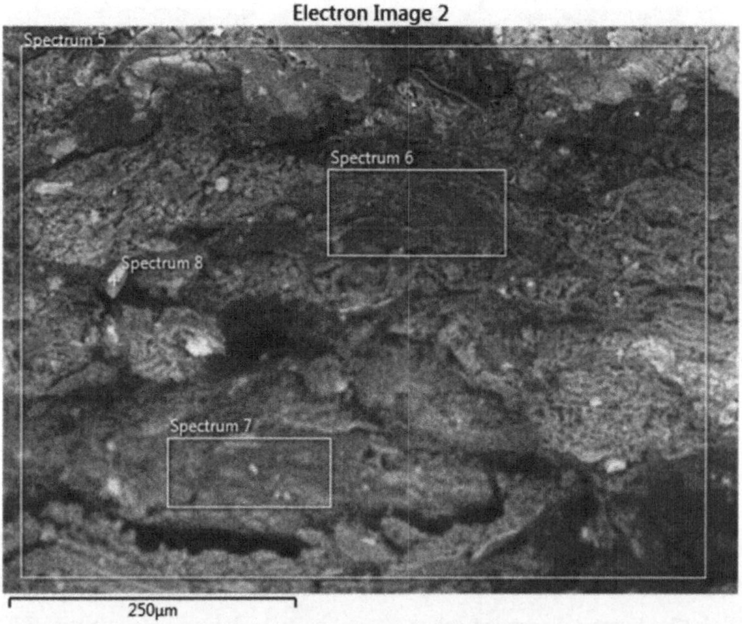

FIGURE 3.25 SEM analysis of composite 3.31: Leaf 90% + PhES-80 – 5% + LG 5%.

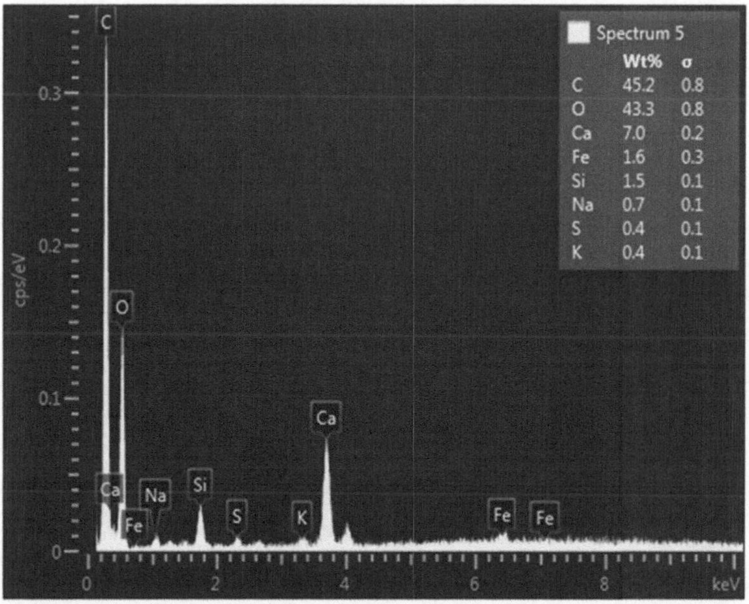

FIGURE 3.26 Energy dispersion micro-X-ray spectral analysis of composite 3.31: Leaf 90% + PhES-80 – 5% + LG 5% (Spectrum 5).

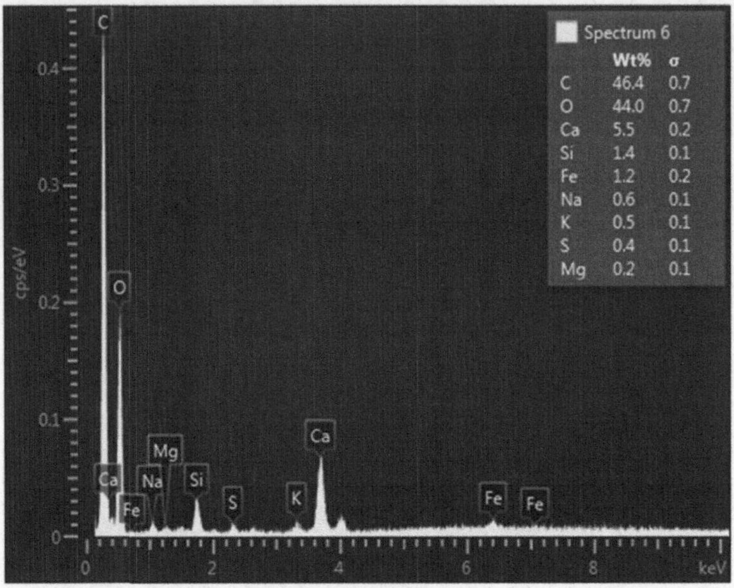

FIGURE 3.27 Energy dispersion micro-X-ray spectral analysis of composite 3.31: Leaf 90% + PhES-80 – 5% + LG 5% (Spectrum 6).

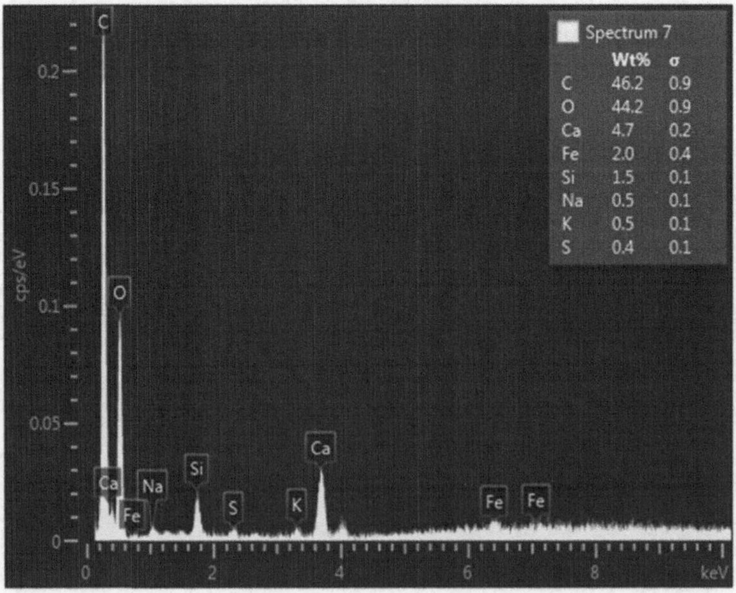

FIGURE 3.28 Energy dispersion micro-X-ray spectral analysis of composite 3.31: Leaf 90% + PhES-80 – 5% + LG 5% (Spectrum 7).

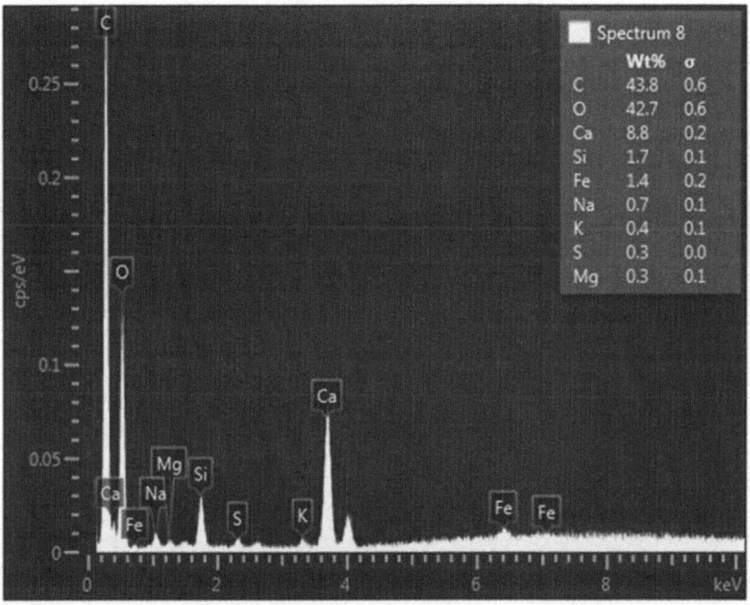

FIGURE 3.29 Energy dispersion micro-X-ray spectral analysis of composite 3.31: Leaf 90% + PhES-80 – 5% + LG 5% (Spectrum 8).

TABLE 3.3 Results of Energy Dispersion Micro-X-Ray Analysis for Composites: Leaves 90% + PhES-80 – 5% + Liquid Glass 5%

Result Type	**Weight (%)**			
Spectrum Label	**Spectrum 8**	**Spectrum 5**	**Spectrum 6**	**Spectrum 7**
C	43.76	45.21	46.37	46.24
O	42.74	43.33	44.01	44.19
Na	0.70	0.71	0.60	0.53
Mg	0.27	–	0.23	–
Si	1.66	1.46	1.35	1.47
S	0.34	0.41	0.35	0.37
K	0.39	0.36	0.46	0.46
Ca	8.77	6.95	5.46	4.73
Fe	1.38	1.57	1.17	2.01
Total	**100.00**	**100.00**	**100.00**	**100.00**

Statistics	**C**	**O**	**Na**	**Mg**	**Si**	**S**	**K**	**Ca**	**Fe**
Max.	46.37	44.19	0.71	0.27	1.66	0.41	0.46	8.77	2.01
Min.	43.76	42.74	0.53	0.23	1.35	0.34	0.36	4.73	1.17
Average	45.39	43.57	0.63	–	1.49	0.37	0.42	6.48	1.53
Standard deviation	1.21	0.67	0.08	–	0.13	0.03	0.05	1.79	0.36

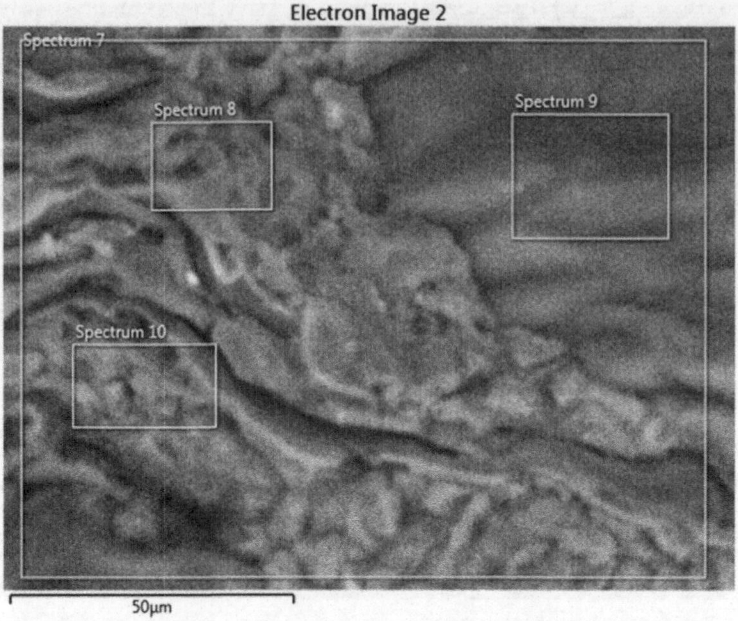

FIGURE 3.30 SEM analysis of composite 3.40: Leaf 90% + PhES-80 – 5% + PE 5%.

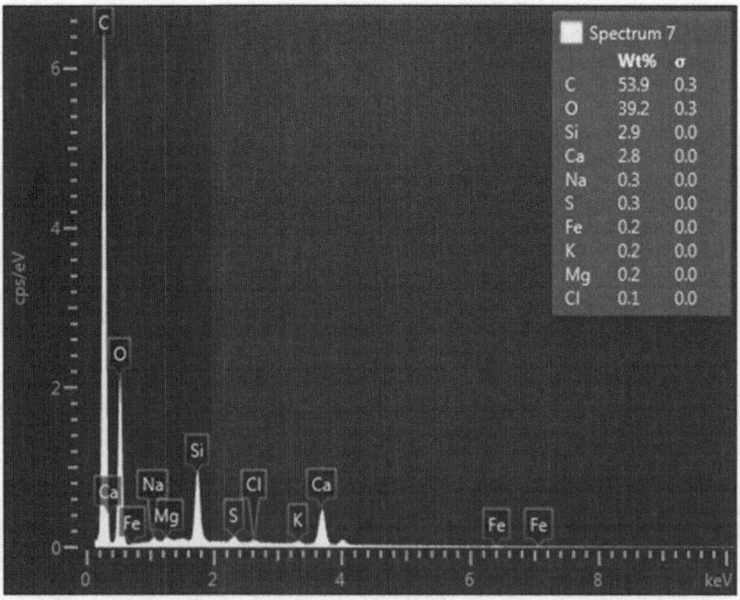

FIGURE 3.31 Energy dispersion micro-X-ray spectral analysis of composite 3.31: Leaf 90% + PhES-80 – 5% + LG 5% (Spectrum 7).

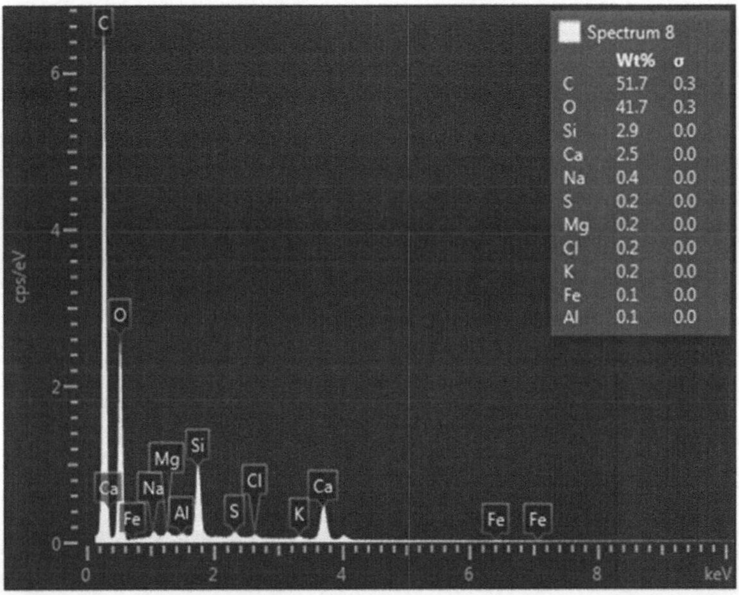

FIGURE 3.32 Energy dispersion micro-X-ray spectral analysis of composite 3.40: Leaf 90% + PhES-80 – 5% + PE 5% (Spectrum 8).

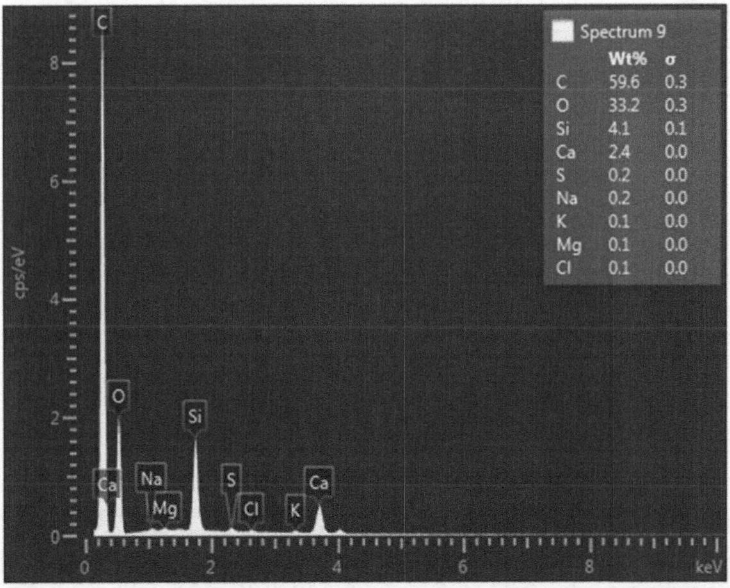

FIGURE 3.33 Energy dispersion micro-X-ray spectral analysis of composite 3.40: Leaf 90% + PhES-80 – 5% + PE 5% (Spectrum 9).

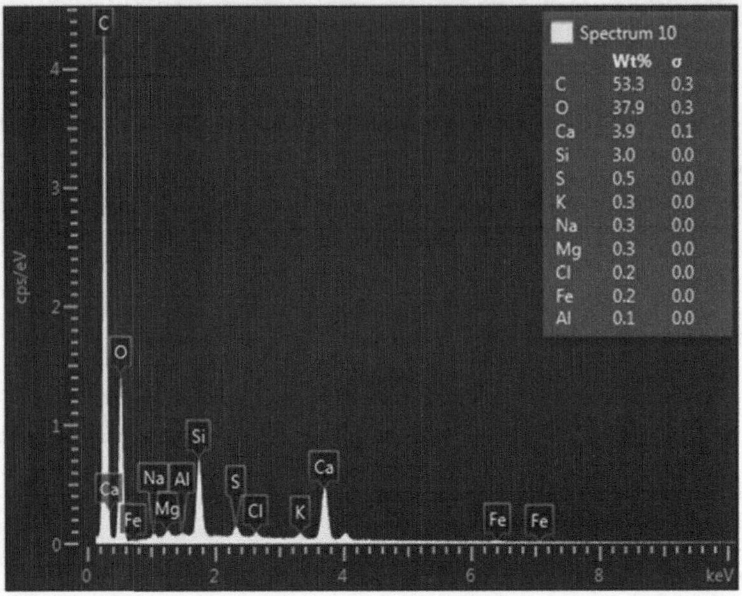

FIGURE 3.34 EDRMS for composite 3.40: Leaf 90% + PhES-80 – 5% + PE 5% (Spectrum 10).

TABLE 3.4 Results of Energy Dispersion Micro-X-Ray Analysis for Composites: Leaves 90% + PE 5%

Result Type	**Weight (%)**			
Spectrum Label	**Spectrum 10**	**Spectrum 7**	**Spectrum 8**	**Spectrum 9**
C	53.31	53.86	51.67	59.60
O	37.91	39.17	41.67	33.16
Na	0.28	0.29	0.35	0.17
Mg	0.27	0.18	0.20	0.12
Al	0.11	–	0.10	–
Si	3.03	2.93	2.85	4.08
S	0.47	0.29	0.23	0.20
Cl	0.24	0.14	0.17	0.12
K	0.32	0.18	0.16	0.14
Ca	3.87	2.75	2.49	2.41
Fe	0.19	0.20	0.11	–
Total	**100.00**	**100.00**	**100.00**	**100.00**

Statistics	**C**	**O**	**Na**	**Mg**	**Al**	**Si**	**S**	**Cl**	**K**	**Ca**	**Fe**
Max.	59.60	41.67	0.35	0.27	0.11	4.08	0.47	0.24	0.32	3.87	0.20
Min.	51.67	33.16	0.17	0.12	0.10	2.85	0.20	0.12	0.14	2.41	0.11
Average	54.61	37.98	0.27	0.19	–	3.22	0.30	0.17	0.20	2.88	–
Standard deviation	3.46	3.58	0.07	0.06	–	0.58	0.12	0.05	0.08	0.68	–

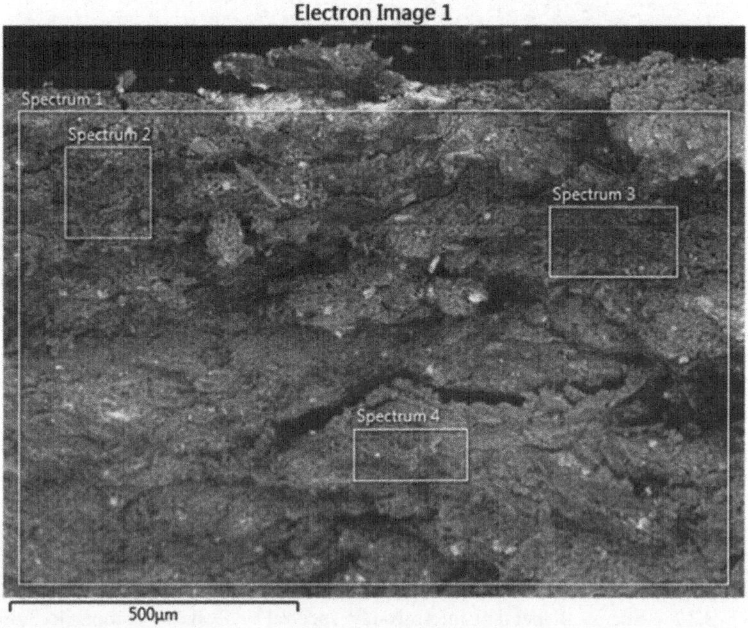

FIGURE 3.35 SEM analysis of composite 3.35: Leaf 90% + PhES-80 – 5% + PE 5% + LG 5%.

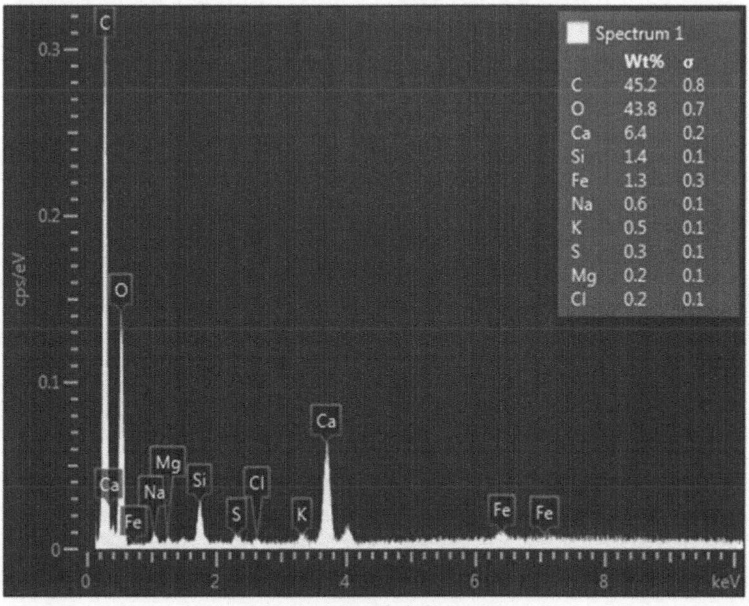

FIGURE 3.36 Energy dispersion micro-X-ray spectral analysis of composite 3.40: Leaf 90% + PhES-80 – 5% + PE 5% + LG 5% (Spectrum 1).

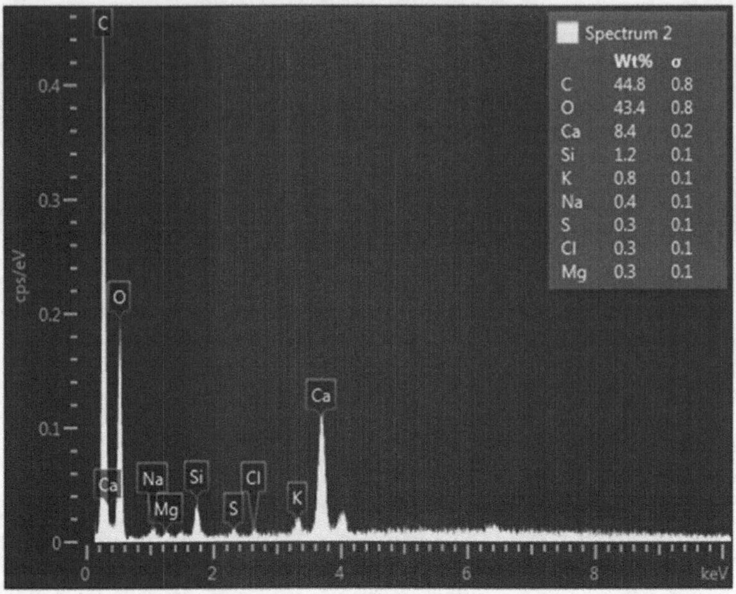

FIGURE 3.37 Energy dispersion micro-X-ray spectral analysis of composite 3.40: Leaf 90% + PhES-80 – 5% + PE 5% + LG 5% (Spectrum 2).

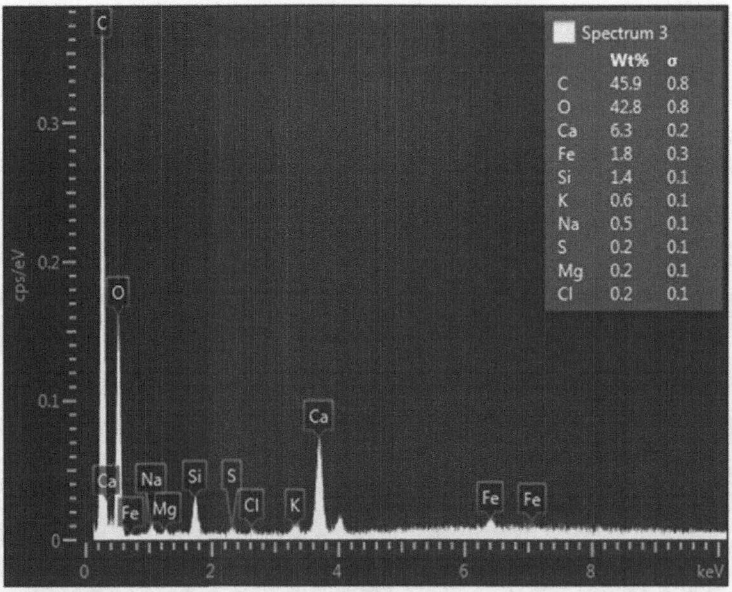

FIGURE 3.38 EDRMS for composite 3.40: Leaf 90% + PhES-80 – 5% + PE 5% + LG 5% (Spectrum 3).

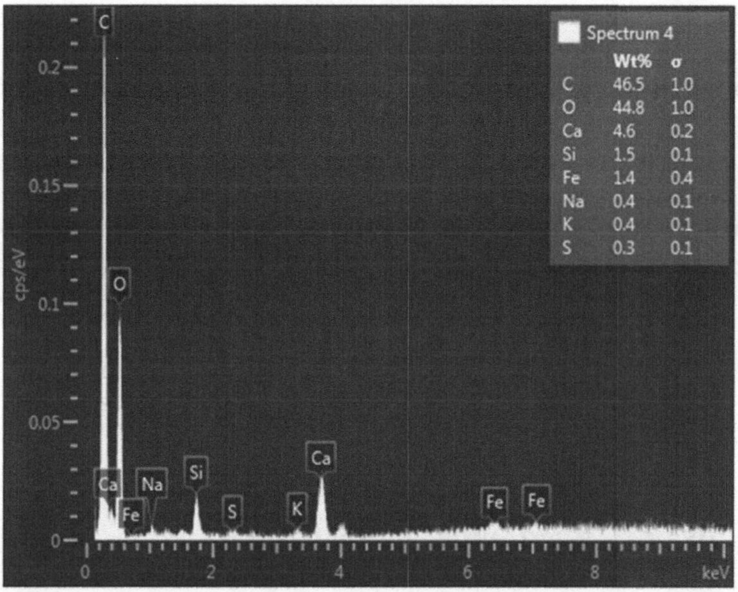

FIGURE 3.39 EDRMS for composite 3.40: Leaf 90% + PhES-80 – 5% + PE 5% + LG 5% (Spectrum 4).

TABLE 3.5 Results of Energy Dispersion Micro-X-Ray Analysis for Composites: Leaves 90% + PE 5%

Result Type	**Weight (%)**			
Spectrum Label	**Spectrum 4**	**Spectrum 1**	**Spectrum 2**	**Spectrum 3**
C	46.46	45.18	44.78	45.93
O	44.83	43.85	43.43	42.84
Na	0.44	0.62	0.45	0.50
Mg	–	0.22	0.28	0.23
Si	1.53	1.37	1.19	1.35
S	0.28	0.33	0.30	0.24
Cl	–	0.18	0.30	0.22
K	0.39	0.51	0.83	0.56
Ca	4.65	6.42	8.44	6.35
Fe	1.42	1.32	–	1.78
Total	**100.00**	**100.00**	**100.00**	**100.00**

Statistics	**C**	**O**	**Na**	**Mg**	**Si**	**S**	**Cl**	**K**	**Ca**	**Fe**
Max.	46.46	44.83	0.62	0.28	1.53	0.33	0.30	0.83	8.44	1.78
Min.	44.78	42.84	0.44	0.22	1.19	0.24	0.18	0.39	4.65	1.32
Average	45.59	43.74	0.50	–	1.36	0.29	–	0.57	6.46	–
Standard deviation	0.75	0.84	0.08	–	0.14	0.04	–	0.19	1.55	–

SEM and energy dispersion micro-X-ray spectral investigations ensure the chemical analysis of the smallest particles of the samples. By use of this methods (using low molecular detector) it may be defined the numeral content of more than 90 elements. On the figures there are X-ray emission spectra charactering for each spectrum of composite with corresponding micro-chemical analysis, which is founded on the transforming of the sample from normal to excitation state. Its principle is based on the phenomena – each element has the uncial atomic structure, which allow give the reflexes in the roentgen spectrum with show of the numeral content of the elements in the sample. They are presented in Tables 3.1–3.6.

It is established that the composites based on leaves contain the following elements: C, O, Si, Ca, Na, K, S, Fe, Mg, Cl, and Al. The content of these elements is changed little in definite region of the spectrum.

3.5 PHYSICAL-MECHANICAL PROPERTIES

Among physical-mechanical properties the strengthening on bending and impact viscosity were investigated. The experimental data are presented in Table 3.6.

The data of this table show that the composites both parameters are differed one from another numerically and depend on the type of binder and its concentration. For example, the mechanical characteristics of samples containing liquid glass, PhES-50 and PhES-80 are in extreme dependence with binder concentration – increasing of the numerical values of these parameters at relatively low concentrations transfers to decreasing of the same ones after more high concentrations. These phenomena may be described in terms of the peculiarities of the composite microstructure. In our case at definite concentrations of binders creates the possibility of appearing of the own phase of this substances, in result of which takes place the formation of the clusters of these small phases and, consequently, creation of the structural inhomogeneous and defects. Increasing of the concentration of these defects leads to weakening of the composites.

It must be noted the positive effect of PE on the composite mechanical properties. The main reason of it is the increasing of the mobility of PE macromolecules at melting, which ensures the intensive diffusion of these macromolecules to the structure of leaves and homogeneous distribution in last. Analogical phenomena take place in case of colophony and wood glue.

TABLE 3.6 Physical-Mechanical Properties of the Composites Based on Leaves and Some Organic-Inorganic Binders

SL. No.	Composite (wt.%)	Pressure	Temperature (°C)	Strength on Bending (Kg/cm^2)	Impact Viscosity (kJ/m^2)
1.	Leaves 97% LG 3%	150	130	12.6	21.2
2.	Leaves 95% + LG 5%	150	130	20.9	21.3
3.	Leaves 90% + LG 10%	150	130	21.5	20.5
4.	Leaves 85% + LG 15%	150	130	14.4	22.4
5.	Leaves 97% + PhES-50 – 3%	250	130	15.5	22.9
6.	Leaves 95% + PhES-50 – 5%	250	130	16.3	22.7
7.	Leaves 90% + PhES-50 – 10%	150	130	11.2	17.7
8.	Leaves 85% + PhES-50 – 15%	150	130	12.0	17.5
9.	Leaves 97% + PhES-80 – 3%	150	130	16.7	17.9
10.	Leaves 95% + PhES-80 – 5%	150	130	17.9	21.0
11.	Leaves 85% + PhES-80 – 10%	150	130	9.6	22.4
12.	Leaves 85% + PhES-80 – 15%	150	130	11.9	20.6
13.	Leaves 95% + PE 5%	150	130	17.9	36.4
14.	Leaves 90% + PE 10%	150	130	16.7	23.5
15.	Leaves 85% + PE 15%	150	130	17.8	24.5
16.	Leaves 80% + PE 20%	150	130	15.6	27.9
17.	Leaves 95% + Colophony 5%	150	130	17.6	31.6
18.	Leaves 90% + Colophony 10%	150	130	19.0	18.6
19.	Leaves 85% + Colophony 15%	150	130	17.0	14.7
20.	Leaves 80% + Colophony 20%	150	130	12.4	24.7
21.	Leaves 95% + Wood glue 5%	150	130	16.5	28

TABLE 3.6 *(Continued)*

SL. No.	Composite (wt.%)	Pressure	Temperature (°C)	Strength on Bending (Kg/cm²)	Impact Viscosity (kJ/m²)
22.	Leaves 90% + Wood glue 10%	150	130	16.9	16.3
23.	Leaves 85% + 15% Wood glue	150	130	15.9	18.0
24.	Leaves 80% + Wood glue 20%	150	130	18.6	25.9
25.	Leaves 94% + PhES-50 – 3% + LG 3% + Hardener	150	130	17.0	15.0
26.	Leaves 90% + PhES-50 – 5% + LG 5%	150	130	13.0	16.3
27.	Leaves 80% + PhES-50 – 10% + LG 10%	150	130	14.3	20.5
28.	Leaves 80% + PhES-50 – 15% + LG 5%	150	130	14.0	27.8
29.	Leaves 80% + PhES-50 – 5% + LG 15%	150	130	13.4	32.0
30.	Leaves 94% + PhES-80 – 3% + LG 3% + Hardener	150	130	13.0	18.0
31.	Leaves 90% + PhES-80 – 5% + LG 5%	150	130	17.0	21.0
32.	Leaves 91% + PhES-50 – 3% + LG 3% + PE 3%	150	130	15.0	23.5
33.	Leaves 85% + PhES-50 – 5% + LG 5% + PE 5%	150	130	16.0	25.0
34.	Leaves 91% + PhES-80 – 3% + LG 3% + PE 3%	150	130	13.4	21.0
35.	Leaves 85% + PhES-80 – 5% + LG 5% + PE 5%	150	130	9.0	18.0
36.	Leaves 94% + PhES-50 – 3% + PE 3%	150	130	16.7	23.0
37.	Leaves 90% + PhES-50 – 5% + PE 5%	150	130	12.0	19.0
38.	Leaves 85% + PhES-50 – 5% + PE 10%	150	130	16.0	20.5
39.	Leaves 94% + PhES-80 – 3% + PE 3%	150	130	17.0	27.0
40.	Leaves 90% + PhES-80 – 5% + PE 5%	150	130	14.0	19.6
41.	Leaves 85% + PhES-80 – 5% + PE 10%	150	130	13.6	21.0

TABLE 3.6 *(Continued)*

SL. No.	Composite (wt.%)	Pressure	Temperature (°C)	Strength on Bending (Kg/cm^2)	Impact Viscosity (kJ/m^2)
42.	Leaves 85% + PhES-80 – 5% + PVA 10%	150	130	13.9	21.3
43.	Leaves 88% + PhES-50 – 3% + LG 3% + PE 3% + $VinSi(OEt)_3$ – 3%	150	130	14.0	23.0
44.	Leaves 80% + PhES-50 – 5% + LG 5% + PE 5% + $VinSi(OEt)_3$ – 5%	150	130	13.0	21.0
45.	Leaves 88% + PhES-80 – 3% + LG 3% + PE 3% + $VinSi(OEt)_3$ – 3%	150	130	13.5	22.0
46.	Leaves 80% + PhES-80 – 5% + LG 5% + PE 5% + $VinSi(OEt)_3$ – 5%	150	130	16.5	25.1
47.	Leaves 90% + PhES-80 – 5% + PE 5%	150	130	13.5	25.1
48.	Leaves 94% + LG 3% + PE 3%	150	130	12.5	27.0
49.	Leaves 90% + LG 5% + PE 5%	150	130	13.0	26.0
50.	Leaves 94% + LG 4% + PhES-80 – 2%	150	130	14.6	33.8
51.	Leaves 94% + LG 2% + PhES-80 – 4%	150	130	10.4	36.7
52.	Leaves 85% + LG 5% + PE 10%	150	130	13.9	36.7
53.	Leaves 85% + LG 10% + PE 5%	150	130	14.4	44.86
54.	Leaves 85% + LG 7.5% + PE 7.5%	150	130	11.1	35.0
55.	Leaves 85% + PhES-50 – 7.5% + PE 7.5%	150	130	11.2	44.3
56.	Leaves 85% + PhES-50 – 10% + PE 5%	150	130	9.7	33.3

For composites contained PE (3.13–3.16) at relatively high concentrations of binder the impact viscosity decreases; same decreasing is observed in case of bending strengthening of this parameter. In the composite Leaf + LG the maximum of impact viscosity is observed for this composite containing 10% of LG., but maximum of impact viscosity of this composite is observed at 15% of LG (bending strengthening at 10% of LG). Comparison of the composites with PhES-50 and PhES-80 shows that they appear to have the same mechanical properties.

The next series of the work has an aim of increasing of composites mechanical properties with use of two and more binders. It must be noted that in separate cases our attempt has success – the mechanical properties were enhanced. In accordance with Table 3.6, the impact viscosity for some composites with two binders is higher than for others. For example this parameter for the following composites has such numerical significant: (i) 21 kJ/m^2 for leaf (90%)+ PhES-80 (5%)+ LG (5%) (composite 3.31); (ii) 19.6 kJ/m^2 for leaf (90%)+ PhES-80 (5%) +PE (5%) (composite 3.40); (iii) 23 kJ/m^2 for leaf (90%)+ PhES-50 (3%)+ PE (3%) (composite 3.36); (iv) 19 kJ/m^2 for leaf (90%)+ PhES-50 (5%)+ PE (5%). For composites contained three types of binders: (a) 23.5 kJ/m^2 for Leaf (90%) + PhES-50 (3%) + LG (3%) +PE (3%) (composite 3.32); and (b) 25 for same composite with 5% of each binder. The composites containing additively fourth binder have higher significance of named parameter – 22–27 kJ/m^2.

3.6 INVESTIGATION OF THE THERMAL CHARACTERISTICS OF THE COMPOSITES BASED ON LEAVES

The thermal stability (TS) of obtained composites based on leaves is expressed by data given in Figures 3.40–3.44. On the basis of analysis of these curves it may be make the following conclusions:

- The softening of the composites with LG monotonically increases with temperature, although with different rate – the higher is binder concentration the lower is this rate (Figure 3.40). However, this rule is damaged for composite contained the binder with 10% concentration. The behavior of this curve is essentially distinguished from analogs. This fact shows on the rather high TS of this composite. Besides of this fact indicate on the rather high compatibility of the ingredients, which leads to the monolithic properties of the composite.

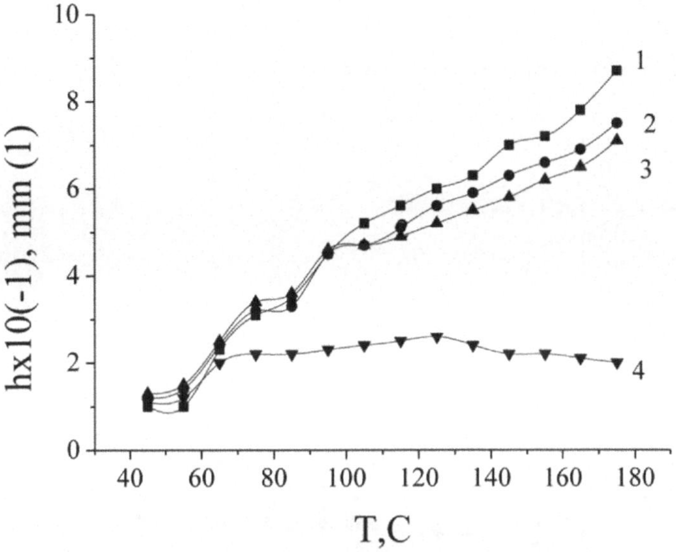

FIGURE 3.40 Dependence of softening on temperature for composites based on leaves and LG, where curves are corresponded to concentrations of LG with 3% (1); 5% (2); 15% (3); and 10% (4).

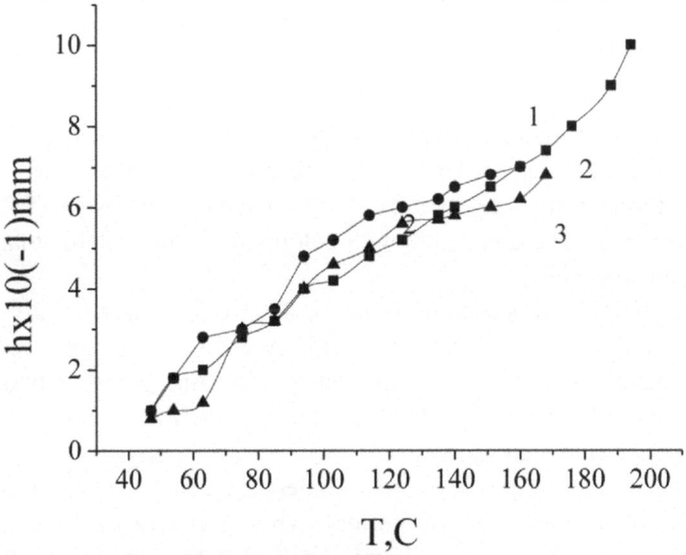

FIGURE 3.41 Dependence of softening on temperature for composites based on leaves and PhES, where curves are corresponded to concentrations of PhES-50 with 3% (1); PhES-80 (3%) (2); and PhES-80 (5%).

- TS of composites contained PhES-50 and PhES-80 (Figure 3.41) is little distinguished one from another, although the composites with PhES-80 is better to some extent.

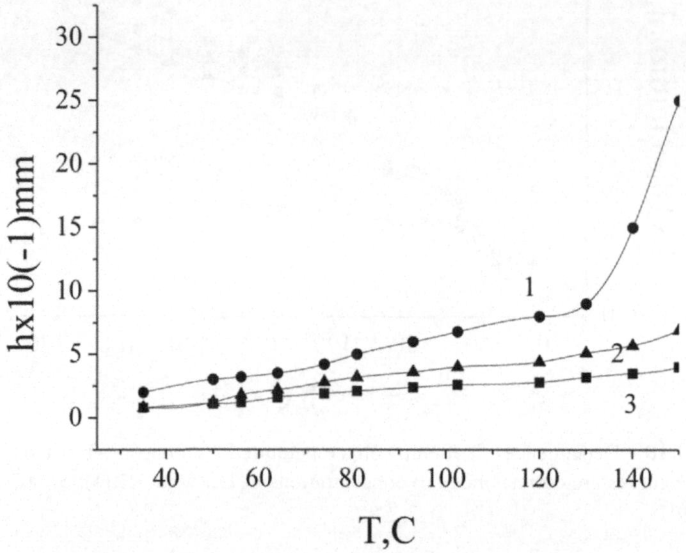

FIGURE 3.42 Dependence of softening on temperature for composites based on leaves and PhES, where curves are corresponded to concentrations of wood glue with 10% (1); Wood glue 15% (2); and Colophony 10% (3).

- TS of the composites with colophony and wood glue at relatively low temperatures is not worse in comparison with other ones, although at more high temperatures (>140C) these composites TS sharply decreases. The composite with colophony is more stable than analog with wood glue.
- The TS of composites presented in Figure 3.43 shows that in general this parameter for composites with three types of binders, in general, is better in comparison with analogs containing one or two type of binders. Here the high thermal stable properties of composites with PhES-es are well shown.
- The TS for composites with 4 binders (Figure 3.44) indicates also on the role of increasing of thermal stable properties of these materials. Curves given on these figures show that for increasing of thermal stable properties of our composites one of best ways is the choice of the number and concentrations of binders.

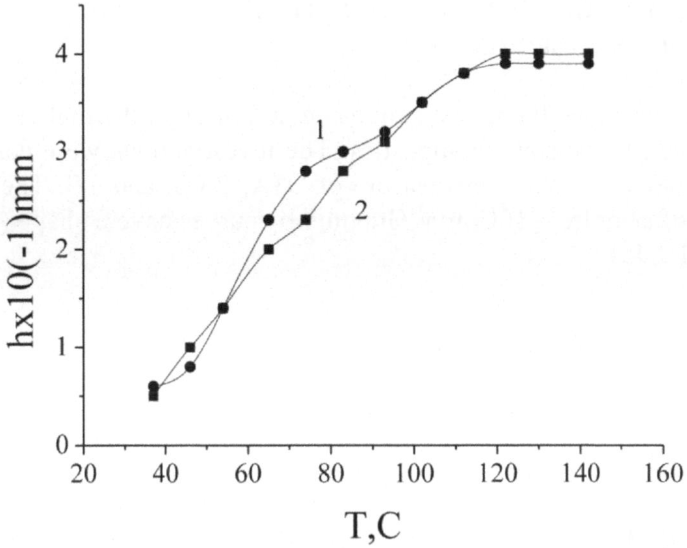

FIGURE 3.43 Dependence of softening on temperature for composites based on leaves with binary binders: 1 – Leaf (90%) + LG (5%) + PE (5%); and 2 – Leaf (90%) + PhES-80 (5%) + PE (5%).

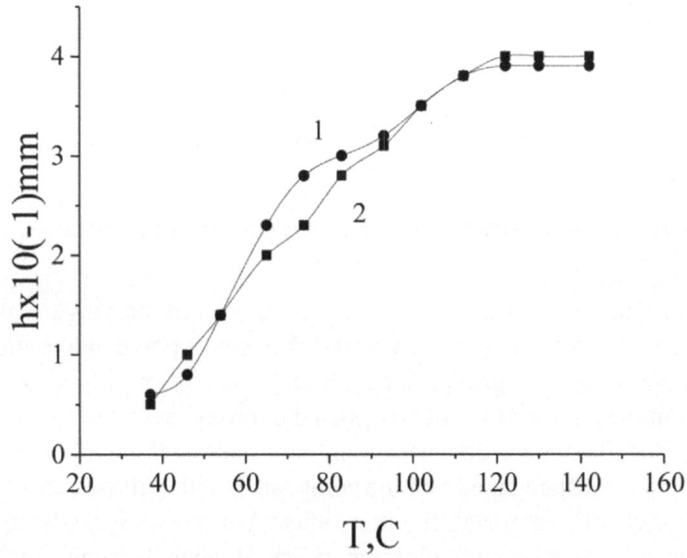

FIGURE 3.44 Dependence of softening on temperature for composites based on leaves with 4 binders: 1 – Leaf (85%) + LG (5%) + PE (5%) + PhES-50 (5%) + $VinSi(EtO)_3$; and 2 – Leaf (85%) + PhES-80 (5%) + PE (5%) + LG (5%) + $VinSi(EtO)_3$.

3.7 THERMOGRAVIMETRIC INVESTIGATIONS OF THE COMPOSITES BASED ON LEAF

There are provided the investigations on definition of thermal gravimetry of obtained by us some composites. The investigations were conducted on the standard device for measure of DTA, DTG, and TG. The rate of heating of samples – 10C/min. Gravimetric curves have S shape (Figures 3.44 and 3.45).

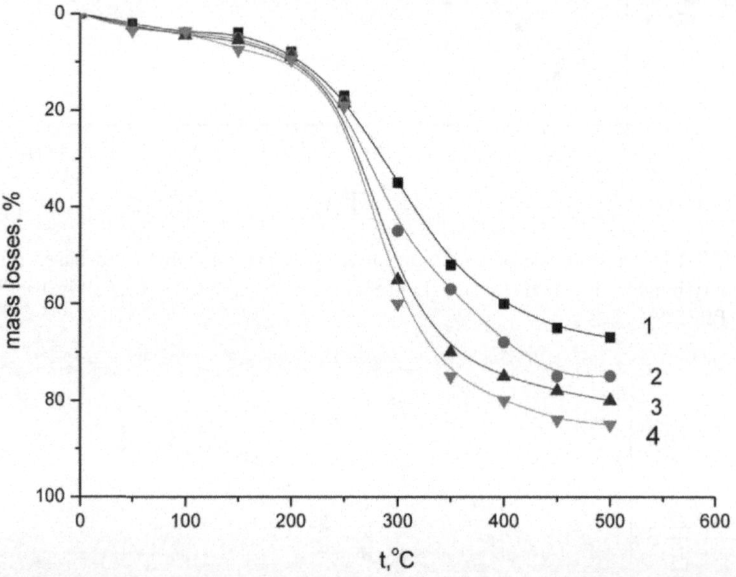

FIGURE 3.45 Thermogravimetric curves for composites on the basis of leaves.

From Figure 3.44, it is clear that the mass loss in the range 100–150°C is about 5%. In this range it is expected that homo-polycondensation reactions on the hydroxyl groups take place. Also, heteropoly condensation reactions in the composite between existed hydroxyl and ethoxy groups may be conducted. The main destruction process goes in the temperature range 250–350°C. It must be noted that the composites with PhES-es are characterized with higher thermal-stability than others, which is due to silsesquioxane fragments of silicones and inhibitor effect of phenyl groups on thermal oxidation processes in the composites.

3.8 WATER ABSORPTION OF COMPOSITES ON THE BASIS OF LEAVES

In Table 3.7, there are presented water absorption after 24 h in water of composites obtained by us. The table data allow made the following conclusions.

The composites with one binder show an extreme dependence of the value of water absorption on the binder concentration. For example, the maximum of the water absorption for composite contained 10% LG corresponds to 70%. This parameter for composites with PhES-50 reaches to 99%, but for composite containing 5% of binder. The maximal absorption of water for composite with 15% of PhES-80 is 80%. These results indicate on formation of the pores systems (with connected each to other by different channels). The study of the water absorption was conducted with definition of dependence of this parameter on the binder concentration, which allow us to find the composites with minimal water absorption. Particularly, with such result (14% absorption) is characterized the composite with 15% of LG, while for composite containing PhES-50 (20%) minimal absorption is equal to 20% and for composite with PhES-80 (20%) – 10%. Best results are obtained for composites with following binders: (i) PE (10%) – 8; (ii) PhES-80 (3%) +LG (3%) + PE (3%) – 8%; (iii) PhES-50 (5%) +PE (10%) – 8%; (iv) PhES-80 (5%)+ PE (5%) – 6%; and (v) PhES-80 (5%) + PE (10%) – 8%. The last results show that PE in all composites effectively increases their hydrophobic properties of the composites, which is, of course, connected with high water stability of PE. The water absorption for the systems with 4 binders is higher in comparison with analogs contained 1, 2, or 3 binders (compare the water absorption values for composites 3.43–3.46).

Therefore, the best results of water absorption in considered systems may be expected in case of skill selection of type and concentration of the binders.

3.9 CONCLUSIONS TO COMPOSITES ON THE BASIS OF LEAVES

On the basis of conducted technological works including preparing of initial blends in mixers with following hot pressing of these blends in spatial press forms under pressures up to 170 kg/cm^2 and temperatures 100–150°C new composite materials are obtained.

There are provided the investigations of the composites on the basis of dry high dispersive leaves and binders (liquid glass, PhES-50 and PhES-80, colophony, and wood glue) in five directions: (i) molecular and physical structural investigations with use FTIR, optical, scanned electron microscope and energy dispersion micro-X-ray analysis; (ii) Physical-mechanical (bending strengthening and impact viscosity) study; (iii) thermal stability (method Vica); (iv) water absorption; and (v) thermal gravimetry.

On the basis of spectroscopic data, it is established that some chemical groups near the leaves surface introduced to the chemical reactions with active chemical groups of binders, in result of which the mechanical properties of the composites enhance.

It is established that the dependence of mechanical properties on the binder concentration has extreme character – at definite concentrations appear the maximums of these parameters. Because of the composites include various types of binders the noted maximums appear at different concentrations of binders in the composites (Table 3.7)

The composites including PE, PhES-50 and PhES-80 separately and in combination, in general, are characterized with relatively high mechanical properties in comparison with other composites. This effect is described in terms of relatively good compatibility of these binders with basis material-dry leaves particles.

Thermal stability of obtained by us composites appear different significances. Mainly the level of this parameter for composites is consistent with data on mechanical characteristics of these materials, which once more confirms the opinion about decisive role of chemical bonds and physical interactions between different ingredients of composites at good compatibility of these elements.

Water absorption of the investigated composites is corresponded to the results considered above. It must be noted that a good hydrophobicity mainly is characteristic for composites including the relatively low amount of the binders.

3.10 COMPOSITES ON THE BASIS OF LEAVES AND *IN-SITU* OBTAINED POLYMERIC MATRIX

One of the techniques used to improve the properties of wood, which has received considerable attention in the past few decades, is the fabrication of wood-polymer composites (WPC) through in-situ formation of polymer from unsaturated monomers within wood pores (vessels, tracheids, capillaries, and ray cells) [19]. The resultant polymer can both strength the mechanical properties of wood and defer or stop wood matrix from being attacked by

TABLE 3.7 Water Absorption of Composites Obtained on the Basis of Leaves and Various Organic-Inorganic Binder

SL. No.	Composite (mass %)	Mass (g)	Volume (cm^3)	Density (g/cm^3)	Mass Composites After 3 h Exposition in Water (g)	Mass of Composites After 24 h Exposition in Water (g)	Water Absorption After 24 h Exposition in Water (%)
1.	Leaves 97% + LG 3%	4.80	3.53	1.36	4.95	5.78	20.4
2.	Leaves 95% + LG 5%	4.89	3.53	1.38	5.17	6.70	36.8
3.	Leaves 90% + LG 10%	4.66	3.18	1.47	5.66	7.83	68.1
4.	Leaves 85% + LG 15%	4.75	3.53	1.34	4.94	5.40	13.9
5.	Leaves 97% + PhES-50 – 3%	4.76	3.53	1.35	5.43	6.88	44.7
6.	Leaves 95% + PhES-50 – 5%	4.53	3.36	1.35	5.07	9.03	99.1
7.	Leaves 90% + PhES-50 – 10%	4.76	3.53	1.35	5.46	8.13	71.0
8.	Leaves 85% + PhES-50 – 15%	4.53	3.35	1.35	4.97	6.92	52.8
9.	Leaves 97% + PhES-80 – 3%	4.80	3.53	1.36	5.02	6.94	44.5
10.	Leaves 95% + PhES-80 – 5%	4.62	3.36	1.38	5.80	7.70	66.6
11.	Leaves 90% + PhES-80 – 10%	4.59	3.18	1.44	4.77	5.66	23.4
12.	Leaves 85% + PhES-80 – 15%	4.50	3.18	1.41	5.54	8.09	79.8
13.	Leaves 95% + PE 5%	3.68	2.64	1.39	3.76	4.54	23.4
14.	Leaves 90% + PE 10%	4.43	3.36	1.32	4.54	4.77	7.95
15.	Leaves 85% + PE 15%	4.62	3.53	1.31	4.91	5.89	27.5
16.	Leaves 80% + PE 20%	3.38	2.64	1.28	3.42	3.79	10.1
17.	Leaves 95% + Colophony 5%	3.76	2.64	1.42	3.81	4.08	8.5
18.	Leaves 90% + Colophony 10%	4.73	3.36	1.41	4.86	5.24	10.8
19.	Leaves 85% + Colophony 15%	4.56	3.18	1.43	4.98	6.45	41.5
20.	Leaves 80% + Colophony 20%	3.40	2.46	1.38	3.50	3.72	9.4
21.	Leaves 95% + Wood glue 5%	3.59	2.82	1.27	3.64	3.77	5.0
22.	Leaves 90% + Wood glue 10%	4.75	3.18	1.49	5.93	8.43	77.5

TABLE 3.7 *(Continued)*

SL. No.	Composite (mass %)	Mass (g)	Volume (cm^3)	Density (g/cm^3)	Mass Composites After 3 h Exposition in Water (g)	Mass of Composites After 24 h Exposition in Water (g)	Water Absorption After 24 h Exposition in Water (%)
23.	Leaves 85% + Wood glue 15%	4.81	3.36	1.44	5.95	6.92	43.7
24.	Leaves 80% + Wood glue 20%	3.73	2.64	1.41	3.96	5.52	47.9
25.	Leaves 94% + PhES-50 – 3% + LG 3% + Hardener	6.42	4.59	1.40	6.61	7.25	12.9
26.	Leaves 90% + PhES-50 – 5% + LG 5% + Hardener	4.45	3.18	1.40	4.57	5.10	14.6
27.	Leaves 80% + PhES-50 – 10% + LG 10%	2.80	1.94	1.44	2.90	3.33	18.9
28.	Leaves 80% + PhES-50 – 15% + LG 5%	3.60	2.64	1.36	3.75	4.14	15.0
29.	Leaves 80% + PhES-50 – 5% + LG 15%	3.68	2.64	1.39	3.79	4.05	10.1
30.	Leaves 94% + PhES-80 – 3% + LG 3% + Hardener	4.61	3.36	1.37	4.71	5.13	11.3
31.	Leaves 90% + PhES-80 – 5% + LG 5%	4.83	3.53	1.37	4.94	5.47	13.3
32.	Leaves 91% + PhES-50 – 3% + LG 3% + PE 3%	4.61	3.36	1.37	4.76	5.15	11.7
33.	Leaves 85% + PhES-50 – 5% + LG 5% + PE 5%	4.64	3.36	1.38	4.85	5.45	17.5
34.	Leaves 91% + PhES-80 – 3% + LG 3% + PE 3%	4.49	3.36	1.34	4.58	4.83	7.6
35.	Leaves 85% + PhES-80 – 5% + LG 5% + PE 5%	4.17	3.00	1.39	4.36	4.96	18.9
36.	Leaves 94% + PhES-50 – 3% + PE 3%	4.28	3.18	1.35	4.37	4.79	11.9
37.	Leaves 90% + PhES-50 – 5% + PE 5%	3.85	2.83	1.36	3.99	4.56	18.4
38.	Leaves 85% + PhES-50 – 5% + PE 10%	4.77	3.70	1.28	4.86	5.14	7.8
39.	Leaves 94% + PhES-80 – 3% + PE 3%	4.77	3.52	1.35	4.95	5.46	14.5

TABLE 3.7 *(Continued)*

SL. No.	Composite (mass %)	Mass (g)	Volume (cm^3)	Density (g/cm^3)	Mass Composites After 3 h Exposition in Water (g)	Mass of Composites After 24 h Exposition in Water (g)	Water Absorption After 24 h Exposition in Water (%)
40.	Leaves 90% + PhES-80 – 5% + PE 5%	4.47	3.52	1.26	4.55	4.79	5.6
41.	Leaves 85% + PhES-80 – 5% + PE 10%	4.52	3.52	1.28	4.64	4.88	7.9
42.	Leaves 85% + PhES-50 – 5% + PVA 10%	4.43	3.34	1.32	4.56	4.85	9.5
43.	Leaves 88% + PhES-50 – 3% + LG 3% + PE 3% + $VinSi(OEt)_3$ –3%	4.75	3.52	1.34	5.00	5.66	19.2
44.	Leaves 80% + PhES-50 – 5% + LG 5% + PE 5% + $VinSi(OEt)_3$ – 5%	4.64	3.52	1.31	4.92	5.47	12.5
45.	Leaves 88% + PhES-80 – 3% + LG 3% + PE 3% + $VinSi(OEt)_3$ – 3%	3.99	3.17	1.26	4.14	4.49	23.1
46.	Leaves 80% + PhES-80 – 5% + LG 5% + PE 5% + $VinSi(OEt)_3$ – 5%	4.63	3.34	1.38	5.09	5.70	23.1
47.	Leaves 90% + LG 5% + PE 5%	4.24	3.17	1.33	4.50	4.85	14.4
48.	Leaves 94% + LG 3% + PE 3%	4.68	3.34	1.40	5.21	5.89	25.9
49.	Leaves 85% + LG 5% + PE 5% + Colophony 5%	4.77	3.52	1.35	5.11	5.98	25.4
50.	Leaves 94% + LG 4% + PhES-80 – 2%	3.52	3.17	1.11	3.73	4.05	15.1
51.	Leaves 94% + LG 2% + PhES-80 – 4%	3.53	2.82	1.25	3.67	3.77	6.8
52.	Leaves 85% + LG 5% + PE 10%	3.82	2.82	1.35	3.90	4.02	5.2
53.	Leaves 85% + LG 10% + PE 5%	3.43	2.64	1.29	3.51	3.68	11.1
54.	Leaves 85% + LG 7.5% + PE 7.5%	3.43	2.64	1.29	3.52	3.63	5.8
55.	Leaves 85% + PhES-50 – 7.5% + PE 7.5%	3.20	2.64	1.21	3.25	3.35	4.7
56.	Leaves 85% + PhES-50 – 10% + PE 5%	3.59	2.64	1.36	3.52	3.76	4.7

water or microorganisms [20–23]. Such multifunctional treatment can help avoid the potential damage of leached preservatives from chemically treated wood on environment [24], strengthen the wood [25], avoid dimensional deformation of compressed wood [26] and color change of heat-treated wood [27]. Thus, such treatment became an environmentally friendly wood modification process. On the consideration of importance of such wood modification and development of research on wood-polymer composite, this part presents an overall review on preparation, performance, and application of WPC.

One of the goals of our project is the obtaining of new constructive materials based on the artificial natural dry leaves at the use of unsaturated monomers with various structure.

For obtaining of new composite materials on the basis of leaves, there was carried out the investigation for obtaining of di and three component systems of leaves with various binders, such as acrylates, diacrylates, allyl compounds and methylhydrosiloxanes, as well as in the presence of various additives. It is shown that on the base of the leaves, and unsaturated type monomers in the presence of peroxides simultaneously during pressing and heating of the mixture it is possible polymerization monomers mixture.

The commonly used peroxides for the polymerization of monomers in wood include dicumyl peroxide (DCP), t-butyl hydroperoxide (TBPB), methyl ethyl ketone peroxide (MEKP), lauroyl peroxide (LPO), isopropyl hydroperoxide (ISO-HPO), cyclohexanone peroxide (CHPO), hydrogen peroxide (HPO), and benzoyl peroxide (BPO). Each of the radicals generated from these peroxides has a different reactivity. The phenyl radical is more reactive than the benzyl radical, and the allyl radical is unreactive. Thus, BPO is one of the most commonly used peroxides initiator. Usually, the amount of peroxide added ranges from 0.2–3% by weight of monomer [28]. Excess peroxide may adversely affect the mechanical properties of the composite because molecular chain scission of the polymer and cellulose occurs when peroxide is too abundant [29, 30]. The composites were obtained after following manipulations:

- Weighing of the ingredients dry powder like leaves and unsaturated monomers by means of the analytical balancer;
- Dry mixing of the ingredients;
- Loading of the blends of the standard press-forms;
- Heating of the blends in the press-forms at definite temperatures (130°C), pressures (150 mmHg) during established time interval (10–15 min); The samples sizes were the same as above.

3.10.1 FTIR INVESTIGATION OF COMPOSITES ON THE BASIS OF LEAVES

The composition of the materials has been studied via Fourier transform infrared spectroscopy. In Figures 3.46–3.50, IR spectra of leave based composites are presented. In Figure 3.46, the peak at 1,263, 1,055 cm^{-1} assigned to the C-O stretching vibration and indicates the esters. The weak absorption band of 796, 700, 617 cm^{-1} indicate the presence chloride, bromide in our plant systems [31]. In the spectrum one can see the absorption bands 1243, 1373, 1449, 1515, 1613–1620, 1731, 2800–2950, 3363 typical for methyl groups, C-H bond absorption (–C/C–/CH_3), CH_2 cellulose – lignin, C=C aromatic, C=C alkene, (C=O etheric bond), C-H methyl, methylene, and phenyl groups, O-H alcoholic group, respectively. It must be noted that the peak at 3,375–3,400 cm^{-1} covers the entire region with a very broad peak. In the IR spectra of composites one can observe additionally absorption bands for used binders (Figures 3.46–3.52).

3.10.2 SEM AND ENERGY DISPERSION MICRO-X-RAY SPECTRAL ANALYSIS OF COMPOSITE

For composite materials made on the basis of powdery leaves and different hardeners scanning electron microscopic (SEM) investigations are provided. Besides of energy dispersive X-ray examinations have been carried out in parallel with the micro-spectral (EDS) examinations. The analysis of leaf epidermis in Merostachys species revealed that epidermal elements on both surfaces vary in shape and size. Leaf epidermis cells on the adaxial surface (Figures 3.53, 3.64, and 3.72) are both long and short, alternately arranged. The surface is smooth, with no ornamentation. Small silica bodies point away from the outer periclinal wall of long cells, arranged in a row. Short cells could not be visualized easily and are represented by suberose cells. The longest axis of these elements is arranged perpendicularly to the long cells and is usually associated with prickles and micro-hairs located at their base.

Bulliform cells are clearly present in the intercostal zones, where three rows of cells were observed [1, 2]. The abaxial surface has a wax layer, either grain-shaped (rough surfaced) or with no defined ornamentation. Like on the adaxial surface, both for the costal and intercostals zones, the epidermis on the abaxial surface consists of long cells, with markedly sinuous anticlinal walls. There are numerous papillae in the long cells. Papillae are either conical or globose structures, occurring in one row, or arranged in two different ways, where epidermal elements may present one or two rows of papillae (Figures 3.54–3.58 and Table 3.8).

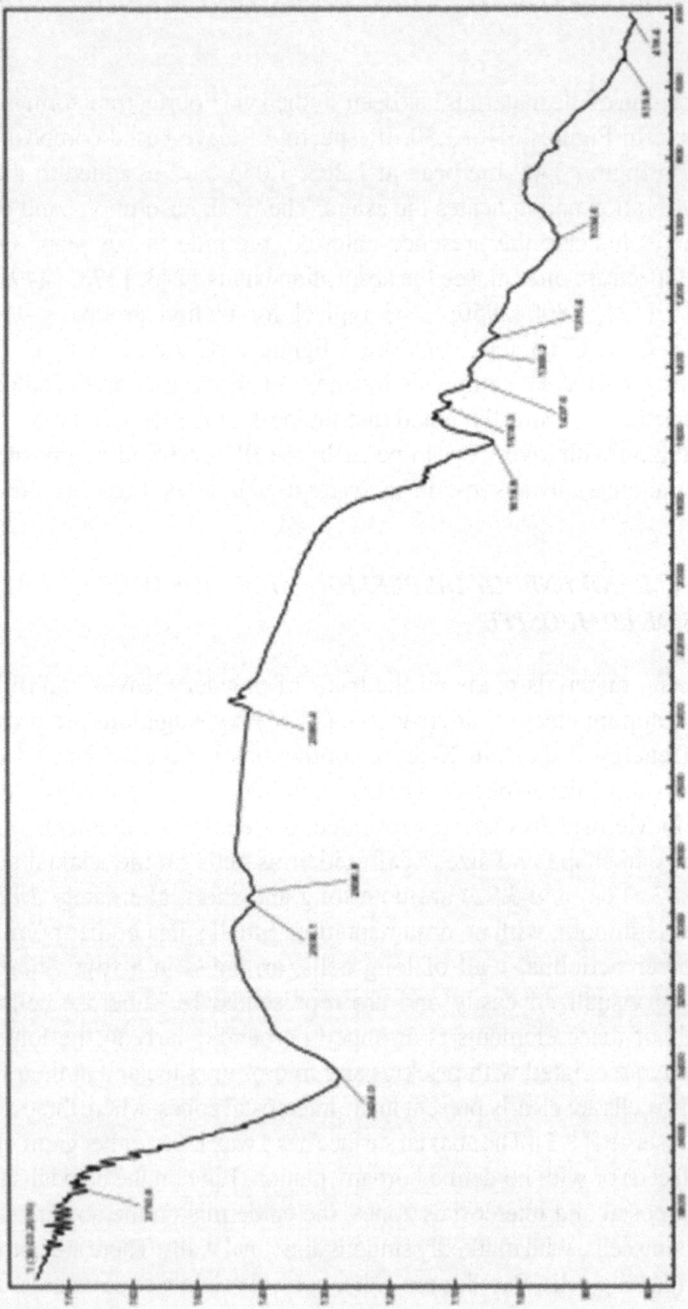

FIGURE 3.46 FTIR spectra of composite leaves (95%) + ethylene methylethyl acrylate (5%) + dicumol peroxide (0.1% mass).

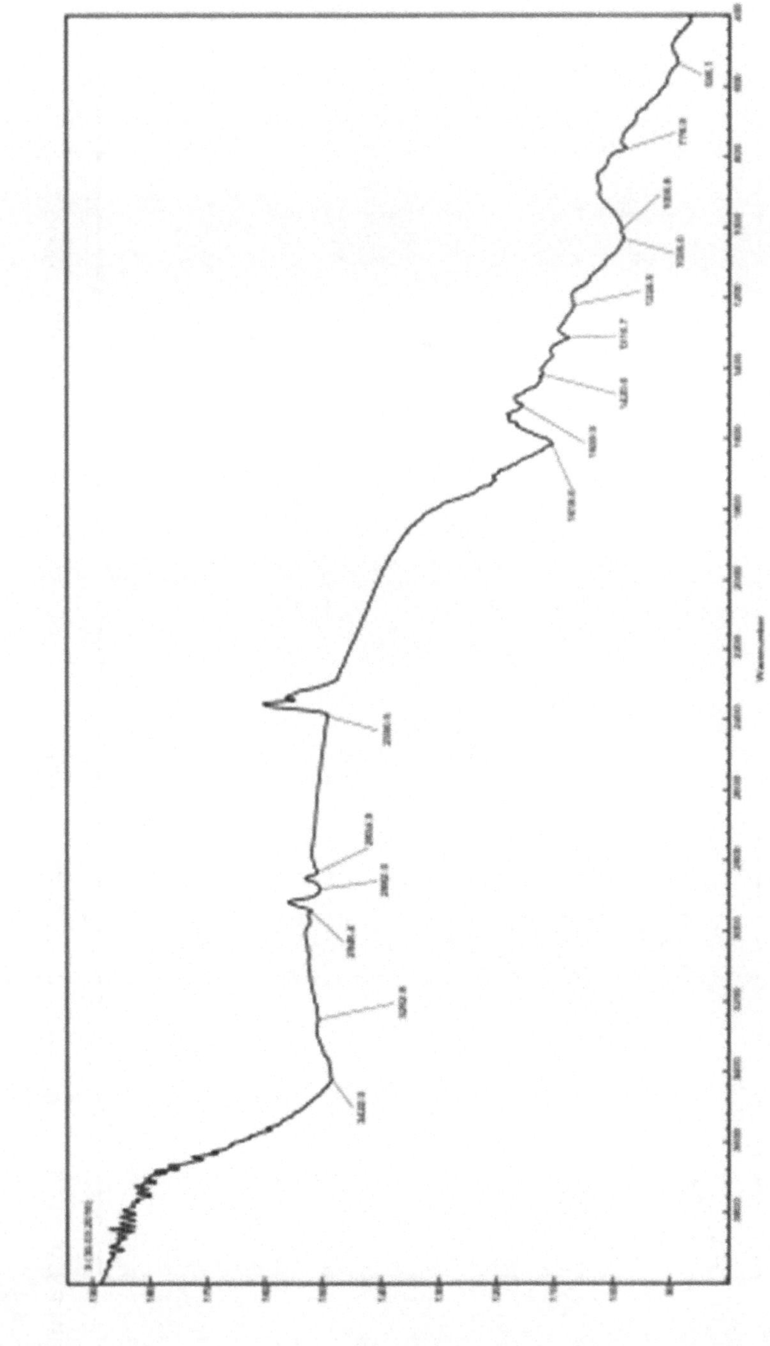

FIGURE 3.47 FTIR spectra of composite leaves (95%) + N-methyl-N-vinylacetamide (5%) + dicumol peroxide (0.1% mass).

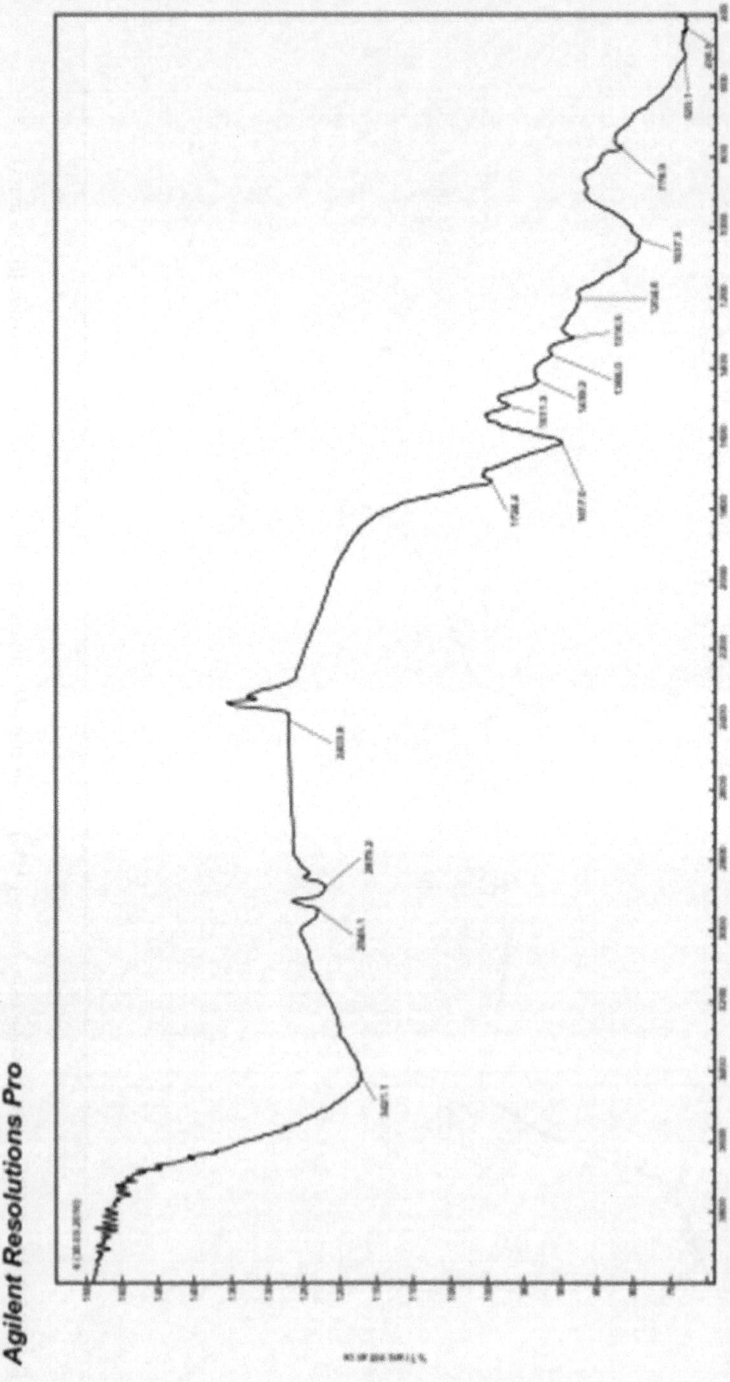

FIGURE 3.48 FTIR spectra of composite leaves (95%) + N-vinyl caprolactam (5%) + dicumol peroxide (0.1% mass).

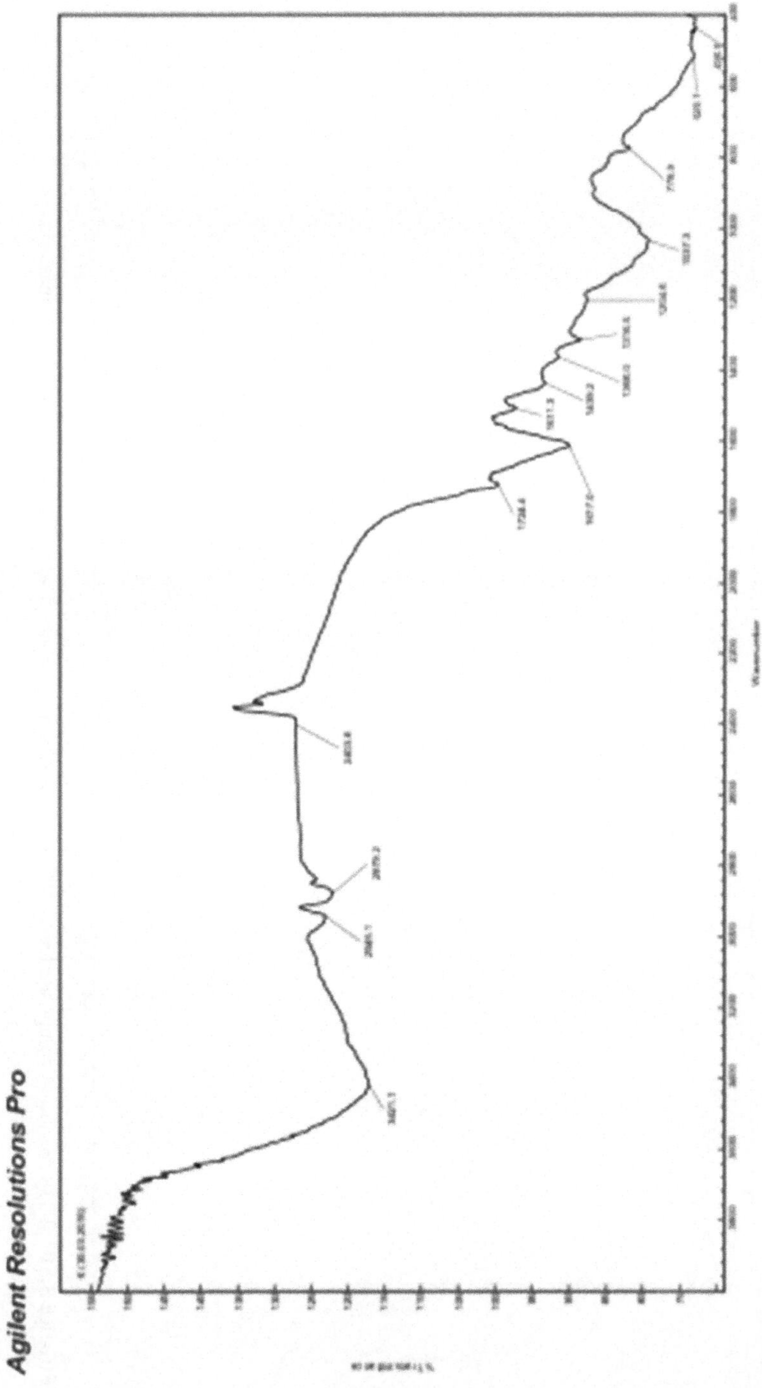

FIGURE 3.49 FTIR spectra of composite leaves (95%) + di(ethylene glycol) 2-ethylhexyl ether acrylate (5%) + dicumol peroxide (0.1% mass).

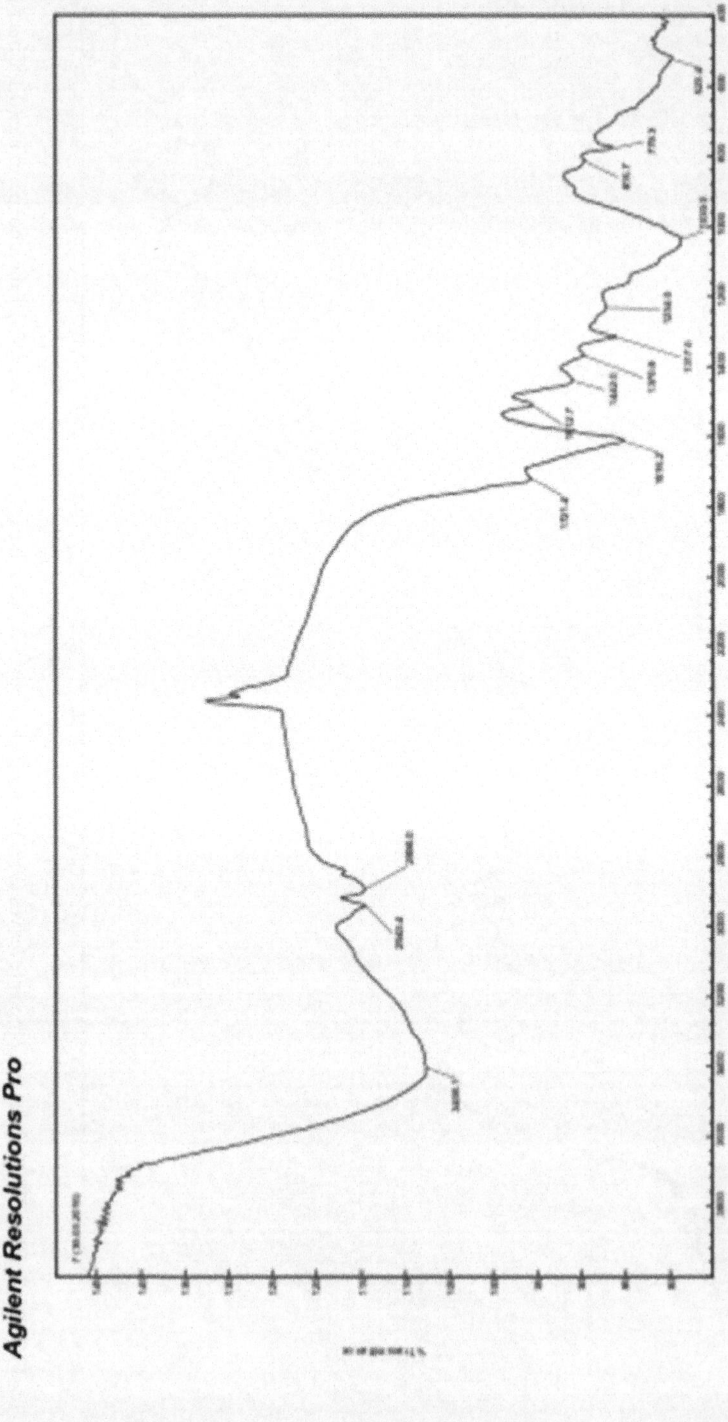

FIGURE 3.50 FTIR spectra of composite leaves (95%) + glycidyl meacrylate (5%) + dicumol peroxide (0.1% mass).

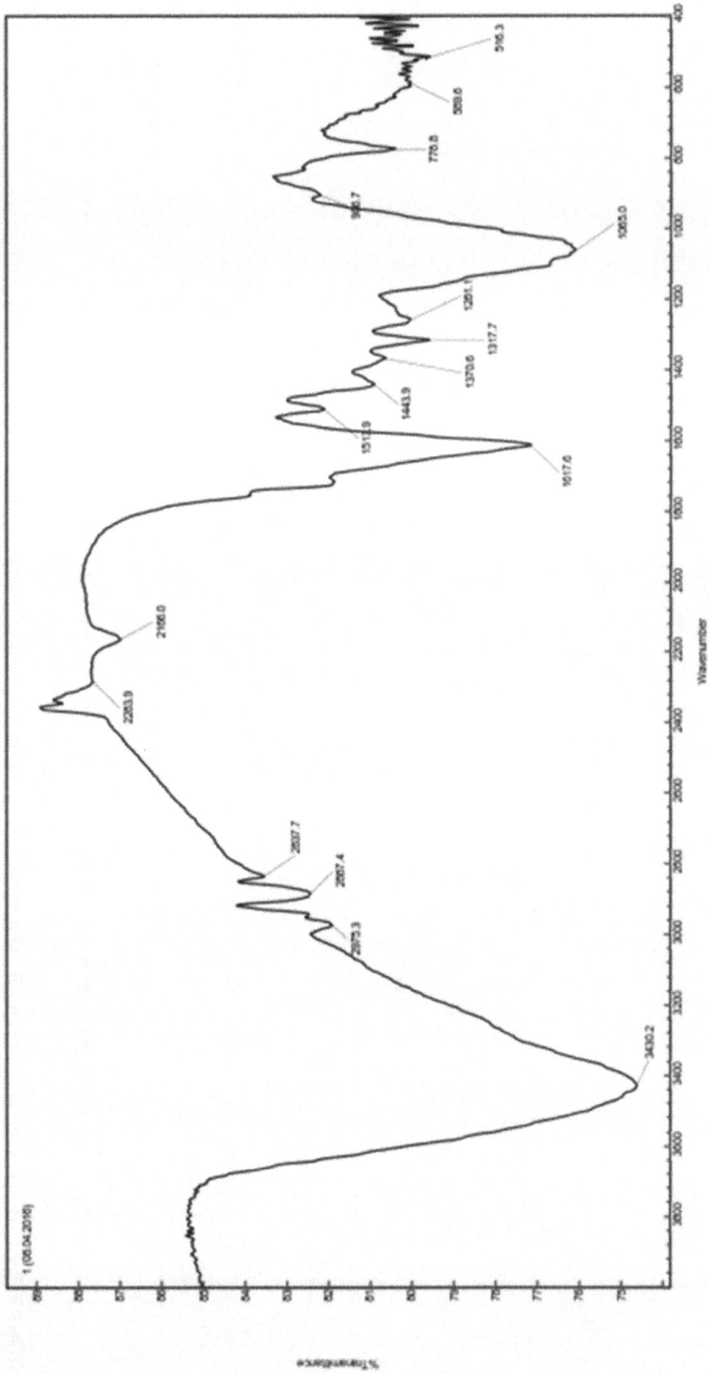

FIGURE 3.51 FTIR spectra of composite leaves (95%) + (PMHSiO + 35 vinyltriethoxy silane) (5%) + platinum catalyst.

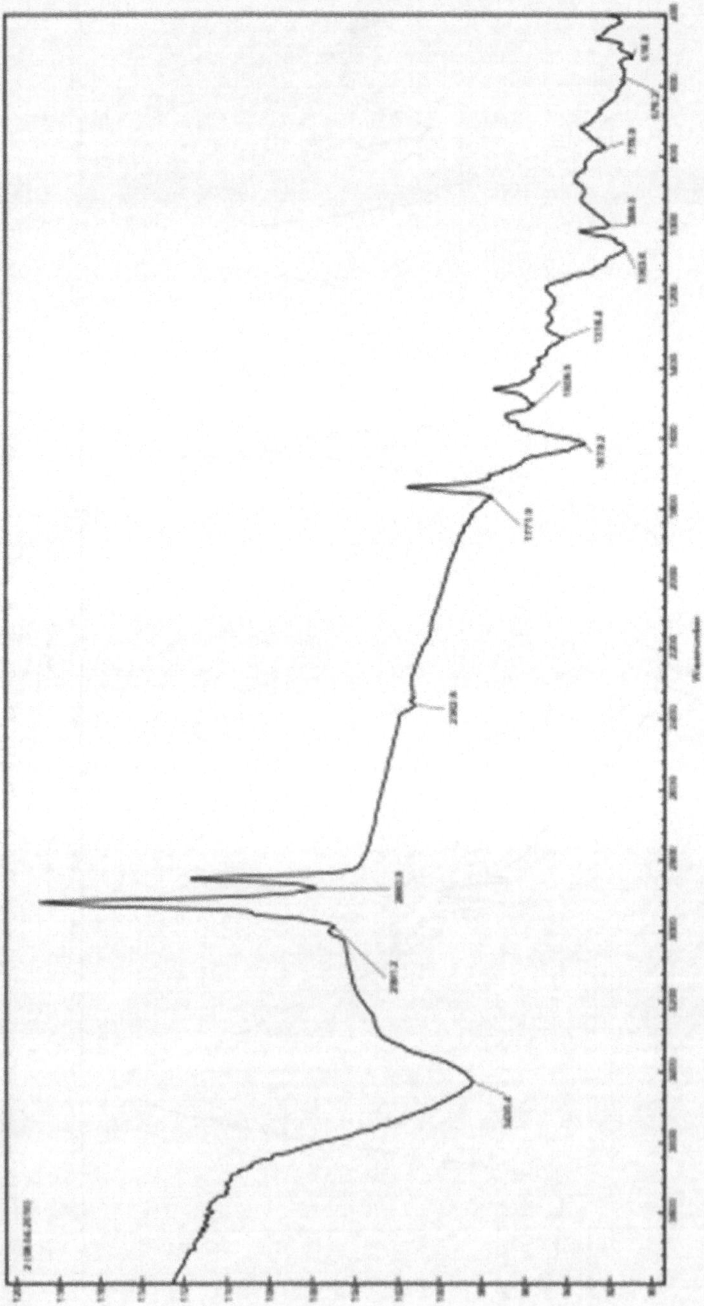

FIGURE 3.52 FTIR spectra of composite leaves (95%) + (PMHSiO+ 35 allyl glycidyl ether) (5%) + platinum catalyst.

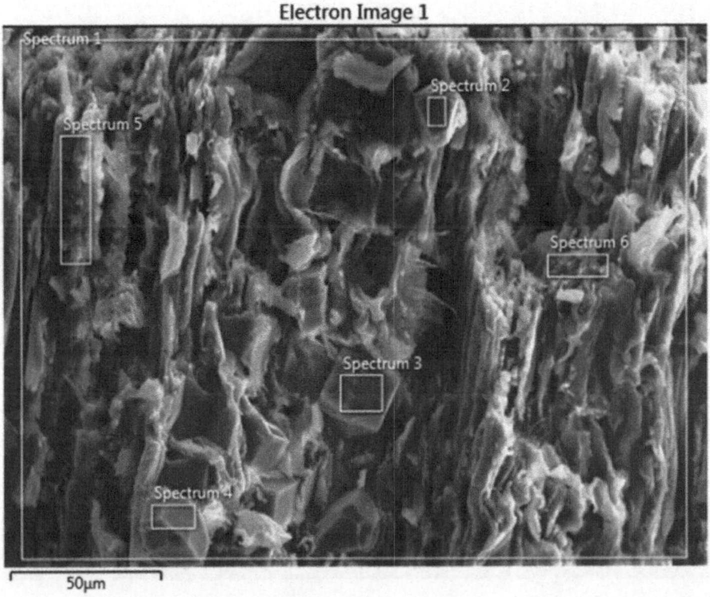

FIGURE 3.53 Scanning electron microscopic micrograms of composite: Leaves (95%) + Di(ethylene glycol) ethyl ether acrylates (5%).

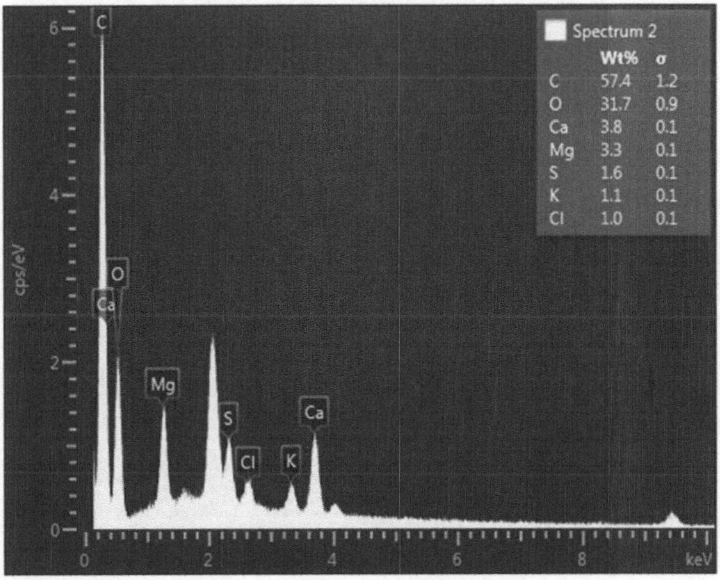

FIGURE 3.54 Energy dispersion micro-X-ray spectral analysis of composite: Leaves (95%) + Di(ethylene glycol) ethyl ether acrylates (5%) (Spectrum 2).

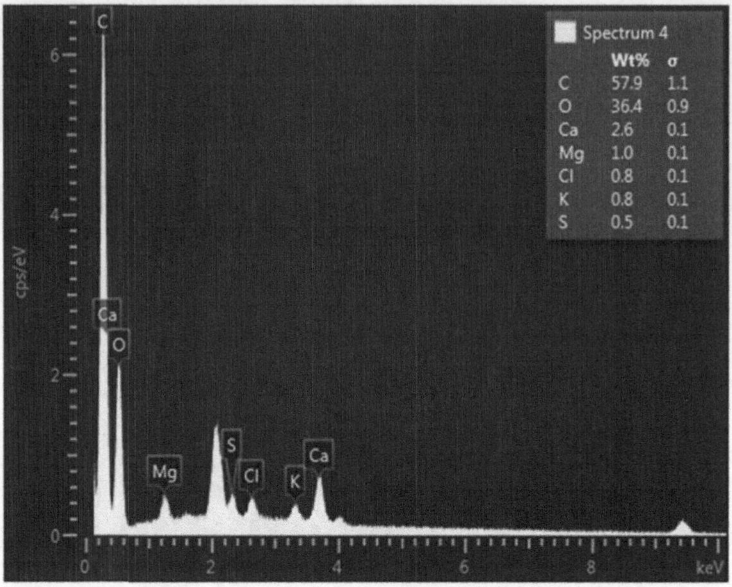

FIGURE 3.55 Energy dispersion micro-X-ray spectral analysis of composite: Leaves (95%) + Di(ethylene glycol) ethyl ether acrylates (5%) (Spectrum 4).

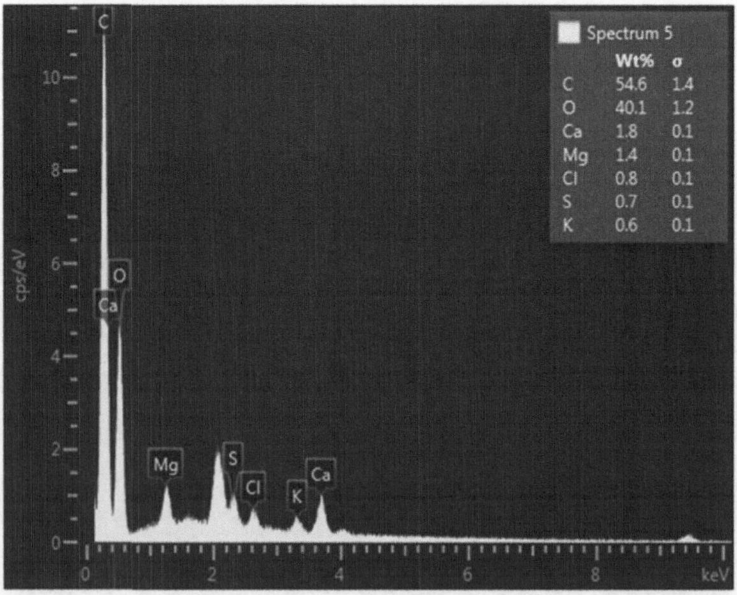

FIGURE 3.56 Energy dispersion micro-X-ray spectral analysis of composite: Leaves (95%) + Di(ethylene glycol) ethyl ether acrylates (5%) (Spectrum 5).

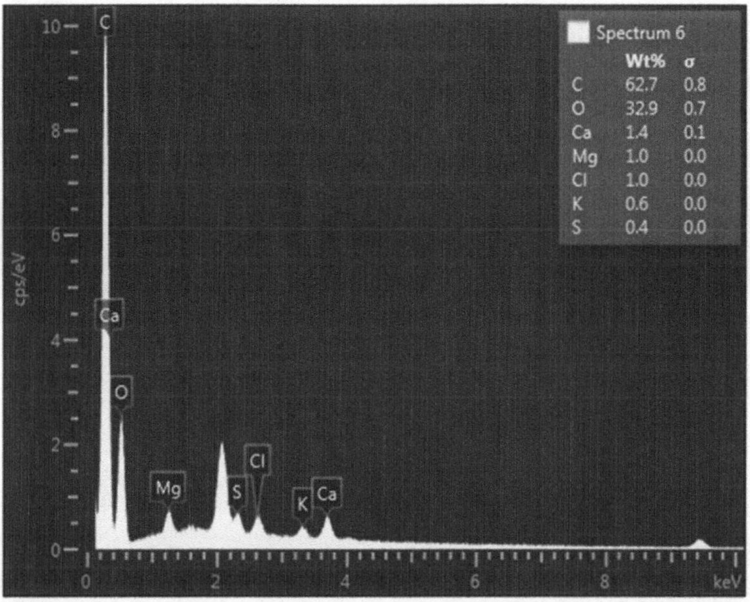

FIGURE 3.57 Energy dispersion micro-X-ray spectral analysis of composite: Leaves (95%) + Di(ethylene glycol) ethyl ether acrylates (5%) (Spectrum 6).

TABLE 3.8 Energy Dispersion Micro-X-Ray Spectral Analysis of Composite: Leaves (95%) + Di(ethylene Glycol) Ethyl Ether Acrylates (5%)

Result Type	**Weight (%)**					
Spectrum Label	**Spectrum 6**	**Spectrum 1**	**Spectrum 2**	**Spectrum 3**	**Spectrum 4**	**Spectrum 5**
C	64.71	38.19	47.44	38.02	67.86	64.59
O	30.89	57.50	41.74	53.69	26.45	30.09
Mg	1.03	0.62	2.30	1.34	1.01	1.42
S	0.44	0.31	2.63	0.47	0.49	0.68
Cl	0.85	0.45	1.02	0.27	0.84	0.75
K	0.69	0.22	1.10	0.40	0.77	0.63
Ca	1.40	2.71	3.77	5.82	2.59	1.84
Total	**100.00**	**100.00**	**100.00**	**100.00**	**100.00**	**100.00**

Statistics	**C**	**O**	**Mg**	**S**	**Cl**	**K**	**Ca**
Max.	61.71	46.50	3.40	1.62	1.03	1.20	5.72
Min.	49.02	32.74	0.62	0.22	0.34	0.22	1.50
Average	53.80	38.73	1.48	0.62	0.71	0.63	3.02
Standard deviation	5.81	6.19	0.93	0.52	0.29	0.28	1.59

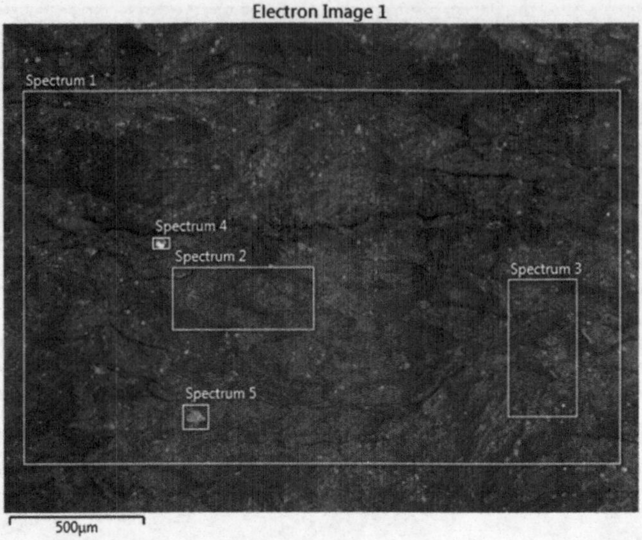

FIGURE 3.58 Scanning electron microscopic micrograms of composite: Leaves (95%) + N-Vinylcaprolactam (5%).

In Figures 3.54–3.57, 3.59–3.63, 3.65–3.71, 3.73–3.76, energy dispersion micro-X-ray spectral analysis of composite on the basis of leaves and various binders is presented. On the figures as well as in Tables 3.6–3.9 the data of

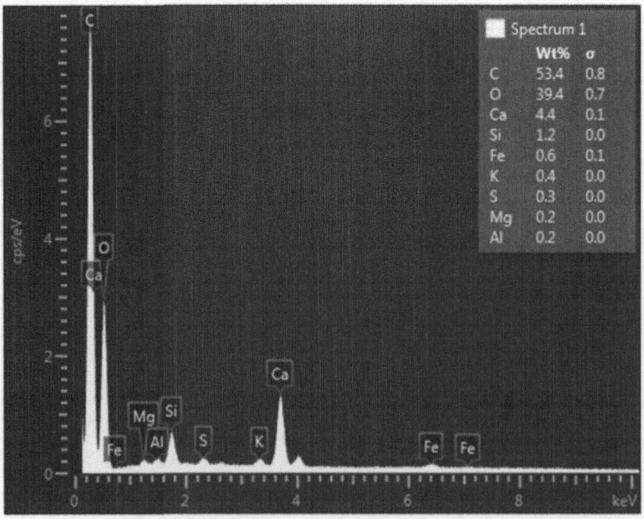

FIGURE 3.59 Energy dispersion micro-X-ray spectral analysis of composite: Leaves (95%) + N-Vinylcaprolactam (5%) (Spectrum 1).

spectral analysis is presented. One can observe that the various of spectral analysis in various part of samples differ from each other. We can conclude that the samples are not homogeneous systems and microelements are distributed in composites anisotropic (Figures 3.59–3.76 and Tables 3.9–3.11).

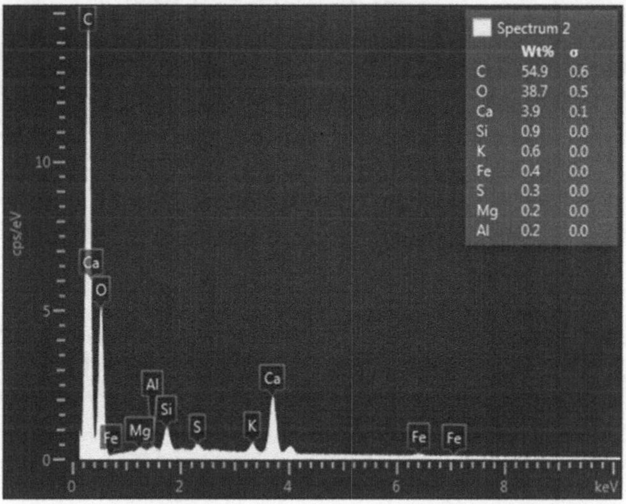

FIGURE 3.60 Energy dispersion micro-X-ray spectral analysis of composite: Leaves (95%) + N-Vinylcaprolactam (5%) (Spectrum 2).

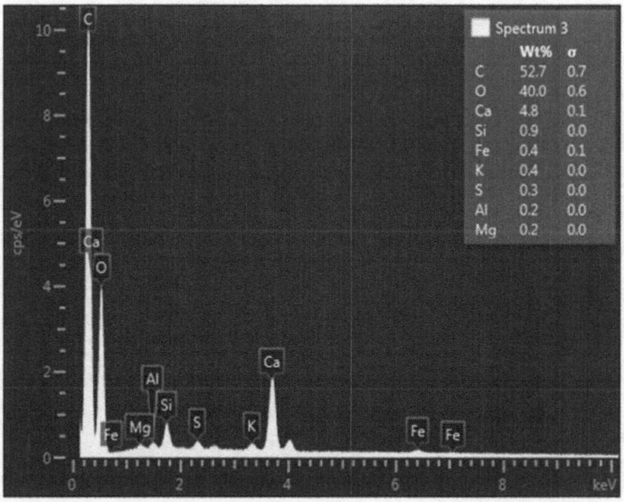

FIGURE 3.61 Energy dispersion micro-X-ray spectral analysis of composite: Leaves (95%) + N-Vinylcaprolactam (5%) (Spectrum 3).

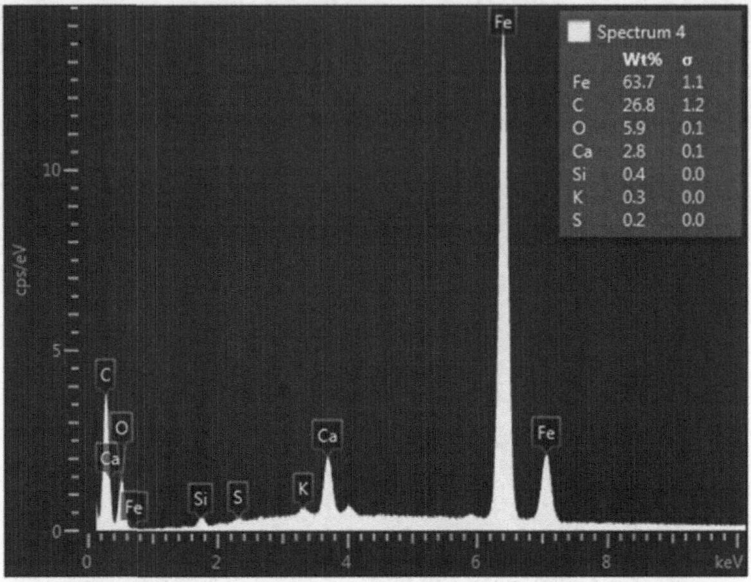

FIGURE 3.62 Energy dispersion micro-X-ray spectral analysis of composite: Leaves (95%) + N-Vinylcaprolactam (5%) (Spectrum 4).

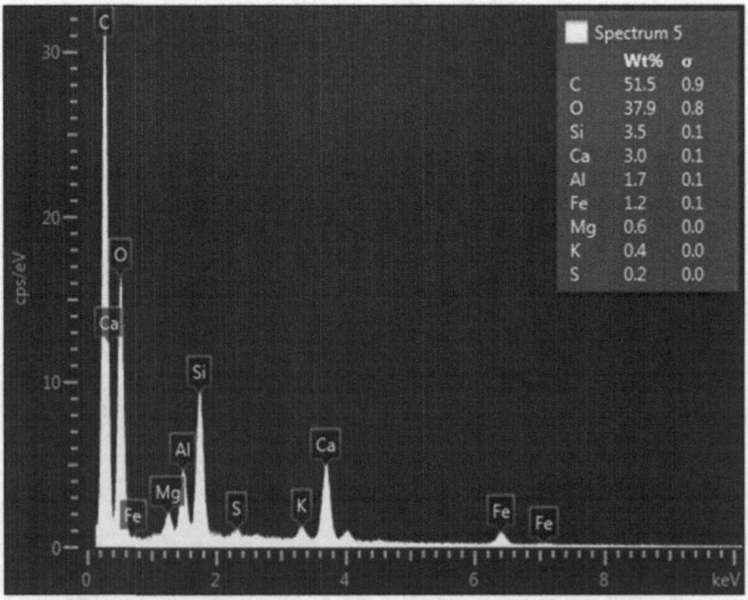

FIGURE 3.63 Energy dispersion micro-X-ray spectral analysis of composite: Leaves (95%) + N-Vinylcaprolactam (5%) (Spectrum 5).

TABLE 3.9 Energy Dispersion Micro-X-Ray Spectral Analysis of Composite: Leaves (95%) + N-Vinylcaprolactam (5%)

Result Type	**Weight (%)**				
Spectrum Label	**Spectrum 1**	**Spectrum 2**	**Spectrum 3**	**Spectrum 4**	**Spectrum 5**
C	53.35	54.91	52.72	26.77	51.49
O	39.40	38.68	39.95	5.89	37.86
Mg	0.23	0.17	0.21	–	0.63
Al	0.22	0.16	0.24	–	1.68
Si	1.17	0.92	0.92	0.40	3.48
S	0.29	0.32	0.33	0.16	0.18
K	0.35	0.57	0.36	0.32	0.43
Ca	4.41	3.91	4.82	2.77	3.02
Fe	0.58	0.35	0.44	63.70	1.23
Total	**100.00**	**100.00**	**100.00**	**100.00**	**100.00**

Statistics	**C**	**O**	**Mg**	**Al**	**Si**	**S**	**K**	**Ca**	**Fe**
Max.	54.91	39.95	0.63	1.68	3.48	0.33	0.57	4.82	63.70
Min.	26.77	5.89	0.17	0.16	0.40	0.16	0.32	2.77	0.35
Average	47.85	32.35	–	–	1.38	0.26	0.41	3.78	13.26
Standard deviation	11.85	14.82	–	–	1.21	0.08	0.10	0.88	28.20

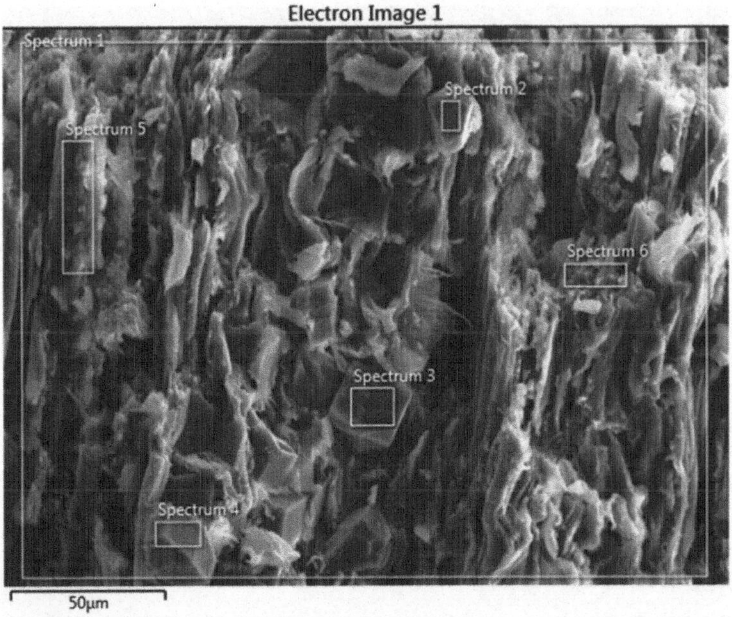

FIGURE 3.64 Scanning electron microscopic micrograms of composite: Leaves (95%) + Glycidyl methacrylate (5%).

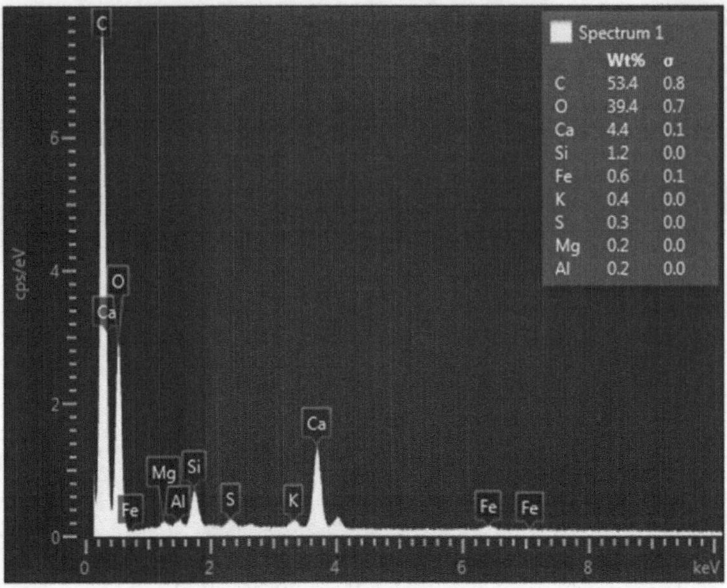

FIGURE 3.65 Energy dispersion micro-X-ray spectral analysis of composite: Leaves (95%) + Glycidyl methacrylate (5%) (Spectrum 1).

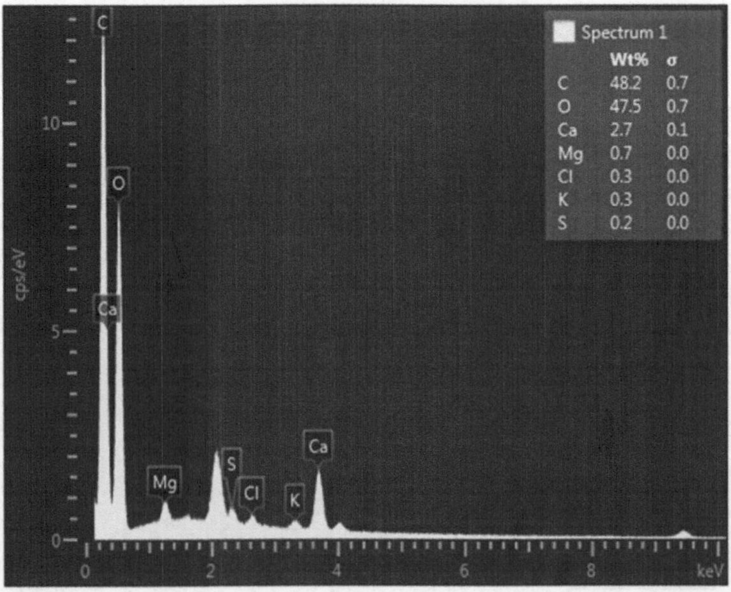

FIGURE 3.66 Energy dispersion micro-X-ray spectral analysis of composite: Leaves (95%) + Glycidyl methacrylate (5%) (Spectrum 1).

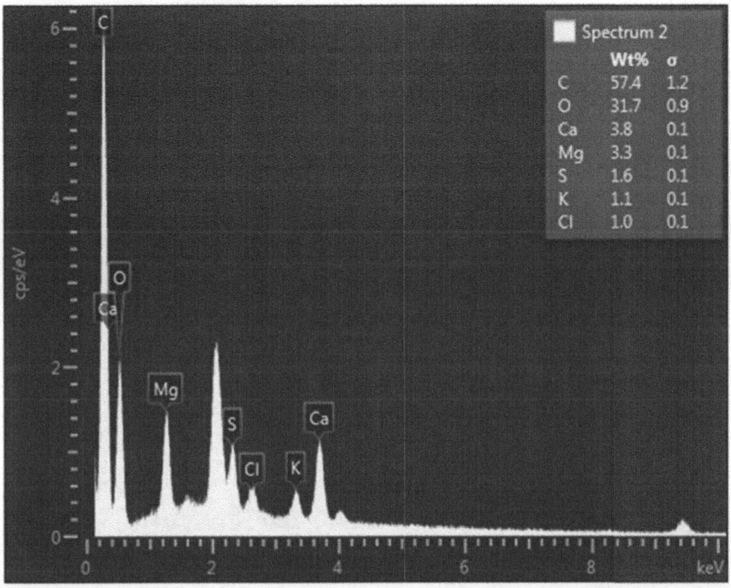

FIGURE 3.67 Energy dispersion micro-X-ray spectral analysis of composite: Leaves (95%) + Glycidyl methacrylate (5%) (Spectrum 2).

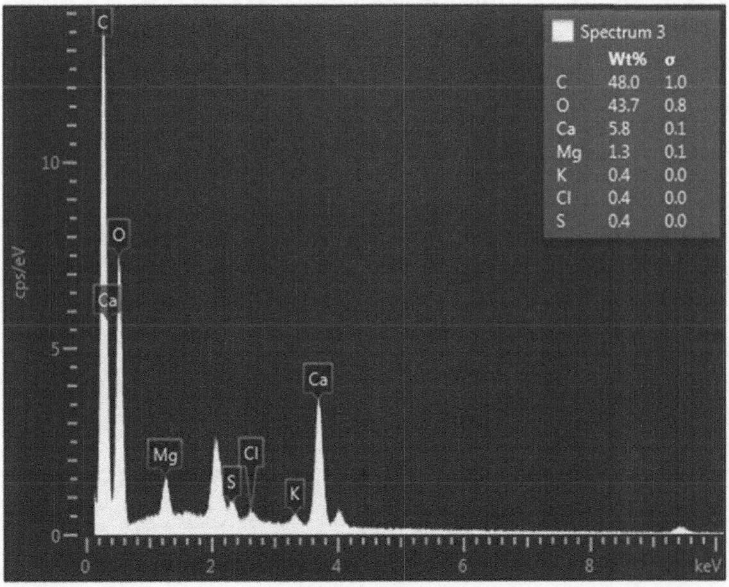

FIGURE 3.68 Energy dispersion micro-X-ray spectral analysis of composite: Leaves (95%) + Glycidyl methacrylate (5%) (Spectrum 3).

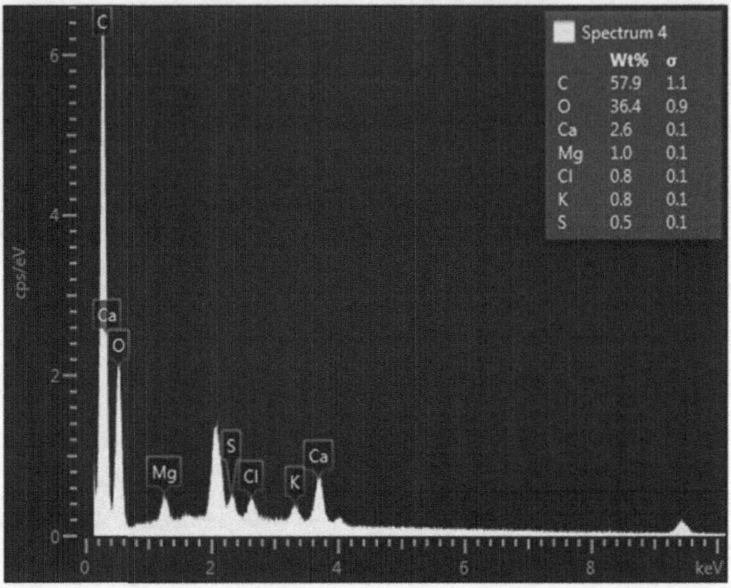

FIGURE 3.69 Energy dispersion micro-X-ray spectral analysis of composite: Leaves (95%) + Glycidyl methacrylate (5%) (Spectrum 4).

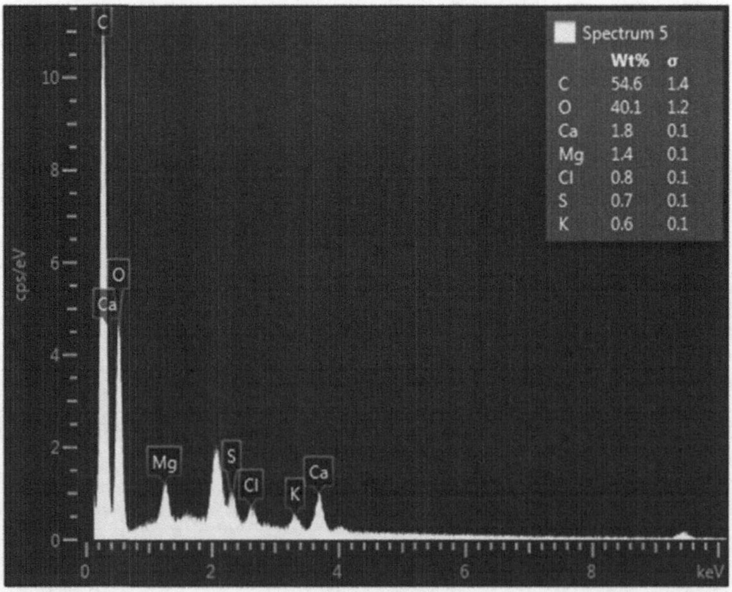

FIGURE 3.70 Energy dispersion micro-X-ray spectral analysis of composite: Leaves (95%) + Glycidyl methacrylate (5%) (Spectrum 5).

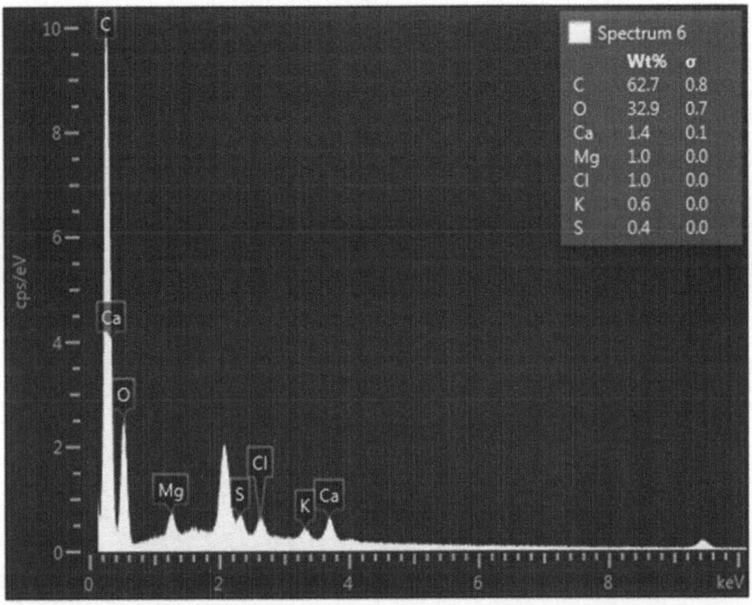

FIGURE 3.71 Energy dispersion micro-X-ray spectral analysis of composite: Leaves (95%) + Glycidyl methacrylate (5%) (Spectrum 2).

TABLE 3.10 Energy Dispersion Micro-X-Ray Spectral Analysis of Composite: Leaves (95%) + Glycidyl Methacrylate (5%)

Result Type	**Weight (%)**					
Spectrum Label	**Spectrum 6**	**Spectrum 1**	**Spectrum 2**	**Spectrum 3**	**Spectrum 4**	**Spectrum 5**
C	62.71	48.19	57.44	48.02	57.86	54.59
O	32.89	47.50	31.74	43.69	36.45	40.09
Mg	1.05	0.72	3.30	1.34	1.01	1.44
S	0.42	0.21	1.63	0.37	0.49	0.66
Cl	0.95	0.35	1.02	0.37	0.84	0.75
K	0.59	0.32	1.10	0.40	0.77	0.63
Ca	1.40	2.71	3.77	5.82	2.59	1.84
Total	**100.00**	**100.00**	**100.00**	**100.00**	**100.00**	**100.00**

Statistics	**C**	**O**	**Mg**	**S**	**Cl**	**K**	**Ca**
Max.	62.71	47.50	3.30	1.63	1.02	1.10	5.82
Min.	48.02	31.74	0.72	0.21	0.35	0.32	1.40
Average	54.80	38.73	1.48	0.63	0.71	0.63	3.02
Standard deviation	5.81	6.19	0.93	0.51	0.29	0.28	1.59

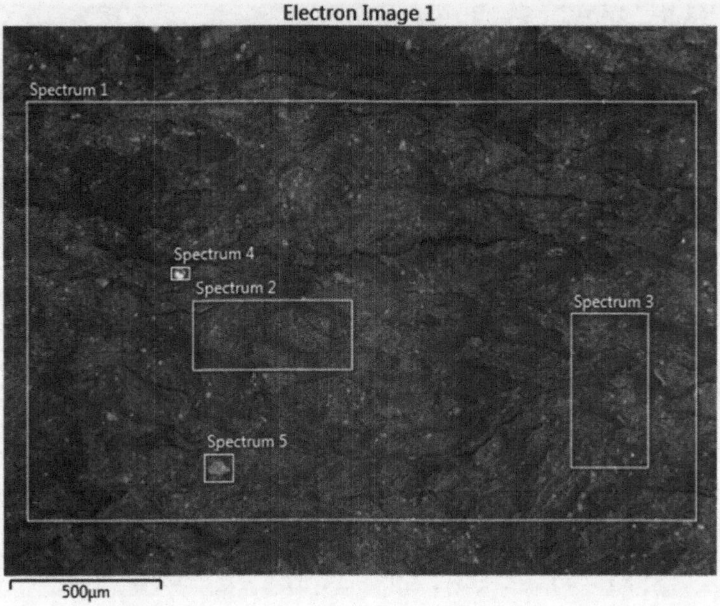

FIGURE 3.72 Scanning electron microscopic micrograms of composite: Leaves (95%) + 2-(Dimethylamino)ethyl acrylates (5%).

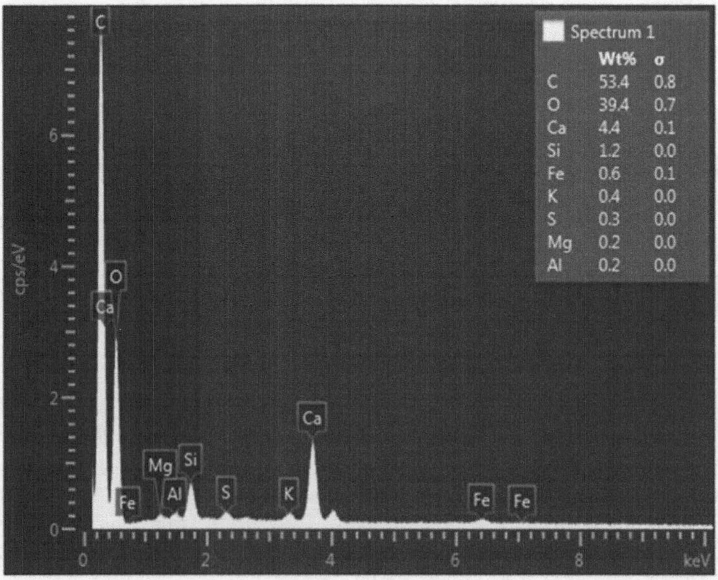

FIGURE 3.73 Energy dispersion micro-X-ray spectral analysis of composite: Leaves (95%) + 2-(Dimethylamino)ethyl acrylates (5%) (Spectrum 1).

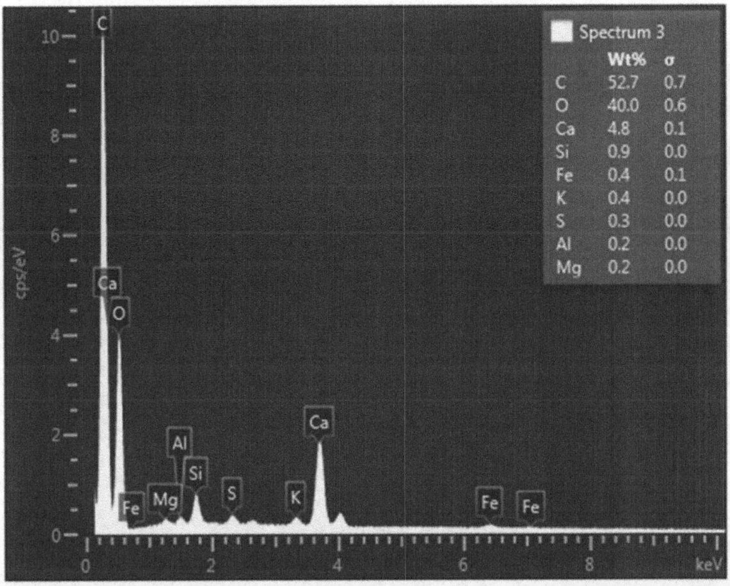

FIGURE 3.74 Energy dispersion micro-X-ray spectral analysis of composite: Leaves (95%) + 2-(Dimethylamino)ethyl acrylates (5%) (Spectrum 3).

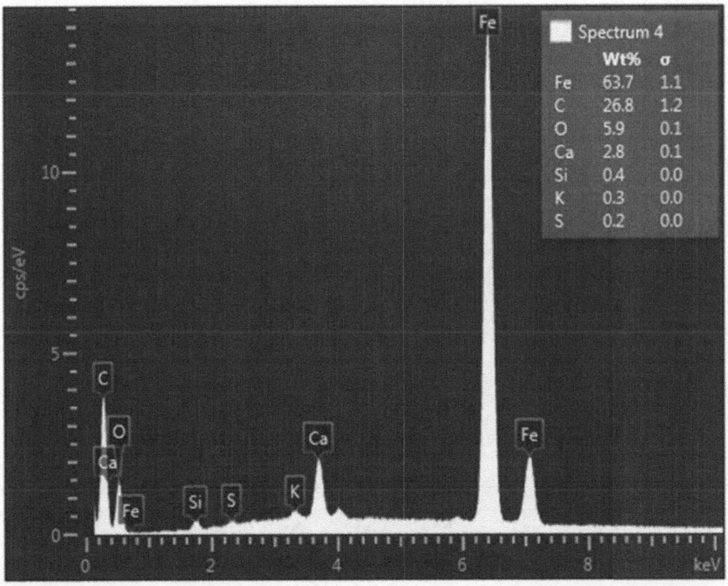

FIGURE 3.75 Energy dispersion micro-X-ray spectral analysis of composite: Leaves (95%) + 2-(Dimethylamino)ethyl acrylates (5%) (Spectrum 4).

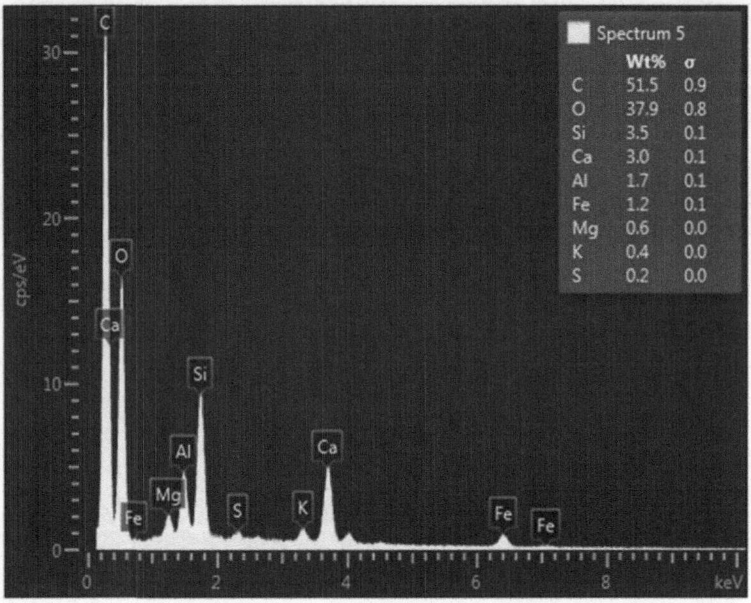

FIGURE 3.76 Energy dispersion micro-X-ray spectral analysis of composite: Leaves (95%) + 2-(Dimethylamino)ethyl acrylates (5%) (Spectrum 5).

TABLE 3.11 Energy Dispersion Micro-X-Ray Spectral Analysis of Composite: Leaves (95%) + 2-(Dimethylamino)Ethyl Acrylates (5%)

Result Type			**Weight (%)**		
Spectrum Label	**Spectrum 1**	**Spectrum 2**	**Spectrum 3**	**Spectrum 4**	**Spectrum 5**
C	55.35	56.91	42.72	26.78	51.59
O	37.40	36.68	49.95	5.88	37.66
Mg	0.21	0.18	0.22	–	0.63
Al	0.24	0.15	0.23	–	1.68
Si	1.18	0.82	0.92	0.40	3.48
S	0.28	0.42	0.33	0.16	0.18
K	0.25	0.57	0.36	0.32	0.43
Ca	4.51	3.91	4.82	2.77	3.02
Fe	0.58	0.35	0.44	63.70	1.23
Total	**100.00**	**100.00**	**100.00**	**100.00**	**100.00**

Statistics	**C**	**O**	**Mg**	**Al**	**Si**	**S**	**K**	**Ca**	**Fe**
Max.	55.91	37.95	0.64	1.68	3.48	0.33	0.57	4.82	63.70
Min.	25.77	3.89	0.16	0.16	0.40	0.16	0.32	2.77	0.35
Average	47.85	32.35	–	–	1.38	0.26	0.41	3.78	13.26
Standard deviation	11.85	14.82	–	–	1.21	0.08	0.10	0.88	28.20

3.10.3 INVESTIGATION OF THE THERMAL STABILITY (TS) OF COMPOSITES WITH USE OF METHOD VICA

In Figures 3.61–3.63, 3.65–3.71, 3.73–3.77, there are presented the temperature dependences of TS of composites based on leaves and different acrylates.

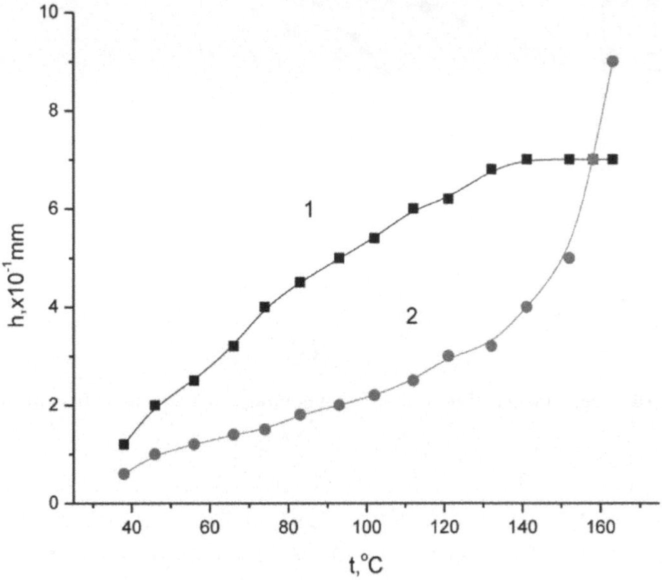

FIGURE 3.77 Dependence of the softening of composites: 3.57-Leaves (95%) + Ethylene-glycol ethyl-ethyl acrylates (5%) – 1; and 3.58-Leaves (95%) + Poly(ethylene glycol) methacrylate (5%) – 2 on the temperature.

In accordance with Figure 3.77, it is shown the essential difference between the composites 3.57 and 3.58. The first dependence indicates on the relatively low TS at temperatures 120–130°C. This parameter becomes constant till 160°C at more high temperatures, although it may be expected that this behavior will be without change at more high temperatures. In case of curve 2 we have reciprocal dependence – the TS lower 120–130°C for this composite is rather best than for first, but at temperatures higher of this interval the TS becomes worst. These results allow us made an opinion about more high movement of the binder molecules in the composite 3.57 in comparison with composite 3.58, although the following increasing of temperature in the composite 3.57 does not leads to any essential changes. In the case of composite 3.58 essential softening begins at temperatures higher 140°C, therefore till this temperature composite 3.58 is stable (Figure 3.78).

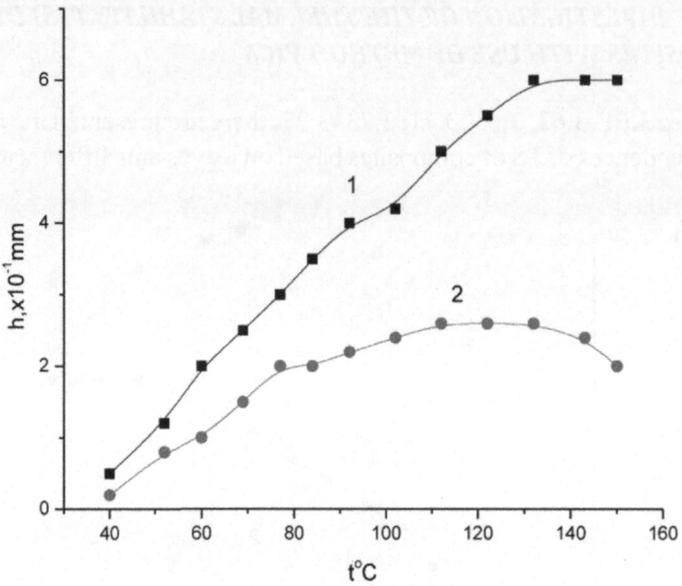

FIGURE 3.78 Dependence of the softening of composites 3.59 and 3.60 on the temperature.

The composites 3.59 and 3.60 analogically have the essential different TS. For the composite 3.60 TS we have in the interval 120–160°C with rather less (about 3 times) temperatures in comparison with composite 3 for this composite stabilization take s place higher 130°C). The obtained result indicates on the high TS of the composite 3.60 (in comparison with composite 3.59), which is, naturally explained with high TS of microstructure of this material and indicates on the stronger interactions between composite ingredients (Figure 3.79).

The curves presented in Figure 3.80 indicate on the rather good TS of composites 5 and 6 at temperatures up to 100°C, although they differ one from another very little. It is evidence shows on the similarity of the morphology of these structures. Higher of 100°C these dependences are differ one from other to some extent, which may be shown on the difference between the temperature dependence of these composite morphology (Figure 3.80).

The character of the curves presented in Figure 3.79 indicates on the similarity of the morphology of the composite microstructures. From the other hand comparison of these curves with analogical ones for composites 3.58 and 3.59 (Figure 3.78) show that a TS of the composites 3.63 and 3.67 is rather low than that for composites 3.58 and 3.59 and to some extent for composites 5 and 6. It shows once more on the relatively high TS of the last composites.

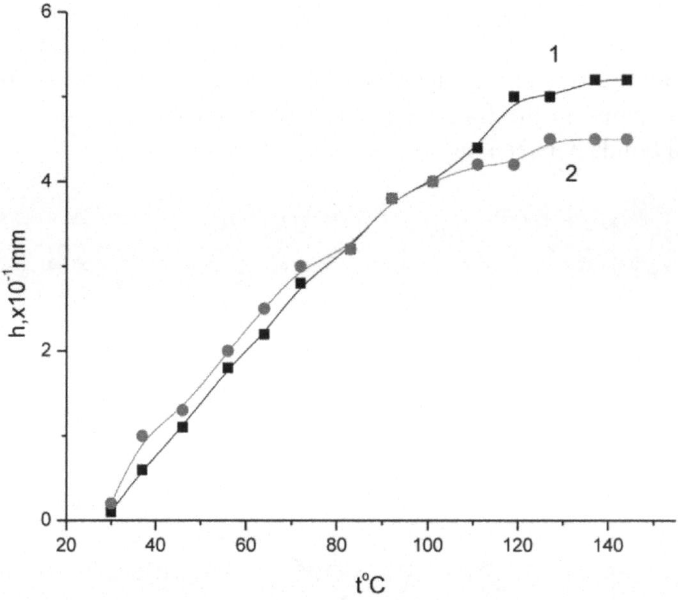

FIGURE 3.79 Dependence of the softening of composites 3.61 and 3.62 on the temperature.

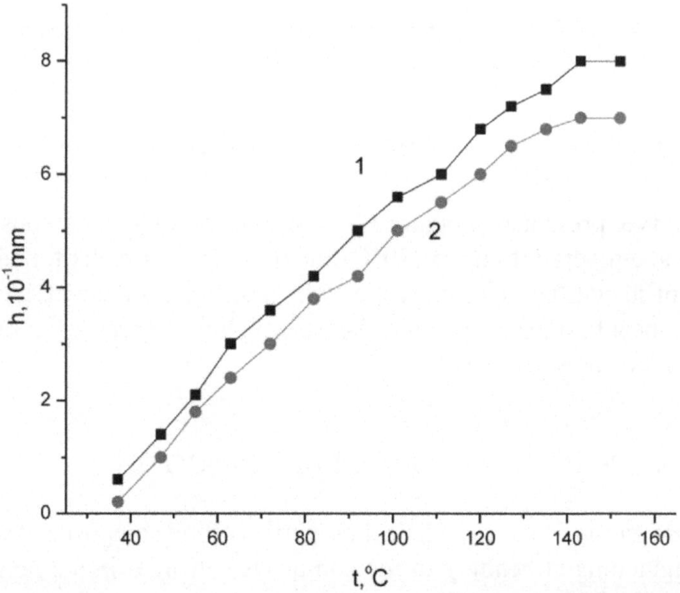

FIGURE 3.80 Dependence of the softening of composites 3.63 curve 1; and 34.67 – 2 on the temperature.

The curves presented in Figure 3.81 show the essential difference between TS of composites 3.68 and 3.69, which indicates on the good TS of the composite 3.69 in the temperature range up to 160°C. This is one of best results, obtained in the frames of the project and shows on the developed structural bonds between phases.

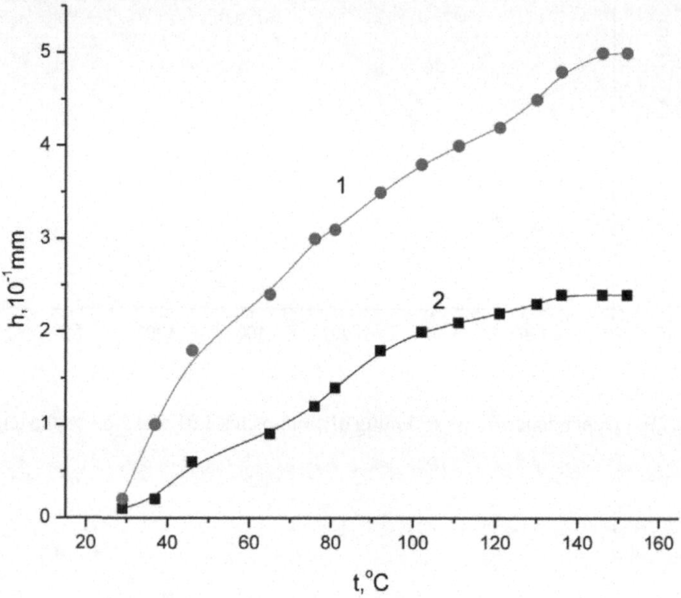

FIGURE 3.81 Temperature dependence of softening for composites 3.68 curve 1; and 3.69 – 2.

The curves presented in Figure 3.82 show on good TS for composites 3.64 and 3.65 at temperatures up to 120°C, although this parameter for composite 3.66 is not so bed. These results are also in good correlation with level of the structural peculiarities of the composites, which are expressed in the extent of interactions between phases.

3.10.4 MECHANICAL PROPERTIES OF COMPOSITES

The investigation of the mechanical properties of the composites shows that the strengthening at bending of the composites changes in the range 6.89–18.4 kg/cm^2 (Table 3.12). The maximum of the strengthening is observed for composites containing methylsiloxane matrix with triethoxysilyl groups in the side chain. The value of impact viscosity changes in the range 17.46–43.88 kJ/m^2 composites has the same order as we have obtained early at investigation of composites contained leaves.

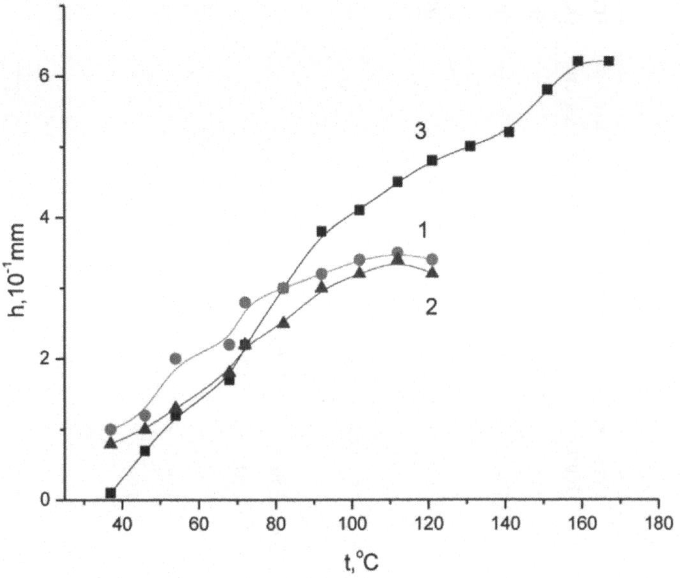

FIGURE 3.82 Temperature dependence of softening for composites 3.64 curve 1; 3.65 – 2; and 3.66 – 3.

3.10.5 INVESTIGATIONS OF THE WATER ABSORPTION OF COMPOSITES BASED ON SAWDUST

It was investigated the water absorption of obtained composites. It was established that in the composites with acrylates the water absorption increases with increasing of length of diethylene glycol fragments on 94%. It seems that water absorption in the composites with ethylene glycol groups during 24 h exposition in the water changes in the interval 16–94%. This phenomenon may be realized in the composites with electrophile groups. Analogical result is obtained for composites with nitrogen containing composites. The works in this direction are continued. The results on water absorption after 3 h exposition in the water of the composites practically is equal to analogs, obtained early for composites with leaves, but with other binders (Table 3.13).

TABLE 3.12 Mechanical Characteristics of the Composites Based on Leaves and Some Unsaturated Monomers

SL. No.	Composite (wt.%)	Pressure (Kg/cm²)	Temperature (T°C)	Strengthening at Bending (Kg/cm²)	Impact Viscosity (kJ/m²)
1.	Leaves (95%) + Ethylene-glycol ethyl-ethyl acrylates (5%)	150	130	14.29	28.44
2.	Leaves (95%) + Poly(ethylene glycol) methacrylate (5%)	150	130	14.13	30.00
3.	Leaves (95%) + N-Methyl-N-vinylacetamide (5%)	150	130	16.52	21.14
4.	Leaves (95%) + Di(ethylene glycol) ethyl ether acrylates (5%)	150	130	12.87	20.00
5.	Leaves (95%) + N-Vinylcaprolactam (5%)	150	130	14.96	19.05
6.	Leaves (95%) + Di(ethylene glycol) 2-ethylhexyl ether acrylates (5%)	150	130	11.16	19.56
7.	Leaves (95%) + Glycidyl methacrylate (5%)	150	130	18.36	22.78
8.	Leaves (95%) + 2 [(Butylamino)carbonyl)oxy] ethyl acrylates (5%)	150	130	8.32	17.46
9.	Leaves (95%) + Diallyl maleate (5%)	150	130	11.43	20.78
10.	Leaves (95%) + Tetramethyl cyclotetrasiloxane (5%)	150	130	10.17	21.33
11.	Leaves (95%) + 2-(Dimethylamino)ethyl acrylates (5%)	150	130	15.52	34.63
12.	Leaves (95%) + (Polymethylhydrosiloxane + 35 Vinyltriethoxysilane) (5%)	150	130	6.89	43.88
13.	Leaves (95%) + (Polymethylhydrosiloxane + 35 Allylglycidyl ether) (5%)	150	130	10.33	18.57

TABLE 3.13 Water Absorption of Samples on the Basis of Leaves and Unsaturated Monomers

SL. No.	Composite (mass %)	Mass (g)	Volume (cm^3)	Density (g/cm^3)	Mass in Water, After 3 h Exposition (g)	Mass in Water, After 24 h Exposition (g)	Water Absorption, After 24 h Exposition (%)
1.	Leaves (95%) + Ethylene glycol methyl ether acrylates (5%)	3.66	2.64	1.39	3.84	4.49	22.68
2.	Leaves (95%) + Poly(ethylene glycol) methacrylate (5%)	3.70	2.64	1.40	3.79	4.41	19.19
3.	Leaves (95%) + N-Methyl-N-vinylacetamide (5%)	3.81	2.64	1.44	3.87	4.02	5.51
4.	Leaves (95%) + Di(ethylene glycol) ethyl ether acrylates (5%)	2.90	2.46	1.18	3.29	3.37	16.21
5.	Leaves (95%) + N-Vinylcaprolactam (5%)	3.59	2.64	1.36	3.65	3.79	5.57
6.	Leaves (95%) + Di(ethylene glycol) 2-ethylhexyl ether acrylates (5%)	3.88	2.64	1.47	4.68	7.53	94.07
7.	Leaves (95%) + Glycidyl methacrylate (5%)	3.89	2.64	1.47	3.97	4.13	6.17
8.	Leaves (95%) + 2 [(Butylamino)carbonyl)oxy] ethyl acrylates (5%)	3.92	2.64	1.48	4.02	4.86	23.98
9.	Leaves (95%) + Diallyl maleate (5%)	3.76	2.64	1.42	3.88	4.89	30.05
10.	Leaves (95%) + Tetramethyl cyclotetrasiloxane (5%)	3.53	2.46	1.43	3.63	4.48	26.9
11.	Leaves (95%) + 2-(Dimethylamino)ethyl acrylates (5%)	4.06	2.82	1.44	4.28	5.15	26.85
12.	Leaves (95%) + (Polymethylhydrosiloxane+ 35 Vinyltriethoxysilane) (5%)	3.84	2.65	1.45	3.91	4.33	12.76
13.	Leaves (95%) + (Polymethylhydrosiloxane+ 35 Allylglycidyl ether) (5%)	3.71	2.65	1.4	3.79	4.57	23.18

3.10.6 CONCLUSIONS TO COMPOSITES ON THE BASIS OF LEAVES AND UNSATURATED MONOMER

1. There are obtained the new composites on the basis of leaves and in-situ obtained of polymeric matrix.
2. With use of method of FTIR there are investigated the chemical structure of obtained composites, on the basis of which it is made the conclusion about formation new chemical bonds between phases–active groups of leaves and in-situ obtained polymers. These bonds enhance the mechanical and other exploitation properties of composites.
3. Investigations of TS with use of method Vica of the composites allow us to conclude that this parameter for new materials is not worse that the analogs obtained on the basis of leaves and other organic/inorganic binders and in separate cases are even better. This result indicates on the existence of intensive chemical bonds between phases.
4. The noted chemical interphases bonds are reflected on mechanical properties and on level of water absorption of the new composites and have the same order than that for analogs given for the composites based on leaves with other binders.

KEYWORDS

- **composite materials**
- **Fourier transform infrared**
- **microstructure**
- **polymeric matrix**
- **sawdust**
- **water absorption**
- **X-ray micro-spectral analysis**

REFERENCES

1. Andrews, L., & Hough, L., (1958). The biosynthesis of polysaccharides: Part I. The composition of plum-leaves polysaccharides. *J. Chem. Soc.*, 4476–4483.

2. Andrews, P., Hough, L., & Stacey, B., (1960). Polysaccharide composition of leaves. *Nature, 185*, 166, 167.
3. Andrews, P., & Hough, L., (1958). The biosynthesis of polysaccharides: Part II. Incorporation of $^{14}CO_2$ into plum-leaves polysaccharides during photosynthesis. *J. Chem. Soc.*, 4483–4488.
4. Neish, A. C., (1958). The biosynthesis of cell wall carbohydrates: Part IV. Further studies on cellulose and xylan in wheat. *J. Biochem. Physiol., 36*, 187–193.
5. Suberkropf, K., Godshalk, G. L., & Klug, M. J., (1976). Changes in the chemical composition of leaves during processing in a woodland stream. *Ecology, 57*, 720–727.
6. Adams, W. W., Winter, K., Schreiber, U., & Schramel, P., (1990). Photosynthesis and chlorophyll fluorescence characteristics in relationship to changes in pigment and element composition of leaves of *Platanus occidentalis* L. during autumnal leaves senescence. *Plant Physiol., 92*(4), 1184–1190.
7. Cote, G. G., (2009). Diversity and distribution of idioblasts producing calcium oxalate crystals in *Dieffenbachia seguine* (*Araceae*). *American Journal of Botany, 96*(7), 1245–1254.
8. Kliosov, A. A., (2007). *Wood-Plastic Composites* (p. 698). John Wiley & Sons, Hoboken, New Jersey.
9. Ellis, W. D., & O'Dell, J. L., (1999). Surface morphology of PECVD fluorocarbon thin films from hexafluoropropylene oxide, 1,1,2,2-tetrafluoroethane and difluoromethane. *J. Appl. Polym. Sci., 73*(12), 2493–2505.
10. Patil, Y. P., Gajre, B., Dusane, D., Chavan, S., & Mishra, S., (2000). Effect of maleic anhydride treatment on steam and water absorption of wood polymer composites prepared from wheat straw, cane bagasse, and teak wood sawdust using Novolac as matrix. *J. Appl. Polym. Sci., 77*(13), 2963–2967.
11. Brember, K. I., & Shneider, M. H., (1985). The significance of the glass transition of lignin in thermomechanical pulping. The significance of the glass transition of lignin in thermomechanical pulping. *Wood Sci. and Technology, 19*, 75–81.
12. Hausen, B. M., Kuhlwein, A., & Schulz, K. H., (1982). Colophony allergy. A contribution to the origin, chemistry, and uses of colophony and modified colophony products, 1. *Derm Beruf Umwelt., 30*(4), 107–115.
13. Muraganatham, S., Anbalagan, G., & Ramamurthy, N., (2009). FTIR and SEM-EDS comparative analysis of medical plants. *Eclipta alba* Hassk and *Eclipta prostrate* Linn. *Rom J. Biophysics, 19*, 285–294.
14. Shirwaikar, A., Malini, S., & Kumari, S. C., (2003). Protective effect of *Pongamia pinnata* flowers against cisplatin and gentamicin induced nephrotoxicity in rats. *Indian J. Exp. Biol., 1*, 58–62.
15. Saikia, C. N., Ali, F., Goswami, T., & Grosh, A. C., (1995). Esterification of high α-cellulose extracted from *Hibiscus cannabinus* L. *Ind. Crops Prod., 4*, 233–239.
16. Mukbaniani, O., Aneli, J., Buzaladze, G., Tatrishvili, T., & Markarashvili, E., (2015). *Composites on the Basis of Renewable Raw Materials* (p. 40). Abstracts of communications of the 6th Nordic Wood Biorefinery Conference, NWBC 2015, Helsinki, Finland.
17. Mukbaniani, O., Aneli, J., Buzaladze, G., Tatrishvili, T., & Markarashvili, E., (2015). Composites on the basis of renewable raw materials. *Proceedings of the 6th Nordic Wood Biorefinery Conference, NWBC 2015* (pp. 467–473). Helsinki, Finland.
18. Mukbaniani, O., Aneli, J., Buzaladze, G., Tatrishvili, T., & Markarashvili, E., (2016). Biocomposite materials on the basis of leaves. *Oxid. Commun.* (In press).

19. Mukbaniani, O., Aneli, J., Buzaladze, G., Tatrishvili, T., & Markarashvili, E., (2015). Composites on the basis of renewable raw materials. In: *3rd International Congress on Energy Efficiency and Energy Related Materials ENEFM* (p. 103). Oludeniz MUGLA/Turkey.
20. Pavla, T., (2011). *Advances in Composite Materials – Analysis of Natural and Manmade Materials* (pp. 229–284). Chapter 9; Publisher In-Tech, 2011, 572. Wood-Polymer Composites – Yongfeng Li.
21. Baysal, E., Sonmez, A., Colak, M., & Toker, H., (2006). Amount of leachant and water absorption levels of wood treated with borates and water repellents. *Bioresource Technology 97*, 2271–2279.
22. Baysal, E., Yalinkilic, M., Altinok, M., Sonmezc, A., Pekerd, H., & Colaka, M., (2007). Some physical, biological, mechanical, and fire properties of wood polymer composite (WPC) pretreated with boric acid and borax mixture. *Construction and Building Materials, 21*(9), 1879–1885.
23. Hashizume, Y., Yoshizawa, S., Nakamura, T., Yajima, H., Ishii, T., & Handa, T., (1988). Dimensional stability of wood-polymer composite (WPC) using the pretreatment of liquid anhydrous ammonia. *Kobunshi Ronbunshu, 45*(8), 617–624.
24. Yalinkilic, M., Gezer, E., Takahashi, M., Demirci, Z., IIhan, R., & Imamura, Y., (1999). Boron addition to non-or low-formaldehyde cross-linking reagents to enhance biological resistance and dimensional stability of wood. *European Journal of Wood and Wood Products, 57*(5), 351–357.
25. Obanda, D., Shupe, T., & Barnes, H., (2008). Reducing leaching of boron-based wood preservatives: A review of research. *Bioresource Technology, 99*(15), 7312–7322.
26. Soulounganga, P., Loubinoux, B., Wozniak, E., Lemor, A., & Gérardin, P., (2004). Improvement of wood properties by impregnation with polyglycerol methacrylate. *European Journal of Wood and Wood Products, 62*(4), 281–285.
27. Ellis, D., & O'Dell, J., (1999). Wood–polymer composites (WPC) made with acrylic monomers, isocyanate, and maleic anhydride. *Journal of Applied Polymer Science, 73*(2), 2493–2505.
28. González-Peña, M., Curling, S., & Hale, M., (2009). On the effect of heat on the chemical composition and dimensions of thermally-modified wood. *Polymer Degradation and Stability, 94*(12), 2184–2193.
29. Ibach, R., & Ellis, W., (2005). Lumen modifications. In: Roger. M. R., (ed.), *Handbook of Wood Chemistry and Wood Composites* (pp. 421–446). CRC Press, Washington, D.C.
30. Rosani Do, C. De. O. A., Doria, M. S. G., Aline, C. De. A., Michelle, L. M., & Mario, G., (2010). Leaf anatomy and micromorphology of six *Posoqueria* Aublet species (*Rubiaceae*). *Rodriguésia, 61*(3), 505–518.
31. Doria, M. S. G., & Léa De, J. N., (2009). Scanning electron microscopy of the leaf epidermis of *Merostachys* Spreng. (*Poaceae: Bambusoideae*). *Acta Bot. Bras., 23*(2), 516–525.

CHAPTER 4

Composite Materials on the Basis of Bamboo

OMAR MUKBANIANI,[1,2] JIMSHER ANELI,[2] and TAMARA TATRISHVILI[1,2]

[1]*Department of Macromolecular Chemistry, Ivane Javakhishvili Tbilisi State Unoiversity, Tbilisi, Georgia*

[2]*Institute of Macromolecular Chemistry and Polymeric Materials, Ivane Javakhishvili Tbilisi State University, Tbilisi, Georgia*

ABSTRACT

The main task of this chapter is an investigation of the composites based on bamboo and binders: liquid glass (LG), phenyletoxysilane of two types–PhES-50 and PhES-80, PE of low density, colophony, and wood glue. All binders besides LG had a powder like view (sizes of the particles–10–100 mcm).

4.1 INTRODUCTION

Ambusa vulgaris, common bamboo, is an open-clump type bamboo species. It is native to Indochina and to the province of Yunnan in southern China, but it has been widely cultivated in many other places and has become naturalized in several [1, 2]. Among bamboo species, it is one of the largest and most easily recognized [3, 4].

There are several differences between bamboo and wood. In bamboo, there are no rays or knots, which give bamboo a far more evenly distributed

Biocomposites: Environmental and Biomedical Applications. Omar Mukbaniani, Tamara Tatrishvili, Neha Kanwar Rawat, & A. K. Haghi (Eds.)

stresses throughout its length. Bamboo is a hollow tube, sometimes with thin walls, and consequently it is more difficult to join bamboo than pieces of wood. Bamboo does not contain the same chemical extractives as wood and can therefore be glued very well [5]. Bamboo's diameter, thickness, and intermodal length have a macroscopically graded structure while the fiber distribution exhibits a microscopically graded architecture, which lead to favorable properties of bamboo [6].

From literature it is known that the bamboo mainly contains sufficient amounts of cellulose, Hemicelluloses, Lignin, Extractives, pentosanes, ash, and as well as silica dioxide.

4.1.1 CHEMICAL COMPOSITION OF THE BAMBOO

- Cellulose: 45–59%;
- Hemicelluloses: 20–25%;
- Lignin: 20–30%;
- Extractives: 2.5–5%.

Ratio of chemical composition among genera and species are slightly different. The average chemical composition is Cellulose 41–44%.

Cellulose is an organic compound with the formula $(C_6H_{10}O_5)_n$, a polysaccharide consisting of a linear chain of several hundreds to many thousands of β(1→4) linked D-glucose units-1 [7–10]. Cellulose is an important structural component of the primary cell wall of green plants, many forms of algae and the oomycetes. Some species of bacteria secrete it to form biofilms [11]. Cellulose is the most abundant organic polymer on Earth [12].

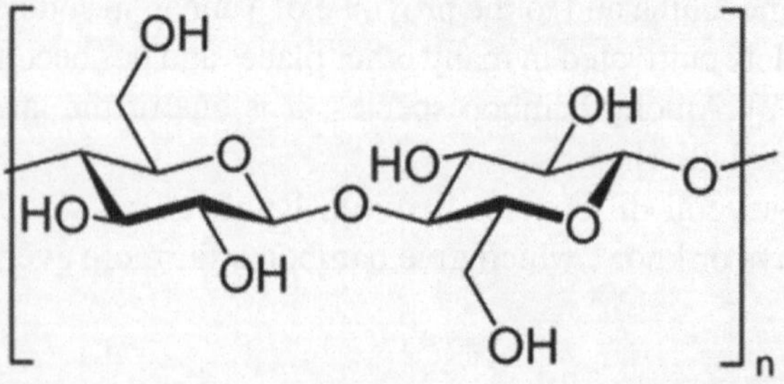

The cellulose content of cotton fiber is 90%, that of wood is 40–50% and that of dried hemp is approximately 45% [13–15].

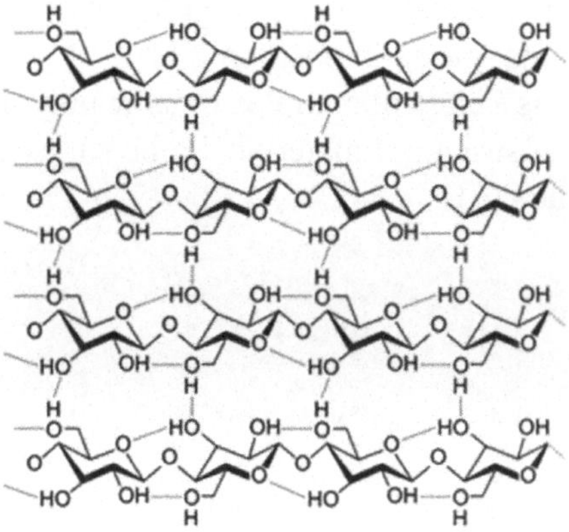

A strand of cellulose (conformation I_α), showing the hydrogen bonds (dashed) within and between cellulose molecules.

A hemicellulose (also known as polyose) is any of several heteropolymers (matrix polysaccharides), such as arabinoxylans, present along with cellulose in almost all plant cell walls [16]. While cellulose is crystalline, strong, and resistant to hydrolysis, hemicellulose has a random, amorphous structure with little strength. It is easily hydrolyzed by dilute acid or base as well as myriad hemicellulose enzymes.

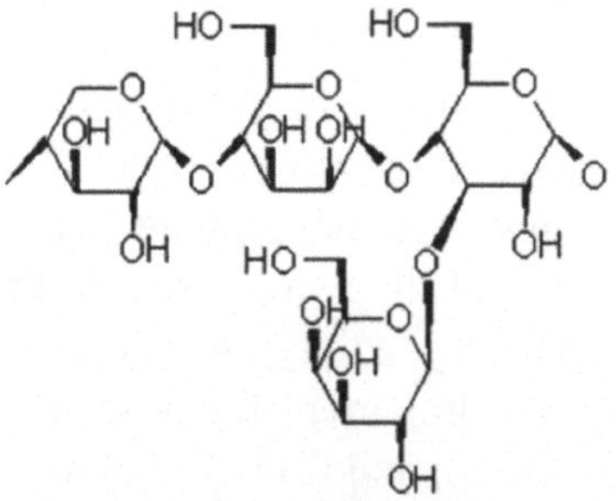

- Xylose - ß(1,4) - Mannose - ß(1,4) - Glucose -
- alpha(1,3) - Galactose

Hemicellulose

Unlike cellulose, hemicellulose (also a polysaccharide) consists of shorter chains – 500–3,000 sugar units as opposed to 7,000–15,000 glucose molecules per polymer seen in cellulose [16]. In addition, hemicellulose is a branched polymer, while cellulose is unbranched.

Pentosans 21–23% – a pentose (aldo- and ketopentose is a monosaccharide with five carbon atoms. Pentoses are organized into two groups.

Aldopentoses have an aldehyde functional group at position 1. Ketopentoses have a ketone functional group in position 2 or 3.

Lignin 26–28% – is a class of complex organic polymers. Lignins are one of the main classes of structural materials in the support tissues of vascular plants and some algae [17].

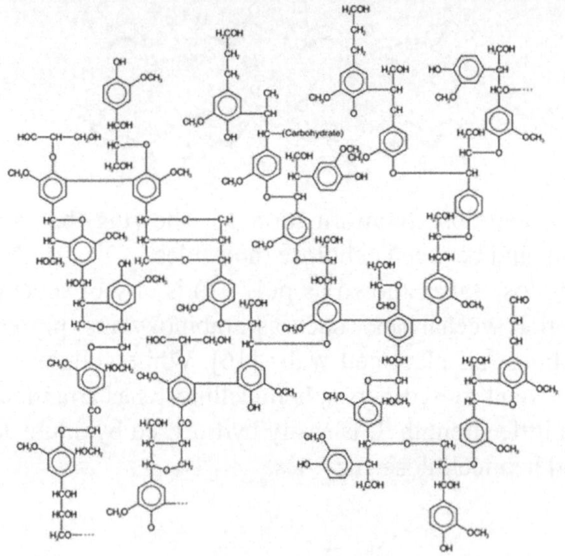

Lignins are particularly important in the formation of cell walls, especially in wood and bark, because they lend rigidity and do not rot easily. Chemically lignin is cross-linked phenol polymers [18].

Ash 1.7–1.9%, Ash is one of the components in the proximate analysis of biological materials, consisting mainly of salty, inorganic constituents. It includes metal salts which are important for processes requiring ions such as Na^+ (Sodium), K^+ (Potassium), and Ca^{2+} (Calcium). It also includes trace minerals which are required for unique molecules, such as chlorophyll and hemoglobin (Table 4.1).

Silica 0.6–0.7% [19]. Silicon dioxide, also known as silica (from the Latin silex), is a chemical compound that is an oxide of silicon with the chemical formula SiO_2. It has been known since ancient times. Silica is most commonly found in nature as quartz, as well as in various living organisms [20, 21]. In many parts of the world, silica is the major constituent of sand. Silica is one of the most complex and most abundant families of materials, existing both as several minerals and being produced synthetically. Notable examples include fused quartz, crystal, fumed silica, silica gel, and aerogels.

Applications range from structural materials to microelectronics to components used in the food industry. Content of bamboo:

TABLE 4.1 Containing of Macro- and Microelements in Bamboo Samples (%) [22]

SL. No.	Elements	Content in Percentage
	Elements of the 1 Group	
1	Ka	14.900
2.	Ca	5.050
3.	Co	10.964×10^{-4}
4.	Fe	0.144
5.	Na	0.878
6.	Zn	371.870×10^{-4}
	Elements of the 2 Group	
7.	Ag	0.275×10^{-4}
8.	Ba	717.00×10^{-4}
9.	Br	109.260×10^{-4}
10.	Cr	1.300×10^{-4}
11.	Sr	296.00×10^{-4}
	Elements of the 3 Group	
12.	As	0.275×10^{-4}
13.	Sb	0.124×10^{-4}
14.	Th	0.214×10^{-4}
15.	U	0.006×10^{-4}

All ingredients of the composites (except of powdery bamboo) were presented in amount 3, 5, 10, 15, and 20 wt.%. The obtained composites may be divided conventionally into the following groups:

1. Bamboo + LG (liquid glass 3–20%);
2. Bamboo + PhES (phenetoxysilsesquioxane 50 and 80 – 3, 5, 10, 15, 20%);
3. Bamboo + PE (5, 10, 15, 20%);
4. Bamboo + Colophony (5, 10, 15, 20%);
5. Bamboo + wood glue (5, 10, 15, 20%);

We were preparing two type form of samples, cylindrical samples of d = 1.5 cm, h = 2 cm (for study of water absorption of samples); and parallelepiped Length 12 cm, height 0.7 cm, width 1.5 cm rectangle shaped designs conducted following parameters: strength on bending, impact viscosity, dependence of the TS and perform determination.

So, all above mentioned binders may react with powdery bamboo, which contain functional groups in cellulose, hemicellulose, pentosanes, lignin, and silicon dioxide with formation of three-dimensional systems.

So, in powdery bamboo by insertion of various binders the following reactions may be proceeds:

$$\equiv SiOH + HO– \rightarrow \equiv SiO– + H_2O$$
$$\equiv SiOH + \equiv SiO– \rightarrow \equiv Si–O–Si\equiv + HO^-$$

Using binders PhES-50 and PhES-80 it is also possible condensation reactions ethoxy groups with hydroxyl groups according to the scheme:

$$–OH + C_2H_5O–Si \equiv \rightarrow –O–Si\equiv + C_2H_5OH$$

which may proceed deeper with obtaining three dimensional systems.

4.2 FTIR INVESTIGATION OF THE COMPOSITES

For samples Fourier transform infrared spectroscopy (Varian 660 FTIR) investigation of composites on the basis of bamboo have been carried out in KBr. The KBr pellets of samples were prepared by mixing (1.5–2.00) mg of samples, finely grounded, with 200 mg KBr (FTIR grade) in a vibratory ball mixer for 20s.

We have investigated the FTIR spectra for pure bamboo, two-, three-, and four-components composites systems.

In Figures 4.1, one can observe absorption bands at 831–833 cm^{-1} characteristic for lignin C-H out of plane deformation Glucomannan (is a water-soluble polysaccharide that is considered a dietary fiber. It is a hemicellulose component in the cell walls of some plant species) and aromatic C–H, C_1–H deformation bands at 897 cm^{-1} of cellulose (β-aromatic linkage),

C-O, and Si-O-Si stretching in cellulose and hemicellulose at 1,039–1,073 cm^{-1}, C-O-C symmetric stretching and aromatic C-H in plane deformation and glucose ring vibration at 1,130–1,132 cm^{-1}, Guaiacyl ring breathing with C-O stretching (lignin and hemicelluloses), esters 1,246–1,249 cm^{-1}, phenol O-H group (cellulose) [22] Syringyl ring breathing and C-O stretching CH_2 wagging, Guaiacyl ring breathing [2] at 1,330–1,334 cm^{-1}, C-H in plane deformation (symmetric) for polysaccharides at 1,375–1,376 cm^{-1}, methylene groups of both holocelluloses and lignin, C-H in plane deformation of CH_2 at 1,430 cm^{-1}, C-H deformation (asymmetric) and aromatic skeletal vibrations at 1,457–1,459 cm^{-1}, C=C of Lignin aromatic skeletal (ring) vibrations at 1,508–1,509 cm^{-1}, C=C of aromatic rings at 1,603–1,604 cm^{-1}, conjugated carbonyl C=O (aryl ketone) at 1,634 cm^{-1}, C=O non conjugated carboxylic acid and their ketones at 1,737 cm^{-1}, C-H, and CH_2 stretching at 2,915–2,934 cm^{-1}, O-H stretching at 3,420–3,434 cm^{-1} [24, 25].

In Figure 4.1, the peak at 1,040–1,073 cm^{-1} assigned to the C-O and Si-O-Si stretching vibration and indicates the esters. The weak absorption band of 796, 700, 617 cm^{-1} indicate the presence of the chloride and bromide in our plant systems [23]. It is interesting, that in composites based on bamboo one can see the absorption bands characteristic for asymmetric valence oscillation of bonds ≡Si-O-Si≡ with maximum near 1,055–1,070 cm^{-1}, which possess to siloxane bond in the cyclotetrasiloxane fragment. Besides of the sorption strips are coincided one to other. In the spectrum one can see the absorption bands 1243–1248, 1376, 1453–1463, 1515, 1613–1620, 1735, 2800–2950, 3421–3434 cm^{-1} typical for methyl groups, C-H bond absorption (–C/C–/CH_3), CH_2 cellulose – lignin, C=C aromatic, C=C alkene, (C=O etheric bond), C-H methyl, methylene, and phenyl groups, O-H alcoholic group, respectively. This O-H stretching indicates also the phenolic compound that has excellent antioxidant properties [24] (Figures 4.2–4.7).

4.3 STUDY OF THE MICROSTRUCTURE OF COMPOSITE MATERIALS ON THE BASIS OF BAMBOO

Bamboo composite microstructure of the samples was studied by NMM-800RF/TRF type of optical microscope.

The target of our work was to investigate the structure of the materials. For that we prepared the sample by following the method: after polishing the surface of the model with a longevity of about 1 cm we transferred it to the emery paper and continued this procedure for 1 h, after which the sample was polished on the coarse calico. After this, the composite was transferred to the

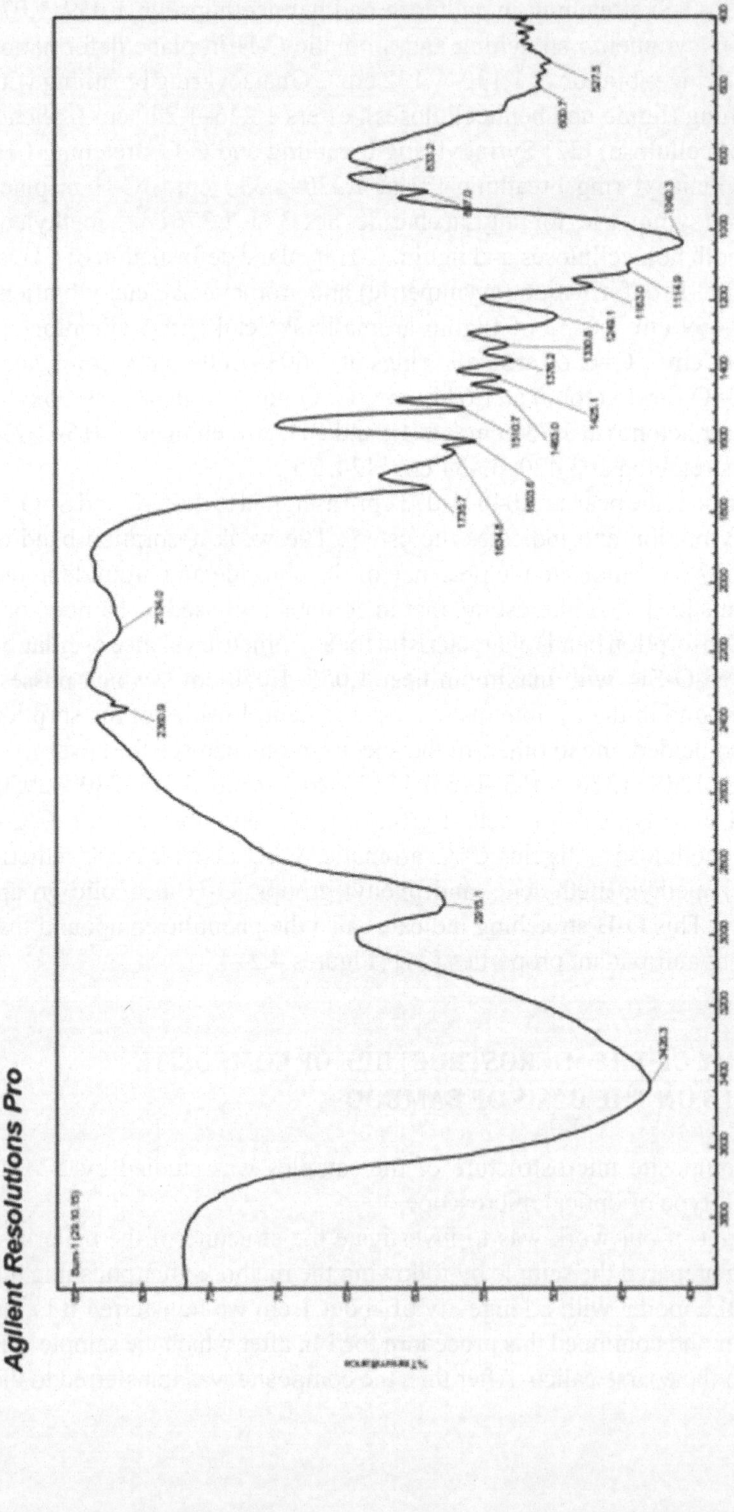

FIGURE 4.1 FTIR spectra of bamboo.

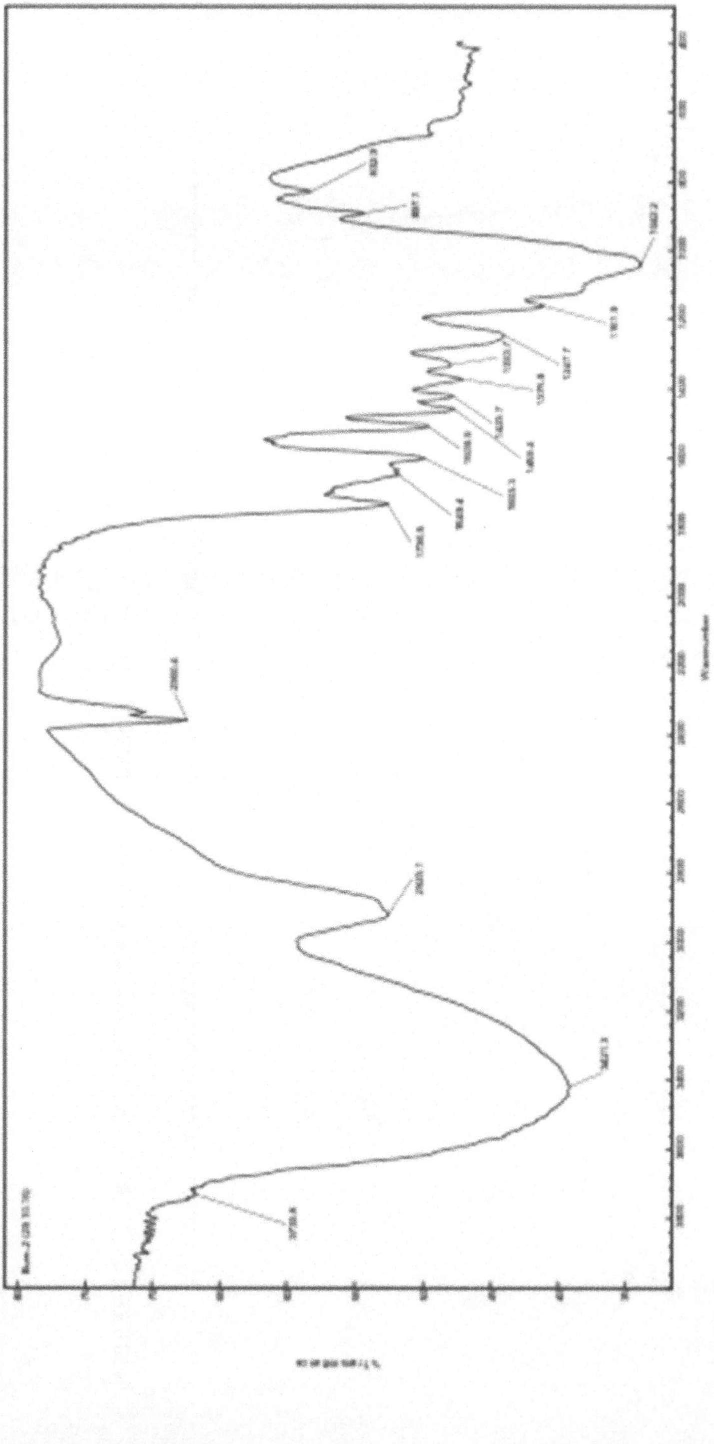

FIGURE 4.2 FTIR spectra of composite 4.2 bamboo 95% + LG 5%.

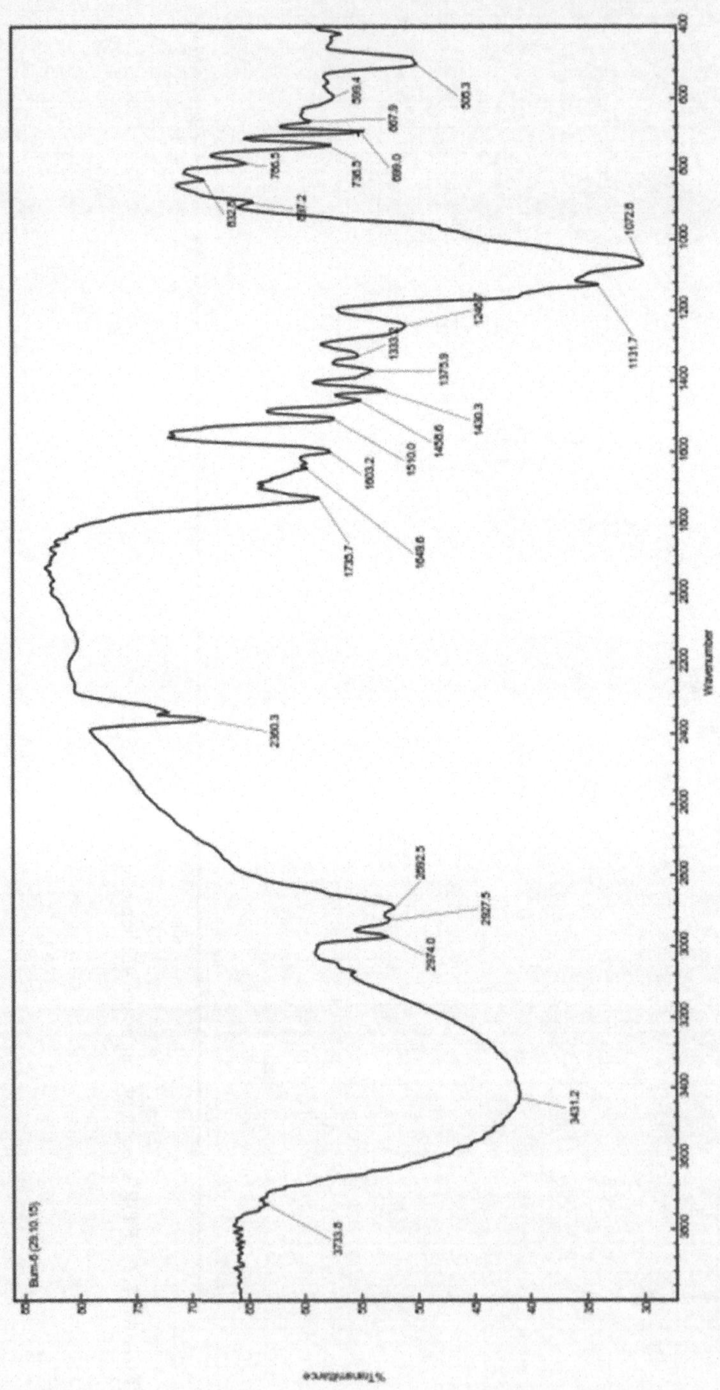

FIGURE 4.3 FTIR spectra of composite 4.6 bamboo 95% + PhES-50 – 5%.

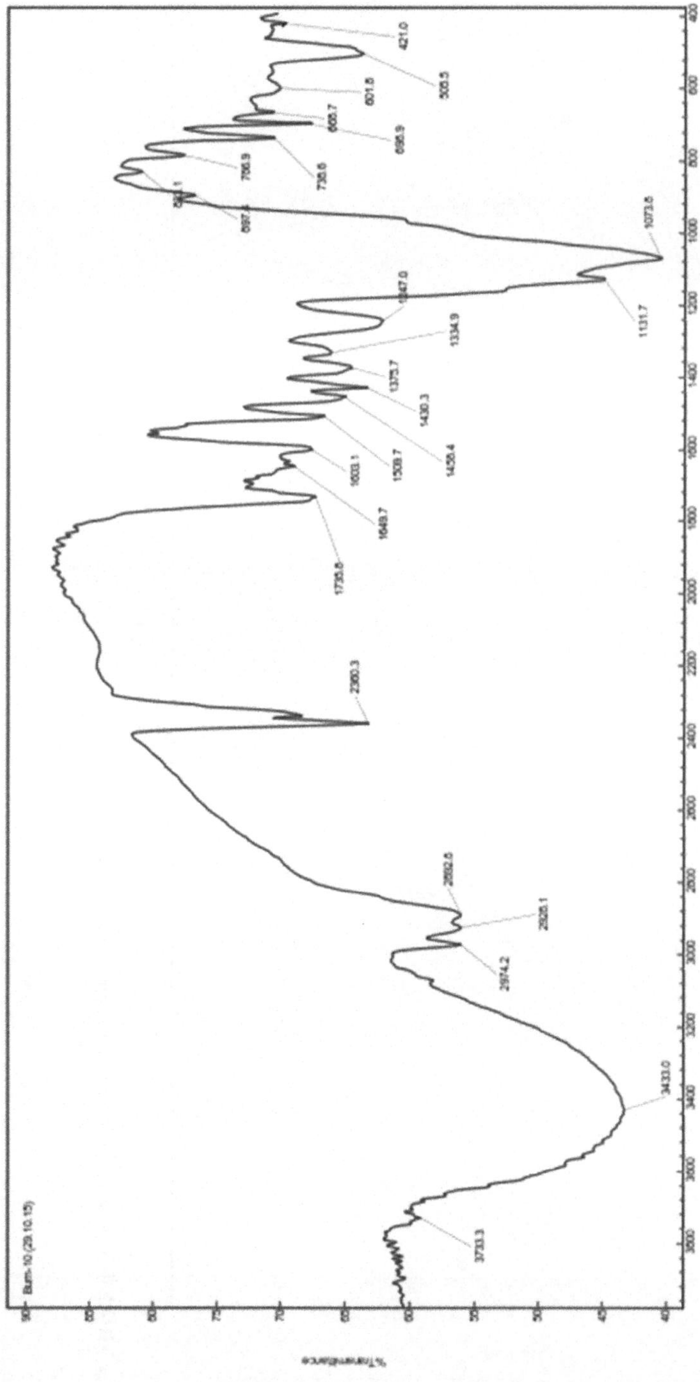

FIGURE 4.4 FTIR spectra of composite 4.10 bamboo 95% + PhES-80 – 5%.

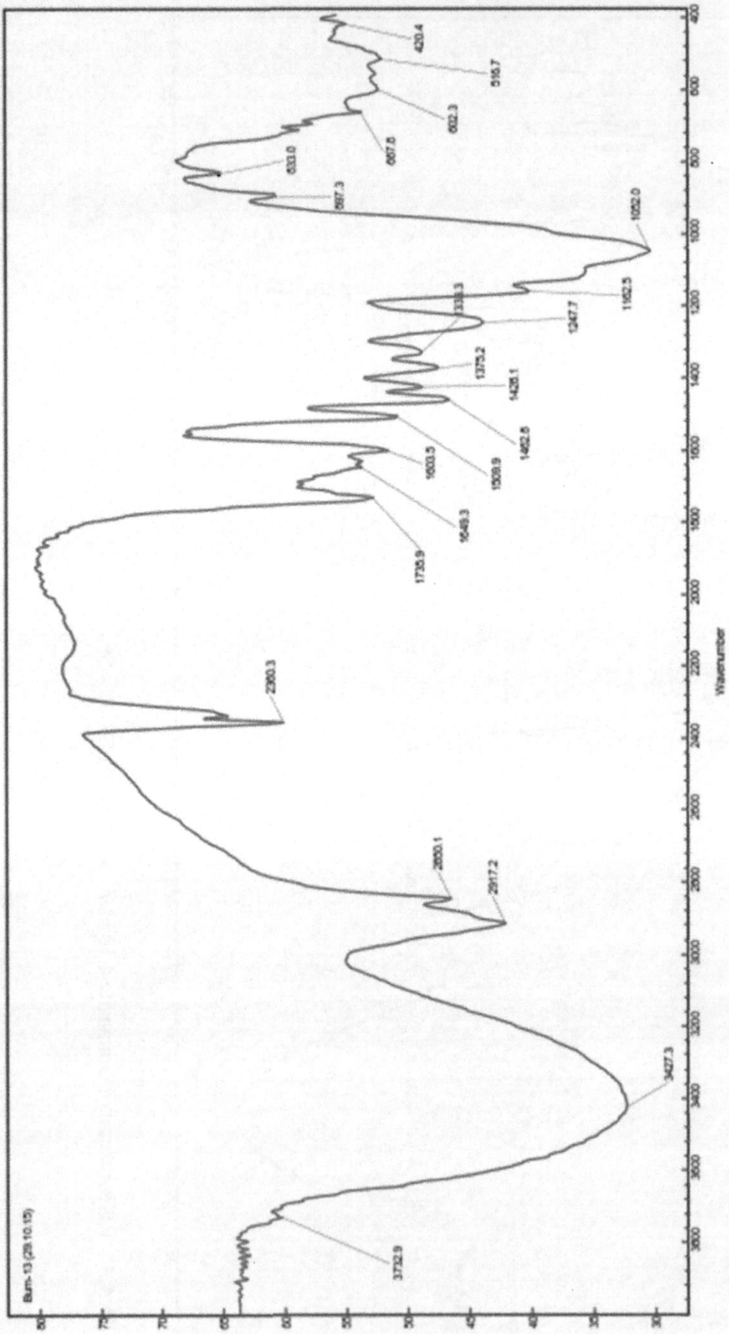

FIGURE 4.5 FTIR spectra of composite 4.15 bamboo 85% + PE 15%.

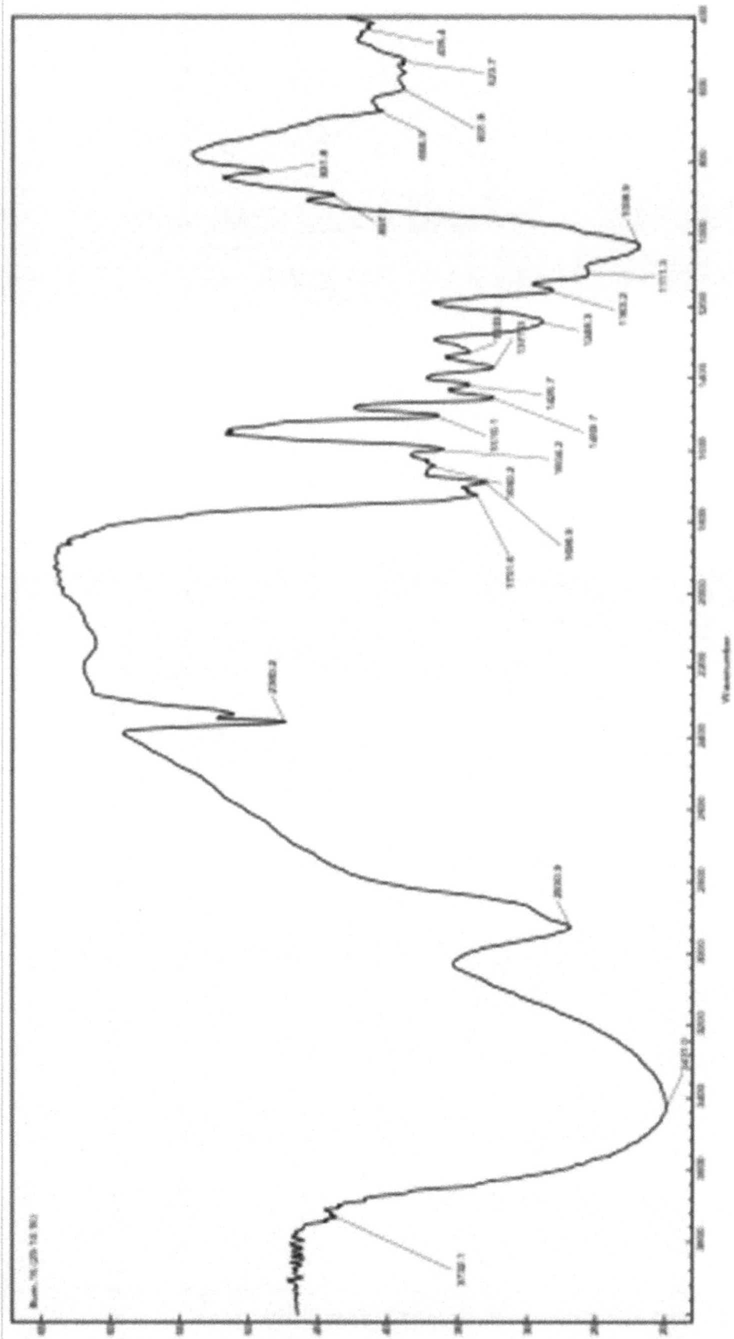

FIGURE 4.6 FTIR spectra of composite 4.18 bamboo 90% + colophony 10%.

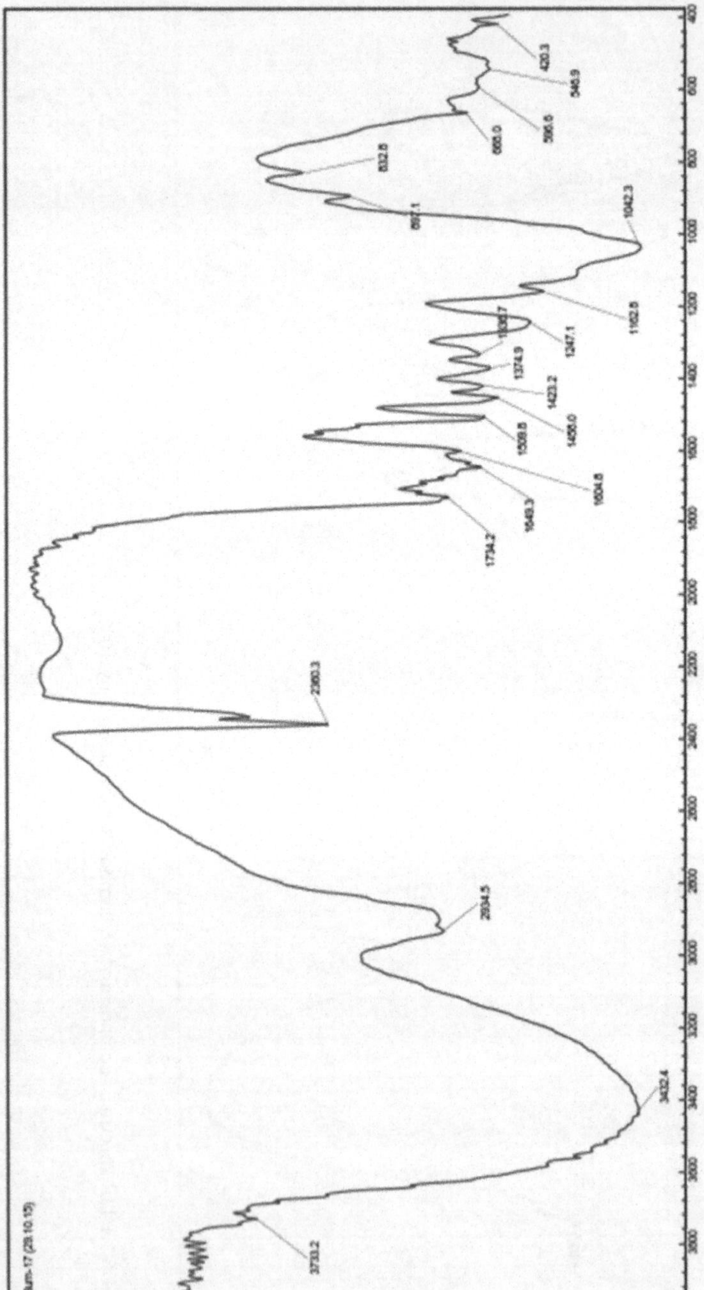

FIGURE 4.7 FTIR spectra of composite 4.22 bamboo 90% + wood glue 10%.

object glass of the optical microscope. The separate parts of the model were studied using the noted optical microscope. The composite obtained under conditions of 150 kg/cm^2 and temperature near 130°C were studied [27, 28].

Particularly, on the figures there are presented the samples of bamboo and the composites based on the bamboo with different concentrations of the binder.

The different color of the patterns caused due to the lighting and temperature of the room. On the figure is indicated the scale of the particles, from figures, we can say that the components are well distributed.

Some samples of the design after grinding remaining scratches can be seen (magnification-x50). The composite (4.6) is a multiplicity component system but does not show a pronounced binder in the form of insertion, which indicates that the optimum temperature of sintering is selected. Gaps and cracks do not observe.

In Figures 4.9–4.15, the optical microscopic data of two and three component containing composite materials is presented, which shows the different colors for different inserts. Inserts do not have a clearly defined border, so the scale is indicated in the figure and changes in the range 400–450 mcm. In Figures 4.8–4.15, one can see fibrous character of supramolecular structure.

In Figure 4.8, is presented optical microscopic data of the bamboo: magnification x50, bamboo particle is quite large and all different sizes, so the scale of the figure, which is about 400 mcm, is within the limits.

FIGURE 4.8 Optical microscopic data of bamboo, magnification is about x50.

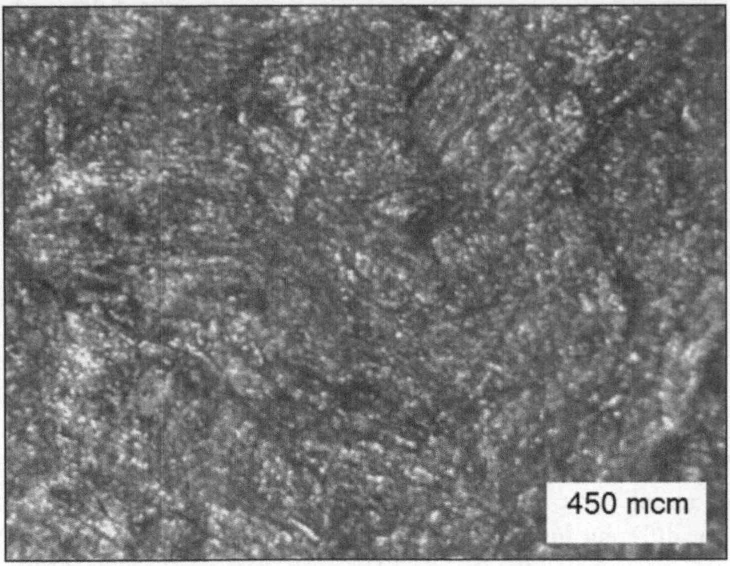

FIGURE 4.9 Optical microscopic data of composite 4.2 (Bamboo 95% – LG 5%), magnification x50.

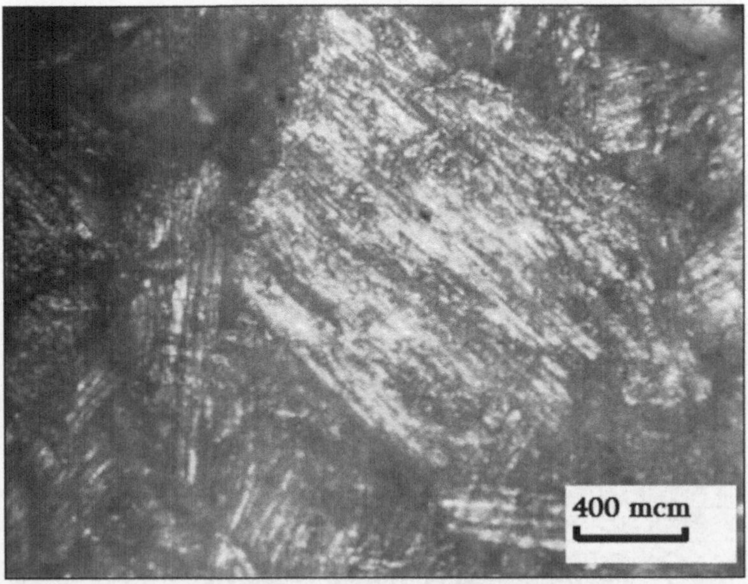

FIGURE 4.10 Optical microscopic data of composite 4.10 (Bamboo 95% – PhES-80 – 5%), magnification x50.

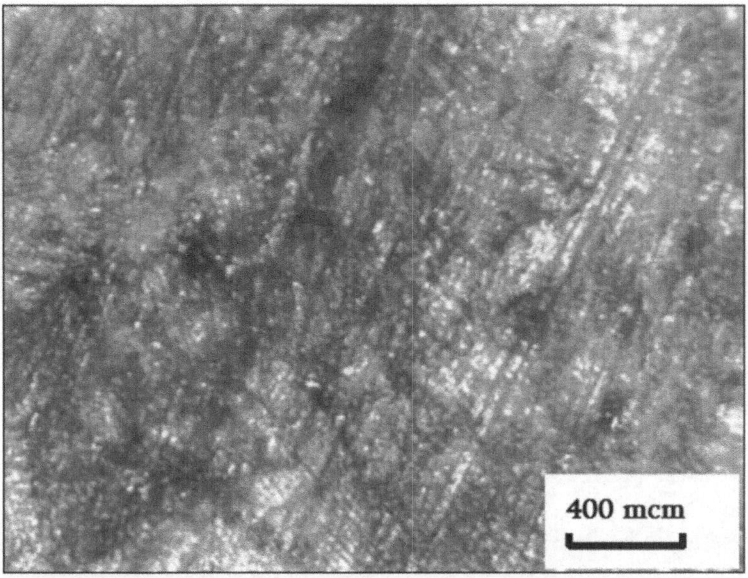

FIGURE 4.11 Optical microscopic data of composite 4.14 (Bamboo 90% – PE 10%), magnification x50.

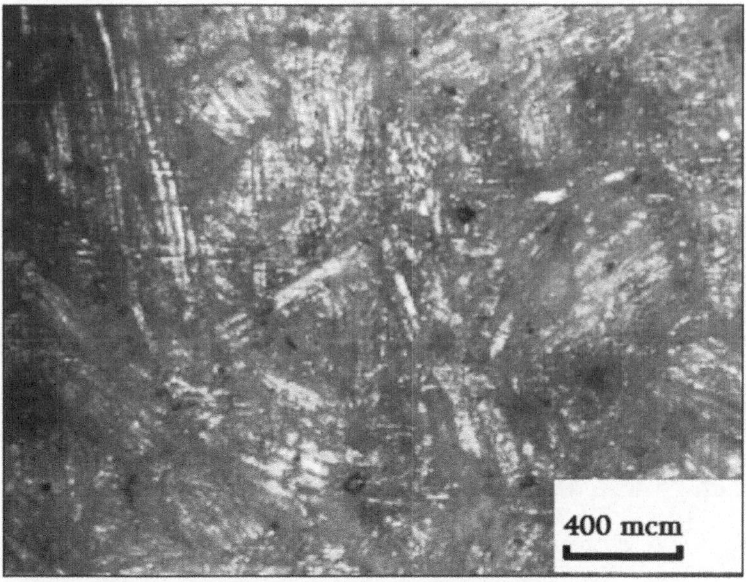

FIGURE 4.12 Optical microscopic data of composite 4.22 (Bamboo 90% – Wood glue 10%), magnification x50.

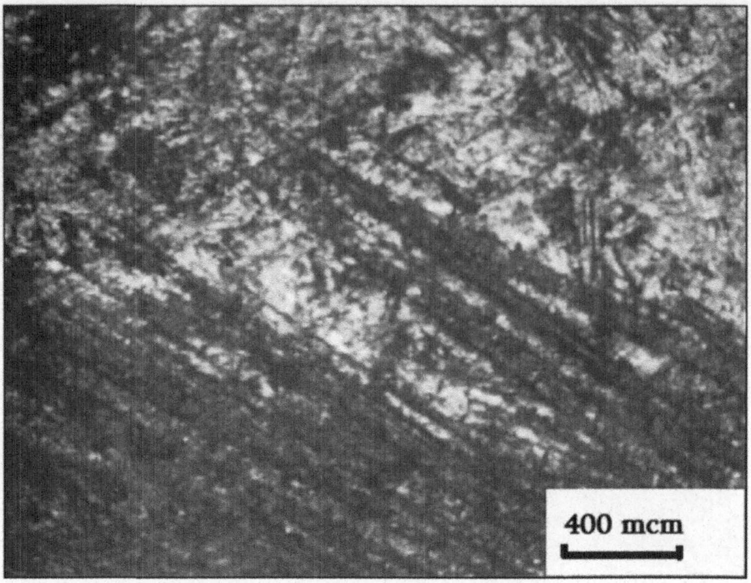

FIGURE 4.13 Optical microscopic data of composite 4.18 (Bamboo 90% – Colophony 10%).

FIGURE 4.14 Optical microscopic data of composite 4.42 (Bamboo 90% – PhES-80 – 5% – PE 5%).

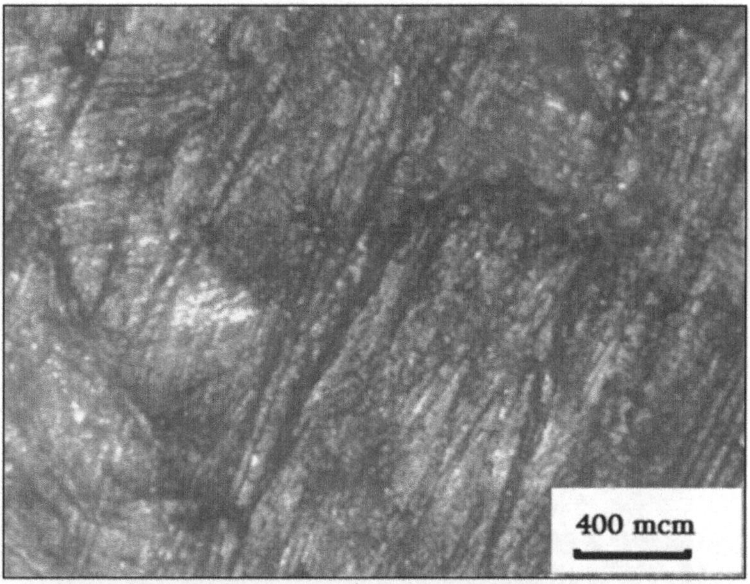

FIGURE 4.15 Optical microscopic data of composite 4.35 (Bamboo 85% – PhES-80 – 5% – LG 5% – PE 5%), magnification x50.

4.4 SCANNING ELECTRON MICROSCOPIC (SEM) AND ENERGY DISPERSION MICRO-X-RAY ANALYSIS OF COMPOSITE MATERIALS ON THE BASIS OF LEAVES

For composite materials made on the basis of bamboo and different hardeners and additives, SEM investigations are provided. In addition, energy-dispersive X-ray examinations have been carried out (EDS).

As can be seen from Figures 4.18–4.20, 4.22–4.24, and 4.26, studies of SEM it is seen the outer epidermis of the leaf, which has a well-organized corrugated structure. The structure includes the sharpened ridge inserts, which have a linear profile. Lemy pointed ridges and grooves in the epidermal cell are connected to a fiber inclusion. Outer surface of the lemy hills includes the knobs of various sizes, which often have a discrete character and usually do not be considered.

Bamboo is divided into 2 major portions, the rhizomes, and the culms. The rhizome is the underground part of the stem and is mostly sympodial or, to a much lesser degree, monopodial.

Interfacial gaps between fibers and polymer are clearly seen in Figures 4.16, 4.21, 4.25, 4.31, and 4.37. Moreover, from these micrographs, the different fiber surfaces were also seen.

Figures 4.16, 4.21, 4.25, 4.31, and 4.37 shows the impact fractured surfaces of the blends obtained from various hardeners. The distribution and aggregation of bamboos particles in the polymer matrix can be observed from Figures 4.16, 4.21, 4.25, 4.31, and 4.37. The figures demonstrated clearly fractured features full of fibrils (Figures 4.16–4.40 and Tables 4.2–4.6).

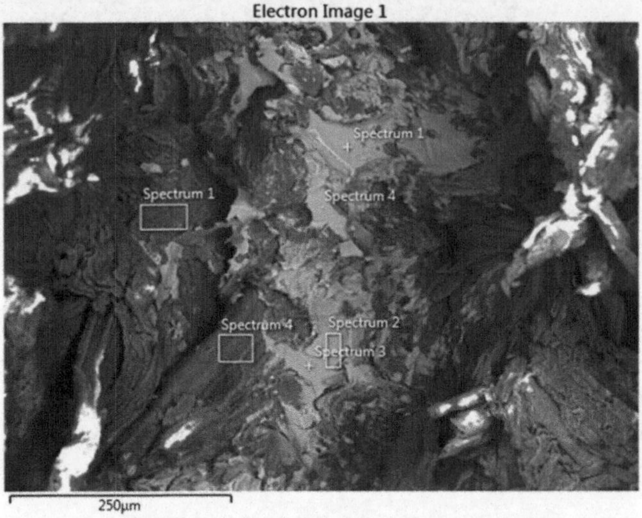

FIGURE 4.16 Scanning electronic microscopic photo for composite 4.2: (Bamboo 95% + LG 5%).

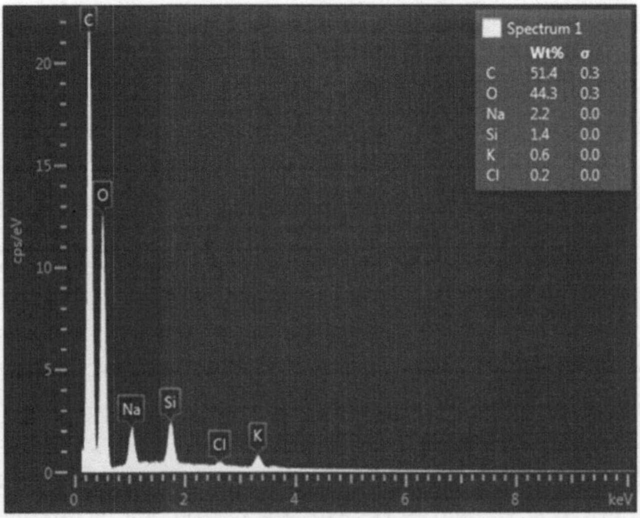

FIGURE 4.17 Energy dispersion micro-X-ray spectral analysis of composite 4.2 (Spectrum 1).

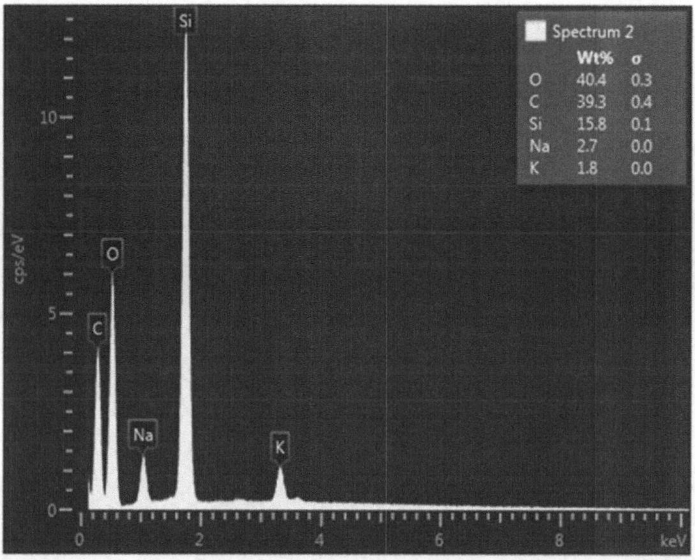

FIGURE 4.18 Energy dispersion micro-X-ray spectral analysis of composite 4.2 (Spectrum 2).

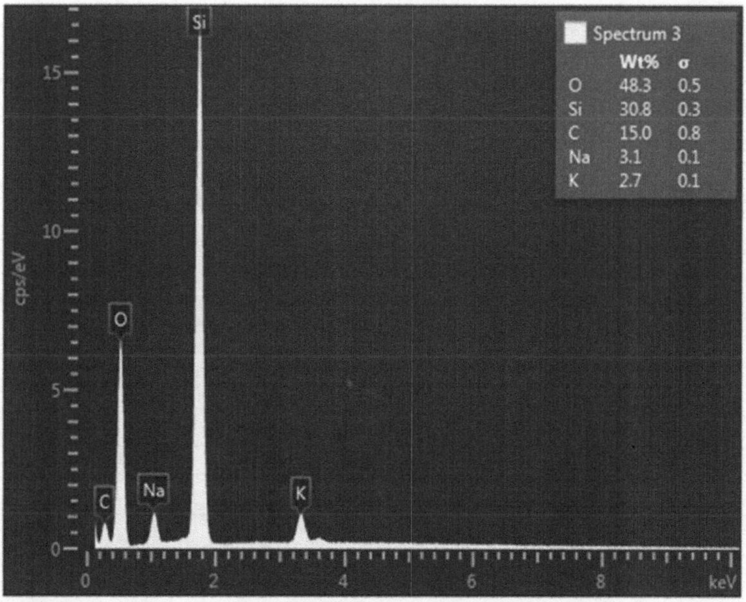

FIGURE 4.19 Energy dispersion micro-X-ray spectral analysis of composite 4.2 (Spectrum 3).

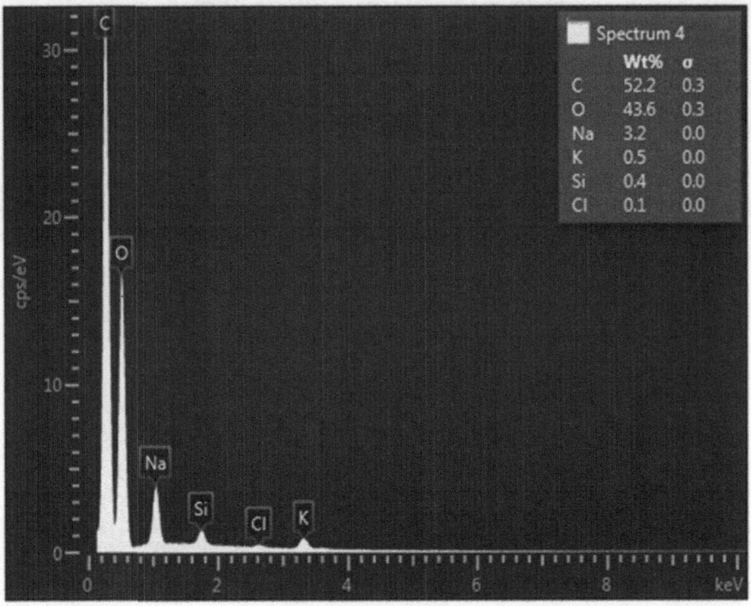

FIGURE 4.20 Energy dispersion micro-X-ray spectral analysis of composite 4.2 (Spectrum 4).

TABLE 4.2 Energy Dispersion Micro-X-Ray Spectral Analysis of Composite 4.2 (Bamboo 95% + Liquid Glass 5%)

Result Type	**Weight (%)**			
Spectrum Label	**Spectrum 1**	**Spectrum 2**	**Spectrum 3**	**Spectrum 4**
C	51.35	39.33	15.03	52.20
O	44.31	40.37	48.34	43.61
Na	2.22	2.68	3.11	3.21
Si	1.37	15.80	30.78	0.42
Cl	0.16	–	–	0.08
K	0.59	1.82	2.74	0.47
Total	**100.00**	**100.00**	**100.00**	**100.00**

Statistics	**C**	**O**	**Na**	**Si**	**Cl**	**K**
Max.	52.20	48.34	3.21	30.78	0.16	2.74
Min.	15.03	40.37	2.22	0.42	0.08	0.47
Average	39.48	44.16	2.80	12.09	–	1.41
Standard deviation	17.33	3.27	0.45	14.31	–	1.08

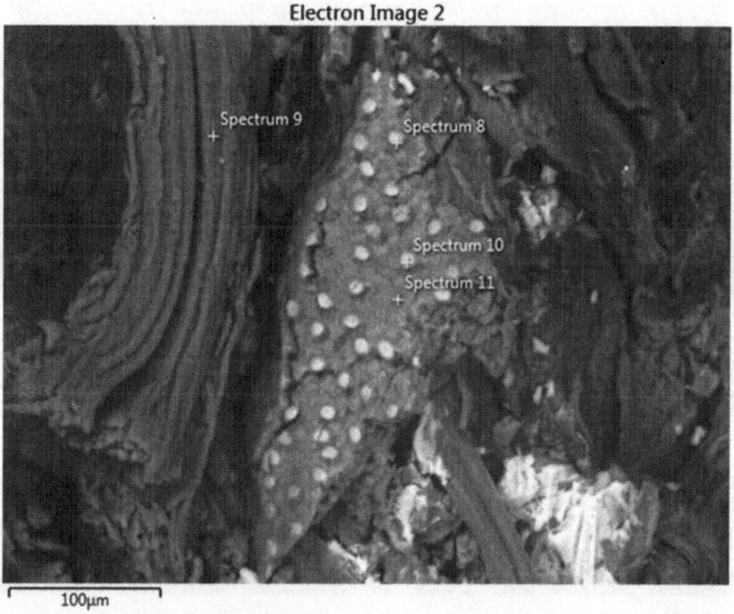

FIGURE 4.21 Scanning electronic microscopic photo for composite 4.10: Bamboo 95% + PhES-80 – 5% (magnification x100).

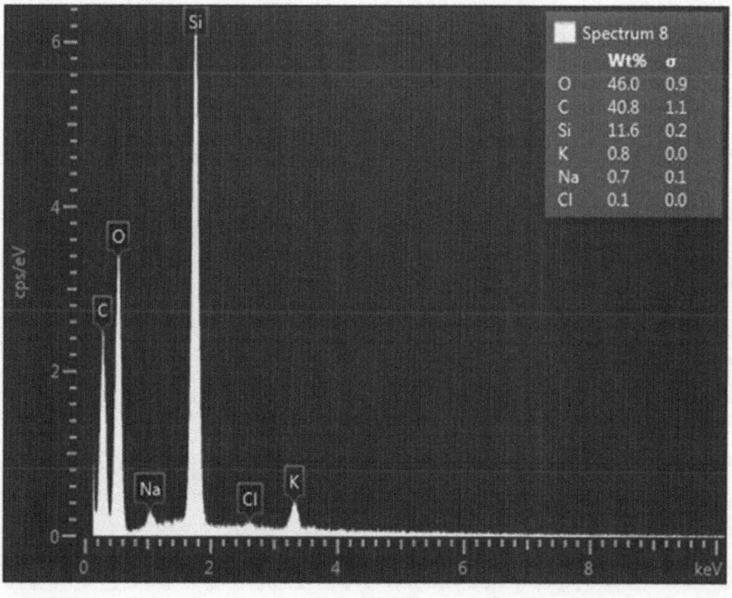

FIGURE 4.22 Energy dispersion micro-X-ray spectral analysis of composite 4.10 (Spectrum 8).

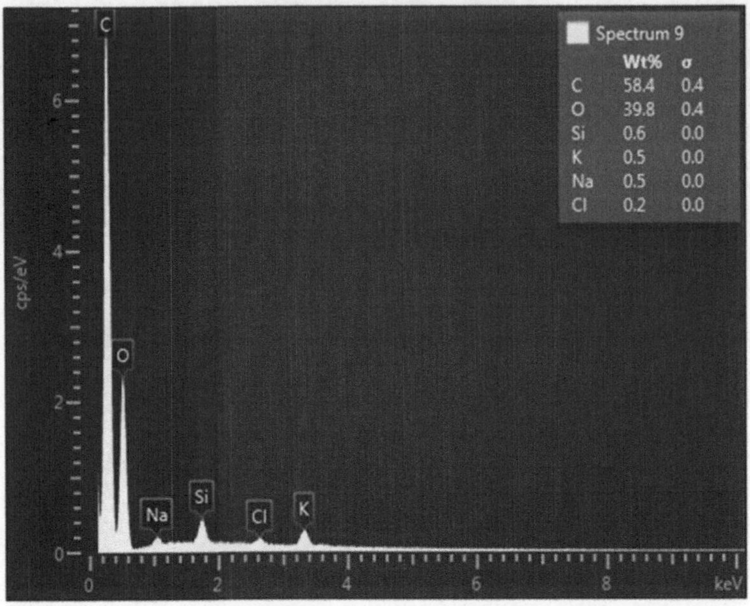

FIGURE 4.23 Energy dispersion micro-X-ray spectral analysis of composite 4.10 (Spectrum 9).

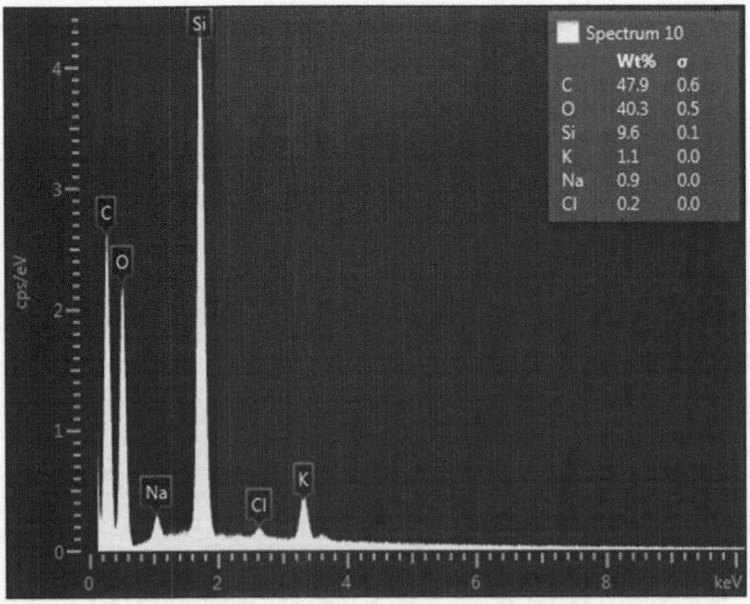

FIGURE 4.24 Energy dispersion micro-X-ray spectral analysis of composite 4.10 (Spectrum 10).

TABLE 4.3 Energy Dispersion Micro-X-Ray Spectral Analysis of Composite 4.10 (Bamboo 95% + PhES-80 – 5%)

Result Type	Weight (%)			
Spectrum Label	**Spectrum 8**	**Spectrum 9**	**Spectrum 10**	**Spectrum 11**
C	39.53	59.43	47.90	40.39
O	46.96	38.84	40.33	42.80
Na	0.81	0.49	0.95	0.52
Si	11.67	0.57	9.55	15.60
Cl	0.16	0.18	0.20	0.09
K	0.87	0.48	1.08	0.61
Total	**100.00**	**100.00**	**100.00**	**100.00**

Statistics	C	O	Na	Si	Cl	K
Max.	58.43	45.96	0.95	15.60	0.20	1.08
Min.	40.39	39.84	0.47	0.59	0.09	0.50
Average	46.88	42.23	0.67	9.32	0.14	0.75
Standard deviation	8.44	2.80	0.22	6.34	0.05	0.25

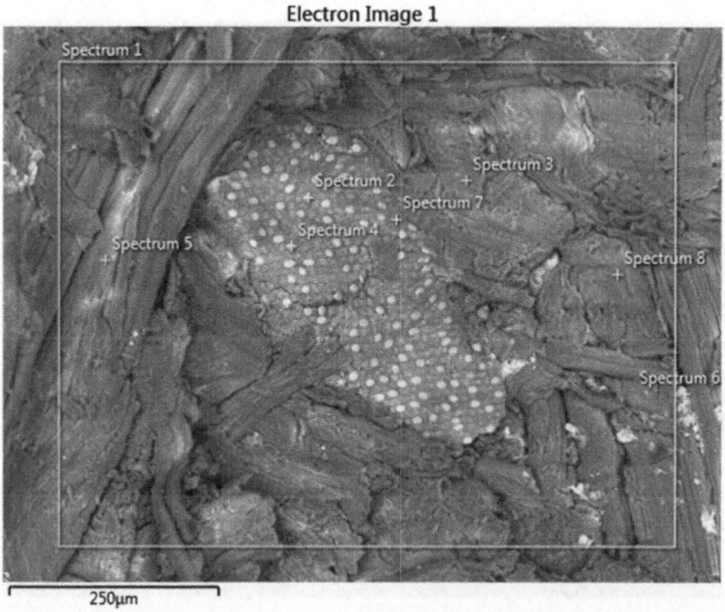

FIGURE 4.25 Scanning electronic microscopic photo for composite 4.14: Bamboo 90% + PE 5% (magnification x250).

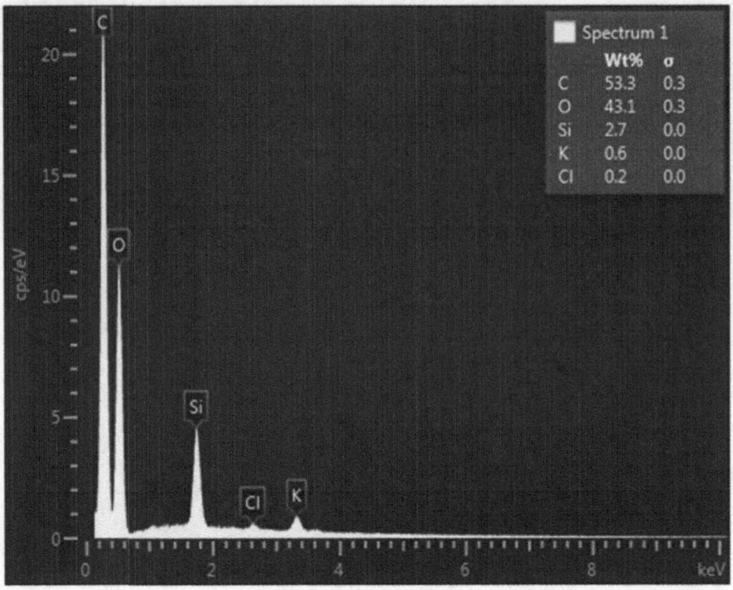

FIGURE 4.26 Energy dispersion micro-X-ray spectral analysis of composite 4.14 (Bamboo 90% + PE 10%) (Spectrum 1).

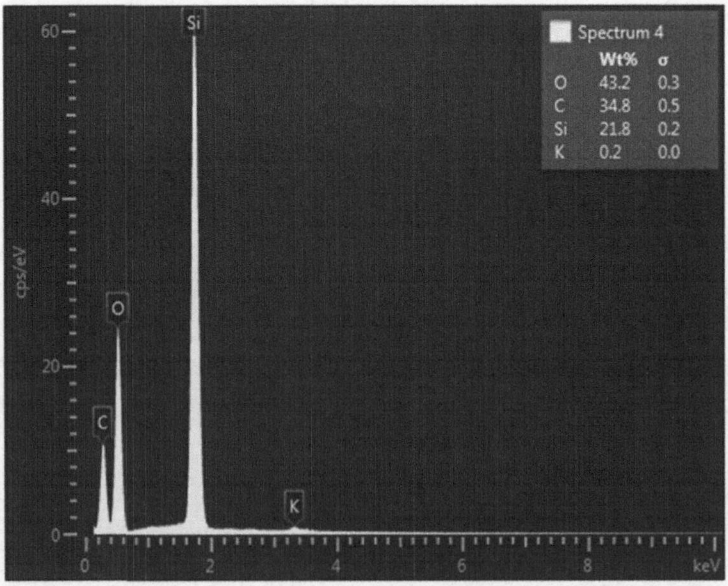

FIGURE 4.27 Energy dispersion micro-X-ray spectral analysis of composite 4.14 (Bamboo 90% + PE 10%) (Spectrum 4).

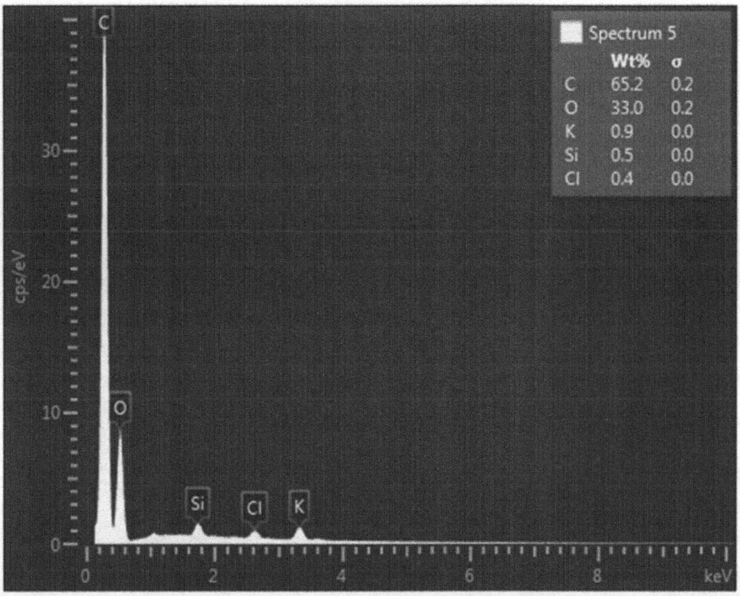

FIGURE 4.28 Energy dispersion micro-X-ray spectral analysis of composite 4.14 (Bamboo 90% + PE 10%) (Spectrum 5).

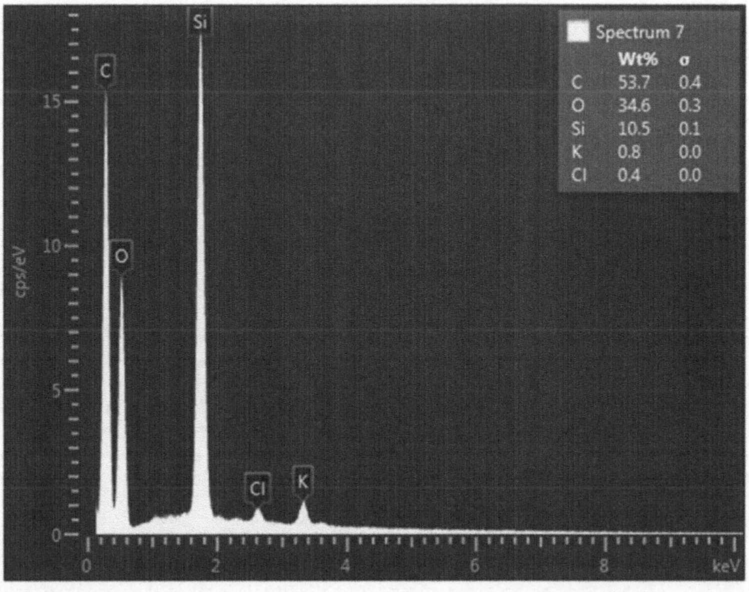

FIGURE 4.29 Energy dispersion micro-X-ray spectral analysis of composite 4.14 (Bamboo 90% + PE 10%) (Spectrum 7).

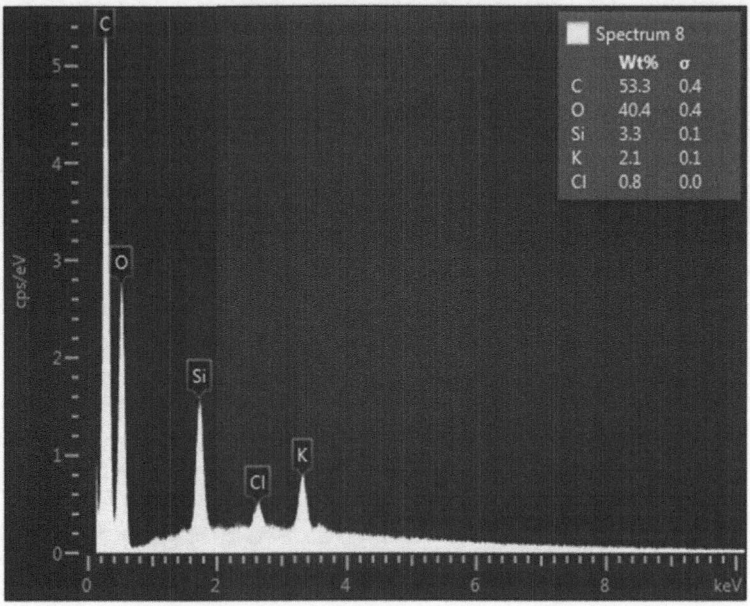

FIGURE 4.30 Energy dispersion micro-X-ray spectral analysis of composite 4.14 (Bamboo 90% + PE 10%) (Spectrum 8).

TABLE 4.4 Energy Dispersion Micro-X-Ray Spectral Analysis of Composite 4.14 (Bamboo 90% + PE 10%)

Result Type	**Weight (%)**							
Spectrum Label	**Spect-rum 1**	**Spect-rum 2**	**Spect-rum 3**	**Spect-rum 4**	**Spect-rum 5**	**Spect-rum 6**	**Spect-rum 7**	**Spect-rum 8**
C	54.32	54.08	60.79	34.81	65.18	28.58	53.67	53.31
O	42.11	32.98	35.41	43.24	33.00	49.05	34.65	40.42
Si	2.75	11.71	2.17	21.77	0.53	21.77	10.50	3.34
Cl	0.17	0.23	0.49	–	0.43	–	0.40	0.80
K	0.65	1.00	1.14	0.17	0.85	0.59	0.78	2.12
Total	**100.00**	**100.00**	**100.00**	**100.00**	**100.00**	**100.00**	**100.00**	**100.00**

Statistics	**C**	**O**	**Si**	**Cl**	**K**
Max.	65.18	49.05	21.77	0.80	2.12
Min.	28.58	31.98	0.53	0.19	0.17
Average	50.72	38.73	9.32	–	0.91
Standard deviation	12.63	6.12	8.65	–	0.57

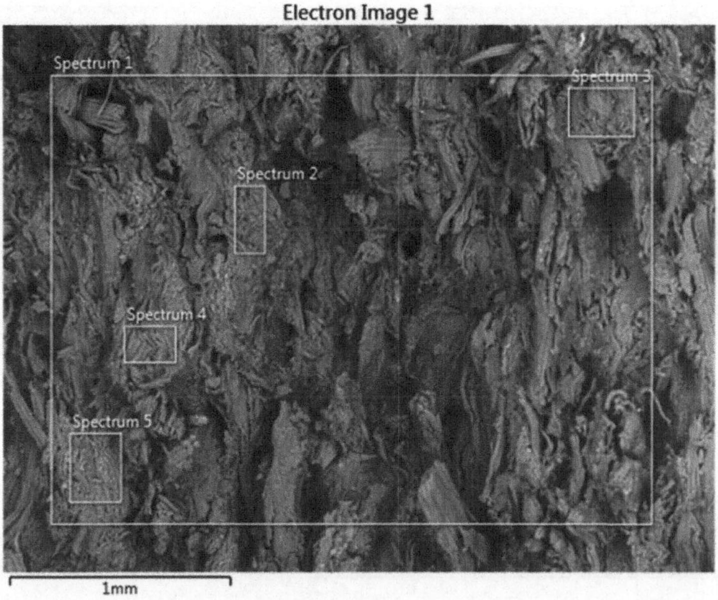

FIGURE 4.31 Scanning electronic microscopic photo for composite 4.18: Bamboo 90% + Colophony 10% (magnification x250).

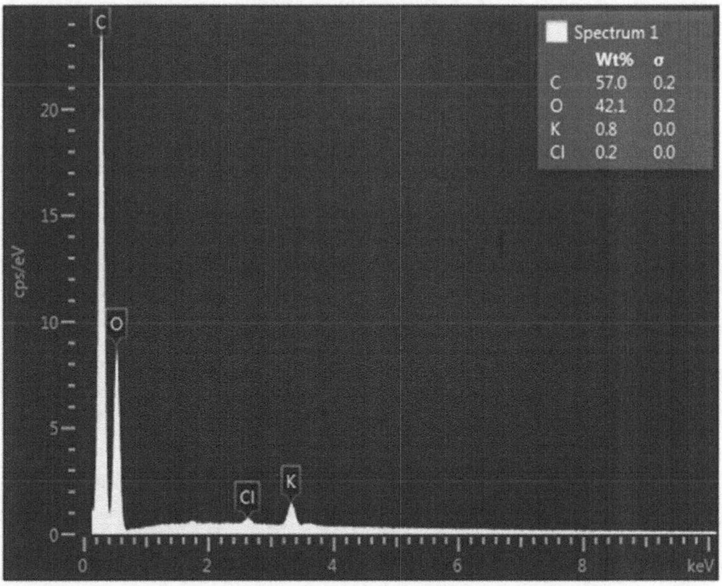

FIGURE 4.32 Energy dispersion micro-X-ray spectral analysis of composite 4.18 (Bamboo 90% + Colophony 10%) (Spectrum 1).

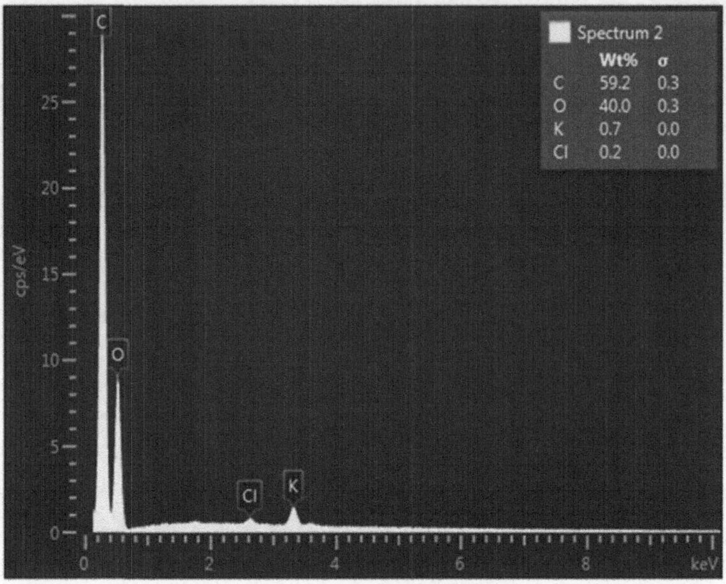

FIGURE 4.33 Energy dispersion micro-X-ray spectral analysis of composite 4.18 (Bamboo 90% + Colophony 10%) (Spectrum 2).

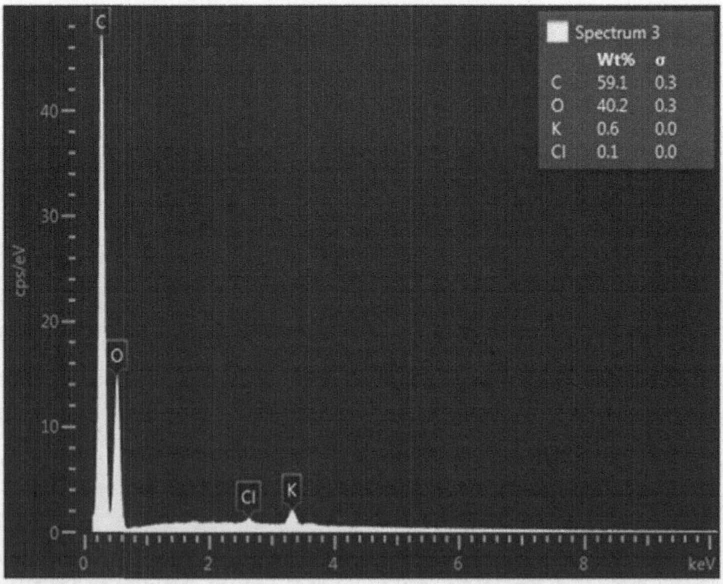

FIGURE 4.34 Energy dispersion micro-X-ray spectral analysis of composite 4.18 (Bamboo 90% + Colophony 10%) (Spectrum 3).

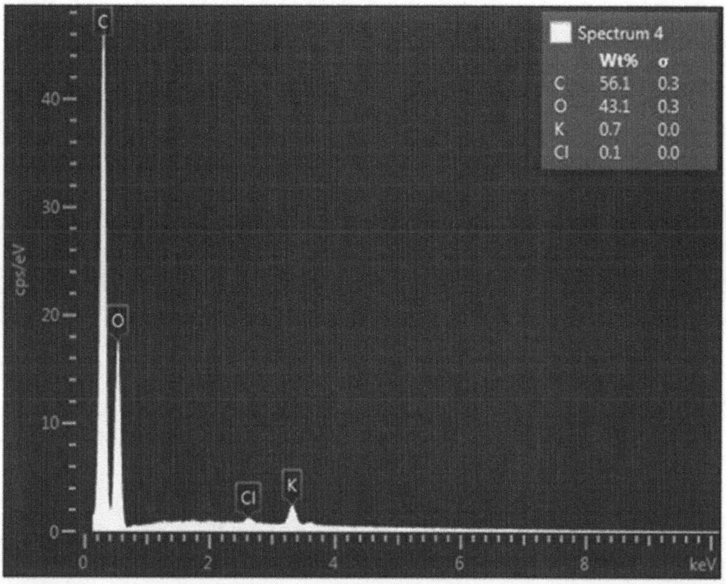

FIGURE 4.35 Energy dispersion micro-X-ray spectral analysis of composite 4.18 (Bamboo 90% + Colophony 10%) (Spectrum 4).

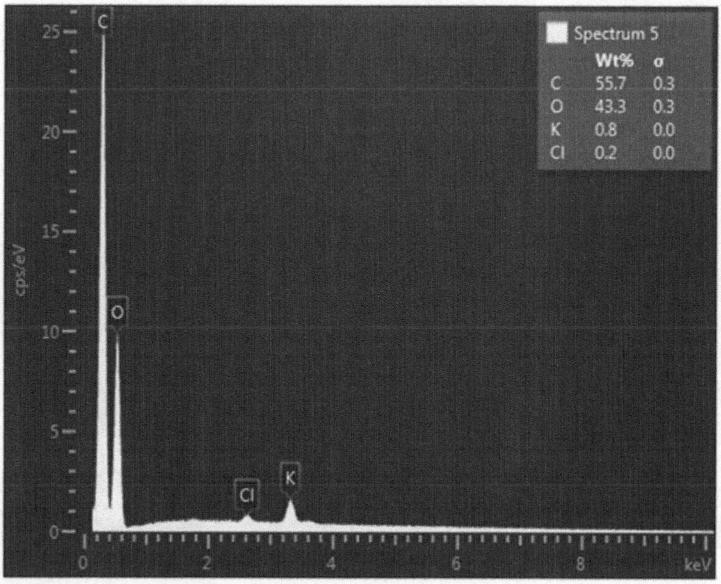

FIGURE 4.36 Energy dispersion micro-X-ray spectral analysis of composite 4.18 (Bamboo 90% + Colophony 10%) (Spectrum 5).

TABLE 4.5 Energy Dispersion Micro-X-Ray Spectral Analysis of Composite 4.18 (Bamboo 90% + Colophony 10%)

Result Type			**Weight (%)**		
Spectrum Label	**Spectrum 1**	**Spectrum 2**	**Spectrum 3**	**Spectrum 4**	**Spectrum 5**
C	56.98	59.20	59.11	56.07	55.70
O	42.09	39.96	40.20	43.11	43.35
Cl	0.17	0.17	0.14	0.15	0.20
K	0.76	0.67	0.56	0.68	0.76
Total	**100.00**	**100.00**	**100.00**	**100.00**	**100.00**

Statistics	**C**	**O**	**Cl**	**K**
Max.	59.20	43.35	0.20	0.76
Min.	55.70	39.96	0.14	0.56
Average	57.41	41.74	0.17	0.69
Standard deviation	1.66	1.59	0.02	0.08

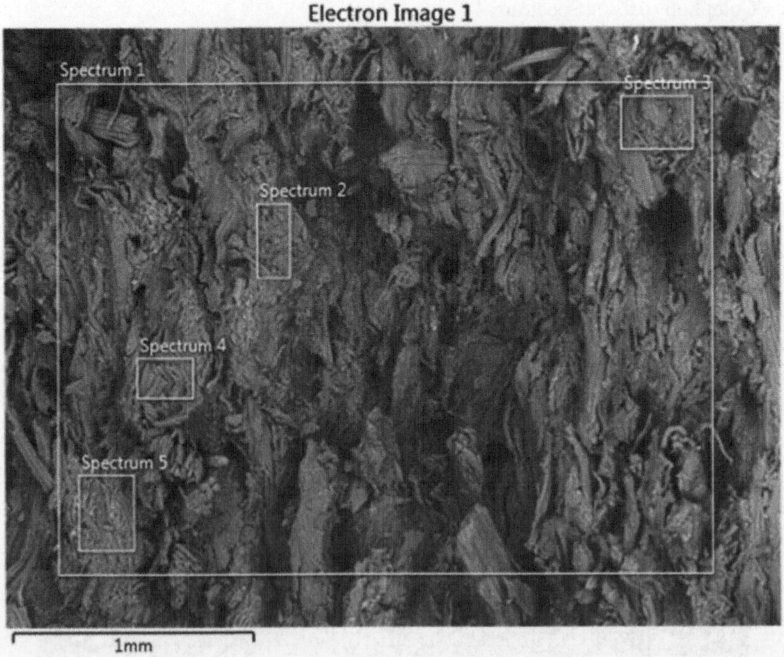

FIGURE 4.37 Scanning electronic microscopic photo for composite 4.22: Bamboo 90% + Wood glue 10% (magnification x250).

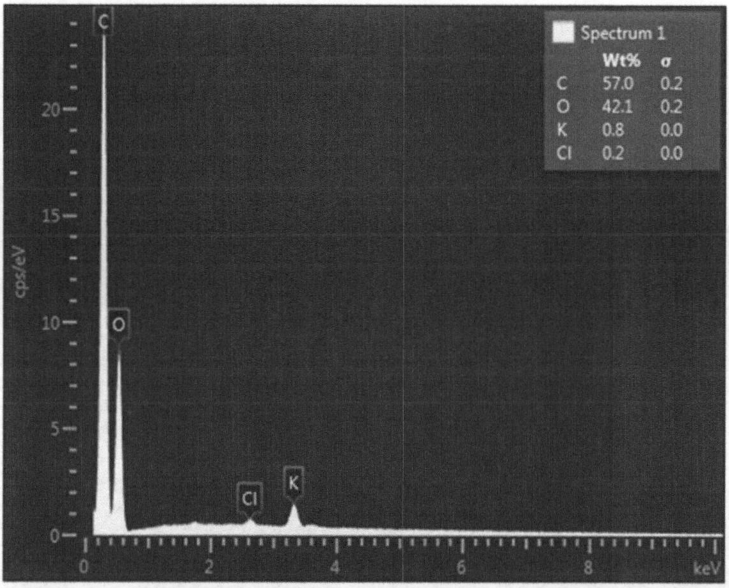

FIGURE 4.38 Energy dispersion micro-X-ray spectral analysis of composite 4.22 (Bamboo 90% + Wood glue 10%) (Spectrum 1).

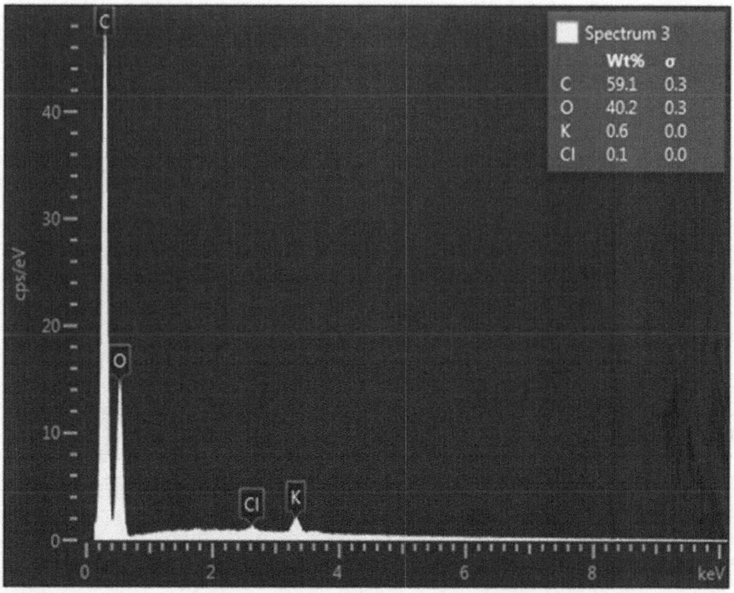

FIGURE 4.39 Energy dispersion micro-X-ray spectral analysis of composite 4.22 (Bamboo 90% + Wood glue 10%) (Spectrum 3).

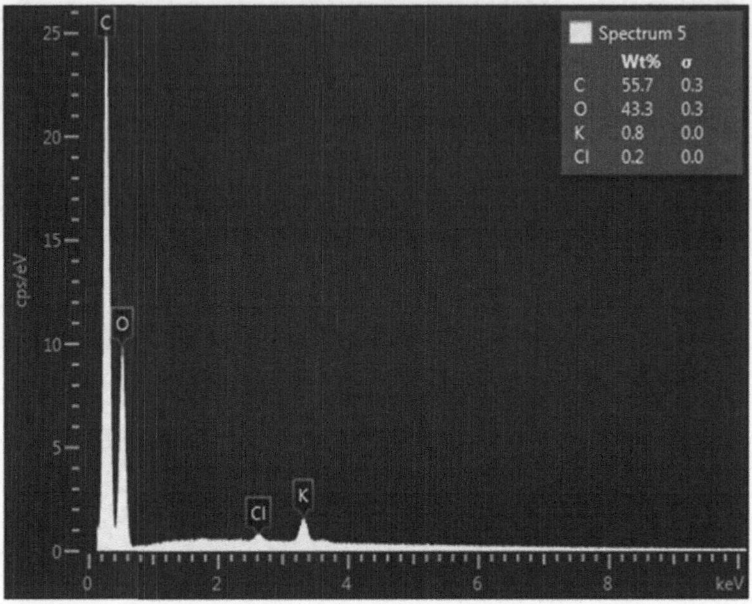

FIGURE 4.40 Energy dispersion micro-X-ray spectral analysis of composite 4.22 (Bamboo 90% + Wood glue 10%) (Spectrum 5).

TABLE 4.6 Energy Dispersion Micro-X-Ray Spectral Analysis of Composite 4.22 (Bamboo 90% + Wood Glue 10%)

Result Type	**Weight (%)**				
Spectrum Label	**Spectrum 1**	**Spectrum 2**	**Spectrum 3**	**Spectrum 4**	**Spectrum 5**
C	58.76	60.17	57.97	56.07	55.70
O	41.09	38.96	41.20	43.11	43.35
Cl	0.15	0.18	0.17	0.15	0.20
K	0.77	0.69	0.66	0.68	0.76
Total	**100.00**	**100.00**	**100.00**	**100.00**	**100.00**

Statistics	**C**	**O**	**Cl**	**K**
Max.	60.17	43.35	0.20	0.77
Min.	55.70	39.96	0.14	0.56
Average	57.41	41.74	0.17	0.69
Standard deviation	1.66	1.59	0.02	0.08

From all these micrograms it is evident that fiber pullout is observed, indicating a poor bonding between the fibers. When the interfacial bonding is partially, the mechanical properties of the composites will be inferior.

SEM and energy dispersive Micro-X-ray spectral studies provide a chemical analysis of samples of the smallest particles. Under this method, using low-molecular-detector can be defined more than 90 element quantitative composition. In Figures 4.18–4.20, 4.22–4.24, 4.26–4.30, 4.32–4.36, 4.38, and 4.39, are given for each composite spectrum characteristic X-ray emission spectrum and corresponding micro-chemical analysis, which is based on the X-ray excitation of the sample. Its fundamental principle is based on the fact that every element has a unique atomic structure, which makes it possible to give us a series of X-ray spectra reflexes, with quantitative analysis of some elements. As well as in Tables 4.2–4.6, for each spectrum are given the characteristic composition. It was proved that composites obtained from leaves contain the following elements: C, O, Si, Ca, Na, K, S, Fe, Mg, Cl, and Al. These content of elements in samples of the spectra changes a little.

4.5 PHYSICAL–MECHANICAL PROPERTIES OF COMPOSITES

First of all, the installation for shredding of dry bamboo was designed. It presents the complex of milling cutter (with diameter about 120 mm). This apparatus allows obtain of milled (to some extent) sawdust containing the particles with sizes from 50 microns up to 1 mm. Liquid glass (LG), PhES-50 and PhES-80, polyethylene (PE), colophony, and wood glue at different concentrations (3–20 wt.%) were used as binders of the composites. The composites were obtained by same manner as in case of other analogical composites based on plant (see materials of early accounts) in the press-forms at external pressure 150 kg/cm^2 and temperature 130°C. Following properties were tested: mechanical strength at bending, toughness, TS.

Following composites on the basis of bamboo were obtained and investigated in the project frames: (a) composites with two ingredients (bamboo + binder 1); (b) composites with three ingredients (bamboo + binder 1 + binder 2). The numerical data of the noted parameters for these composites are tabulated in Table 4.7.

In accordance with Table 4.7, data the mechanical strength of type of shock viscosity for composites containing liquid glass are characterized with small maximum at 10 wt.% of binder, although the mechanical strength at bending in the presented range of concentrations such maximum is absent. In this case extreme magnitude of mechanical strength at bending may be at more high concentrations of this binder. With maximums of the strength at bending the composites containing PhES-50 or PhES-80 are characterized at concentrations of binders 5 and 10 wt.%, respectively. For the same composites shock viscosity maximums correspond to concentration 10 wt.%. The main reason of such dependences found on the degree of homogeneity of distribution of a binder substance in the composites. At relatively high content of binders, it takes place the creation of own structural phase of the binders as clusters, which in general lead to weakening of the composite's mechanical properties. With relatively high mechanical properties are characterized the composites with colophony and wood glue, which is due to homogeneity of distribution of binder in composites because of good wettability of these substances initiated by melting of these substances in the process of the heating of press vessels under pressure. The mechanical properties of composites with double binders depend on the proportion of them in composites. For example, the shock viscosity for composite containing PhES-50 and liquid glass (10 and 10 wt.%, respectively) reach to 64.8 kj/cm^2. It is evident that by finding of optimal proportion of the components it can be formed the structure with more high mechanical properties. The attempt of obtaining of the composites with more high mechanical properties by way of introduction to the composites third binder was not successful.

4.6 TGA INVESTIGATION OF THE COMPOSITES ON THE BASIS OF LEAVES AND DIFFERENT BINDERS

Thermogravimetric investigations were carried out on a "Paulic-Pauluc-Erday" Derivatograph model MOM-102). The test conditions were temperature rise rate of ≈10 deg/min in an open area. In Figures 4.41–4.47, there are presented TG, DTA, and DTG curves of composites. From which it is evident that with an increase of temperature the mass losses of composites rise (Figures 4.41–4.47).

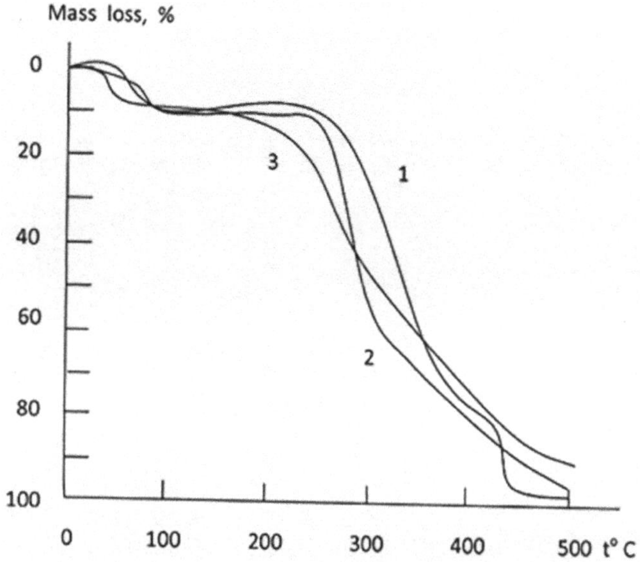

FIGURE 4.41 Mass loss of two component composites based on bamboo with binders: Liquid glass (5%) – 1; PhES-50 (5%) – 2; and PhES-80 (5%) – 3 (composites 4.2 and 4.10 accordingly).

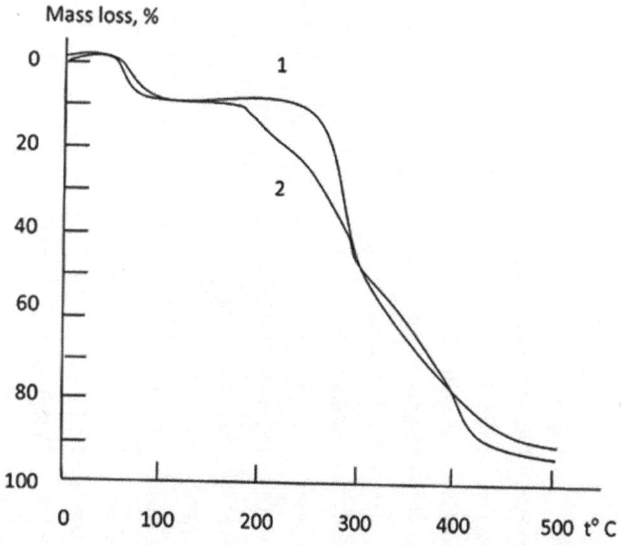

FIGURE 4.42 Mass loss of two component composites based on bamboo with binders: PE (10%) – 1; and PhES-80 (15%) – 2 (composites 4.14 and 4.12 accordingly).

TABLE 4.7 Physical-Mechanical Properties of the Composites Based on Bamboo

SL. No.	Composite (Mass %)	Pressure (kg/cm^2)	Temperature (T°C)	Strength on Bending (kg/cm^2)	Impact Viscosity (kJ/M^2)
1.	Bamboo 97% + Liquid glass 3%	150	130	19.7	56.7
2.	Bamboo 95% + Liquid glass 5%	150	130	20.4	58.5
3.	Bamboo 90% + Liquid glass 10%	150	130	22.7	61.1
4.	Bamboo 85% + Liquid glass 15%	150	130	26.4	57.9
5.	Bamboo 97% + PhES-50 – 3%	250	130	11.8	47.2
6.	Bamboo 95% + PhES-50 – 5%	250	130	22.5	48.8
7.	Bamboo 90% + PhES-50 – 10%	150	130	13.0	52.2
8.	Bamboo 85% + PhES-50 – 15%	150	130	18.1	50.5
9.	Bamboo 97% + PhES-80 – 3%	150	130	11.8	50.4
10.	Bamboo 95% + PhES-80 – 5%	150	130	16.9	48
11.	Bamboo 90% + PhES-80 – 10%	150	130	21.7	53.9
12.	Bamboo 85% + PhES-80 – 15%	150	130	17.8	53.6
13.	Bamboo 95% + PE 5%	150	130	19.9	53.9
14.	Bamboo 90% + PE 10%	150	130	24.5	56.9
15.	Bamboo 85% + PE 15%	150	130	24.1	60.5
16.	Bamboo 80% + PE 20%	150	130	19.4	54.3
17.	Bamboo 95% + Colophony 5%	150	130	20.4	61.8
18.	Bamboo 90% + Colophony 10%	150	130	28.8	54.8
19.	Bamboo 85% + Colophony 15%	150	130	25.1	56.6
20.	Bamboo 80% + Colophony 20%	150	130	18.4	51.1

TABLE 4.7 *(Continued)*

SL. No.	Composite (Mass %)	Pressure (kg/cm^2)	Temperature (T°C)	Strength on Bending (kg/cm^2)	Impact Viscosity (kJ/M^2)
21.	Bamboo 95% + Wood glue 5%	150	130	19.9	60.3
22.	Bamboo 90% + Wood glue 10%	150	130	28.9	64.5
23.	Bamboo 85% + Wood glue 15%	150	130	34.2	62.8
24.	Bamboo 80% + Wood glue 20%	150	130	17.6	49.1
25.	Bamboo 94% + PhES-50 – 3% + Liquid glass 3% + Hardener	150	130	31.8	55.5
26.	Bamboo 90% + PhES-50 – 5% + Liquid glass 5% + Hardener	150	130	17.3	59.1
27.	Bamboo 80% + PhES-50 – 10% + Liquid glass 10%	150	130	18.6	64.8
28.	Bamboo 80% + PhES-50 – 15% + Liquid glass 5%	150	130	17.6	49.1
29.	Bamboo 80% + PhES-50 – 5% + Liquid glass 15%	150	130	18.6	50.0
30.	Bamboo 94% + PhES-80 – 3% + Liquid glass 3% reinforcement	150	130	17.6	60.0
31.	Bamboo 90% + PhES-80 – 5% + Liquid glass 5%	150	130	20.9	51.6
32.	Bamboo 91% + PhES-50 – 3% + Liquid glass 3% + PE 3%	150	130	16.3	48.8
33.	Bamboo 85% + PhES-50 – 5% + Liquid glass 5% + PE 5%	150	130	15.9	47.6
34.	Bamboo 91% + PhES-80 – 3% + Liquid glass 3% + PE 3%	150	130	16.5	44.4
35.	Bamboo 85% + PhES-80 – 5% + Liquid glass 5% + PE 5%	150	130	12.4	47.7
36.	Bamboo 94% + PhES-50 – 3% + PE 3%	150	130	20.4	53.7
37.	Bamboo 90% + PhES-50 – 5% + PE 5%	150	130	21.5	51.3
38.	Bamboo 85% + PhES-50 – 5% + PE 10%	150	130	20.7	59.3
39.	Bamboo 85% + PhES-50 – 10% + PE 5%	150	130	11.8	48.9
40.	Bamboo 85% + PhES-50 – 7.5% + PE 7.5%	150	130	16.9	55.9

TABLE 4.7 *(Continued)*

SL. No.	Composite (Mass %)	Pressure (kg/cm^2)	Temperature (T°C)	Strength on Bending (kg/cm^2)	Impact Viscosity (kJ/M^2)
41.	Bamboo 94% + PhES-80 – 3% + PE 3%	150	130	16.1	59.2
42.	Bamboo 90% + PhES-80 – 5% + PE 5%	–	–	15.6	54.3
43.	Bamboo 85% + PhES-80 – 5% + PE 10%	150	130	16.1	51.4
44.	Bamboo 85% + PhES-50 – 5% + Polyvinylacetate 10%	150	130	16.5	59.3
45.	Bamboo 88% + PhES-50 – 3% + Liquid glass 3% + PE 3% + $VinSi(OEt)_3$ – 3%	150	130	13.4	33.7
46.	Bamboo 80% + PhES-50 – 5% + Liquid glass 5% + PE 5% + $VinSi(OEt)_3$ – 5%	150	130	15.7	37.9
47.	Bamboo 88% + PhES-80 – 3% + Liquid glass 3% + PE 3% + $VinSi(OEt)_3$ – 3%	150	130	16.2	35.3
48.	Bamboo 80% + PhES-80 – 5% + Liquid glass 5% + PE 5% + $VinSi(OEt)_3$ – 5%	150	130	12.10	32.5
49.	Bamboo 90% + Liquid glass 5% + PE 5%	150	130	18.1	68.5
50.	Bamboo 94% + Liquid glass 3% + PE 3%	150	130	16.7	48.3
51.	Bamboo 90% + Liquid glass 5% + PE 5%	150	130	11.8	50.2
52.	Bamboo 94% + Liquid glass 4% + PhES-80 – 2%	150	130	7.2	36.5
53.	Bamboo 94% + Liquid glass 2% + PhES-80 – 4%	150	130	15.3	49.04
54.	Bamboo 85% + Liquid glass 5% + PE 10%	150	130	20.2	61.8
55.	Bamboo 85% + Liquid glass 10% + PE 5%	150	130	14.4	62.6
56.	Bamboo 85% + Liquid glass 7.5% + PE 7.5%	150	130	15.3	59.8

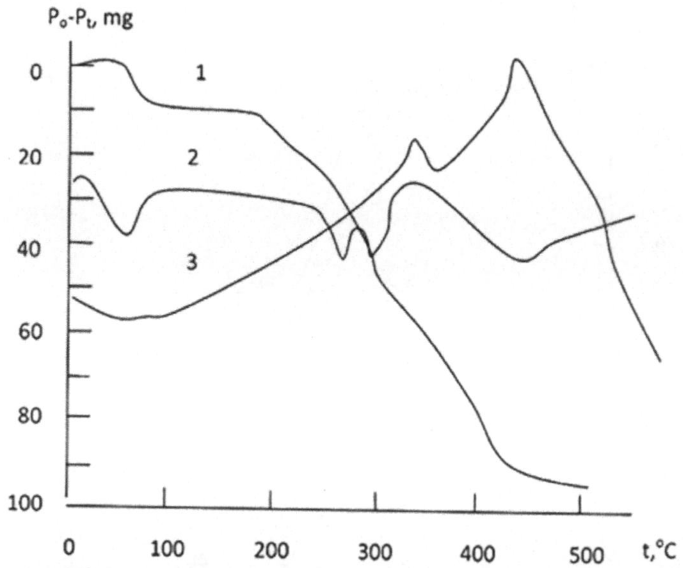

FIGURE 4.43 Thermogravimetric curves (TG-1; DTG-2; and DTA-3) for two components composite: Bamboo (85%) + PhES-80 (15%) (composites 4.13).

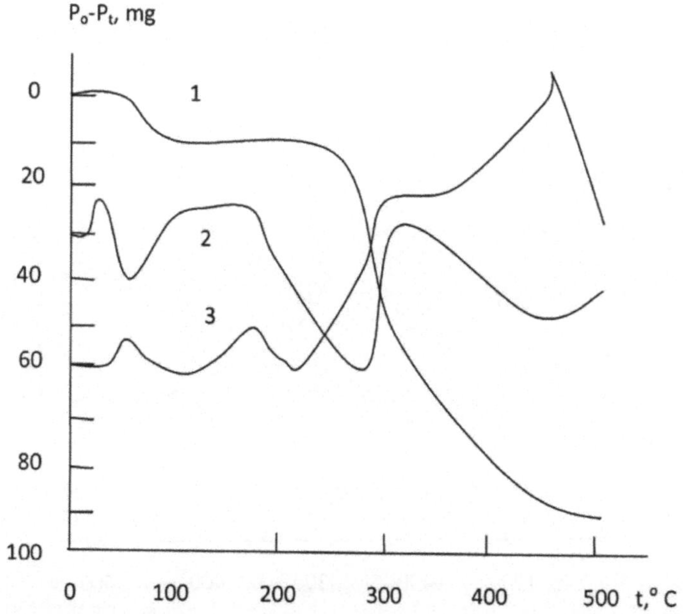

FIGURE 4.44 Thermogravimetric curves (TG-1; DTG-2; and DTA-3) for two component composites: Bamboo (90%) + PE (10%).

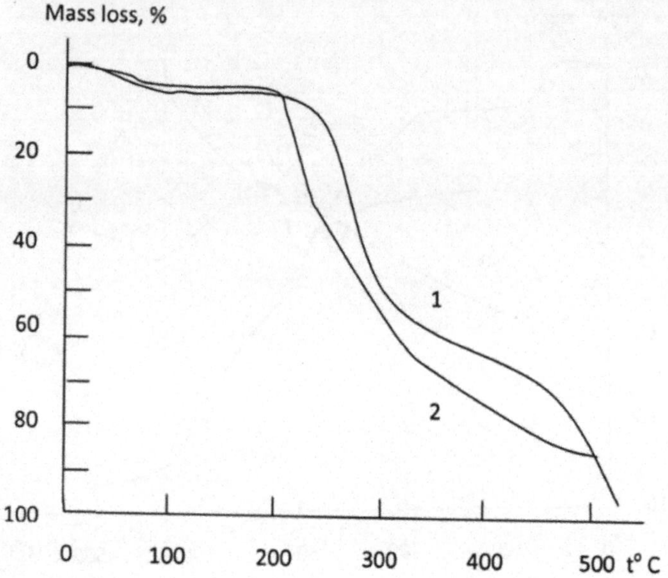

FIGURE 4.45 Mass loss of three component composites based on bamboo with binders: Bamboo (90%) + Liquid glass (5%) + Polyethylene (5%) – 1; Bamboo (90%) + Liquid glass (5%) + PhES-80 (5%) – 2 (composites 4.51 and 4.31 accordingly).

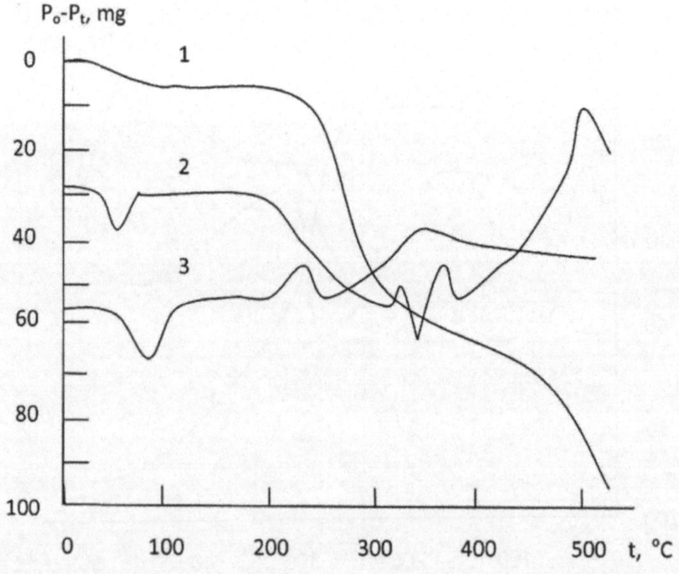

FIGURE 4.46 Thermogravimetric curves (TG-1; DTG-2; and DTA-3) for three component composites: Bamboo (90%) + Liquid glass (5%) + PE (5%).

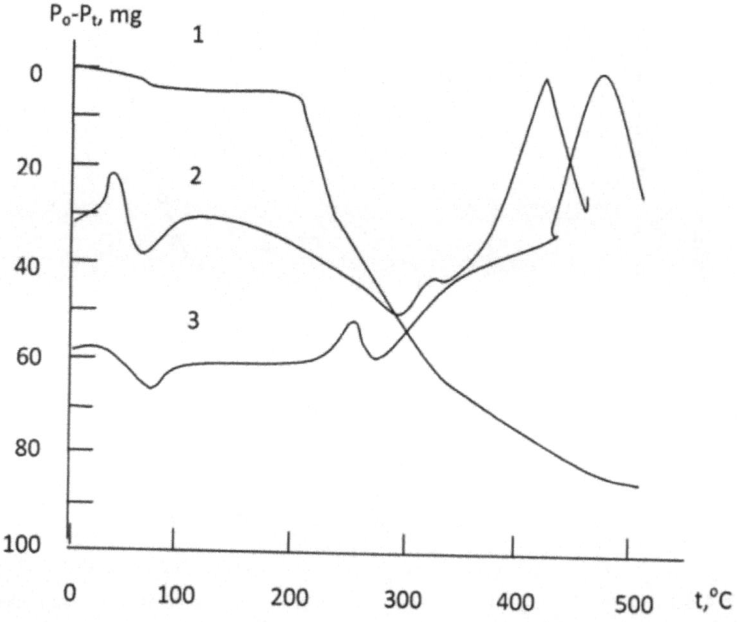

FIGURE 4.47 Thermogravimetric curves (TG-1; DTG-2; and DTA-3) for four component composites: Bamboo (85%) + Liquid glass (5%) + PhES-80 (5%) + PE (5%).

As it is seen from thermogravimetric curves about 10% mass losses one can observe in the temperature range up to 200–210°C in this temperature interval (endothermic peaks on DTA and DTG curves) condensation processes may be proceeds with unreacted hydroxyl groups as well as with hydroxyl and ethoxy groups. The main destruction process take place from 230–280°C temperature and after 500–600°C the full destruction of composite materials take place. It has been found that the used bindery does not affect on the thermal oxidative stability of the composites.

4.7 THERMAL STABILITY (TS) OF THE COMPOSITES ON THE BASIS OF BAMBOO

The dependences of the softening of composites on temperature are presented in Figures 4.48–4.52. The curves presented on them show that the TS of composites essentially depend on a structural factor. In general, the composites with relatively high physical-mechanical properties are characterized with higher TS.

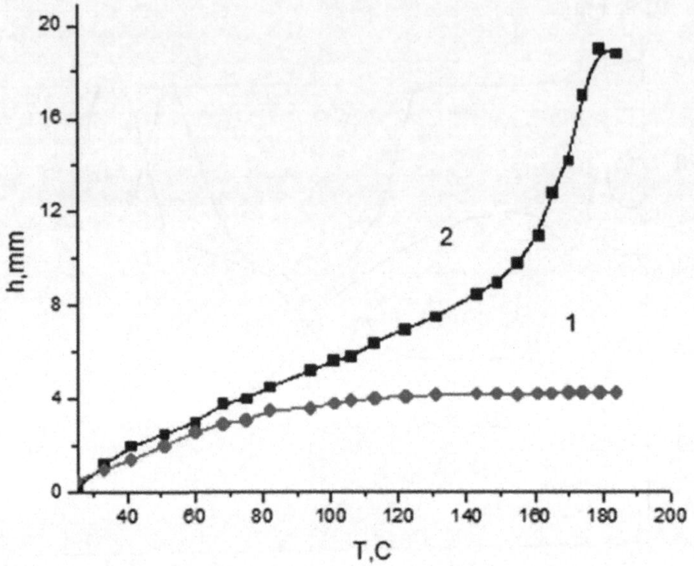

FIGURE 4.48 Dependence of the softening temperature for composites: 1 – Bamboo (90%) + Liquid glass (10%); and Bamboo (90%) + PhES-80 (10%).

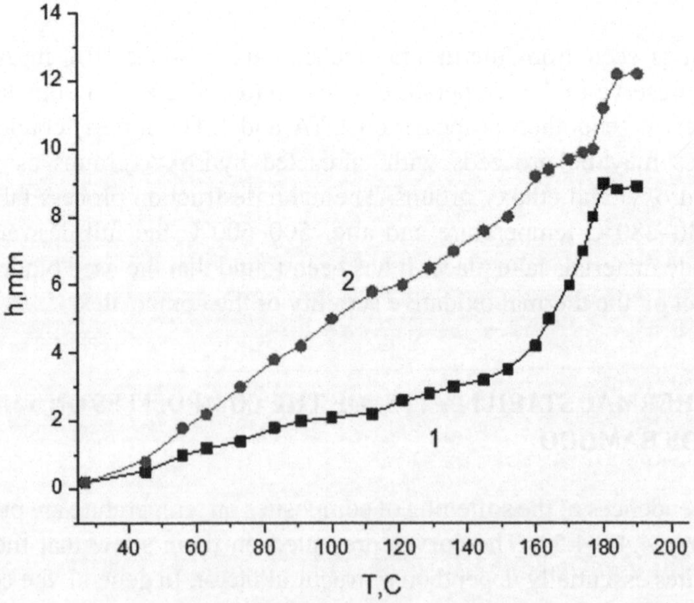

FIGURE 4.49 Dependence of softening on temperature for composites: 1 – Bamboo 90% + PE (10 wt.%); and 2 – Bamboo 90% + PE (15 wt.%) (composites 4.14 and 4.15).

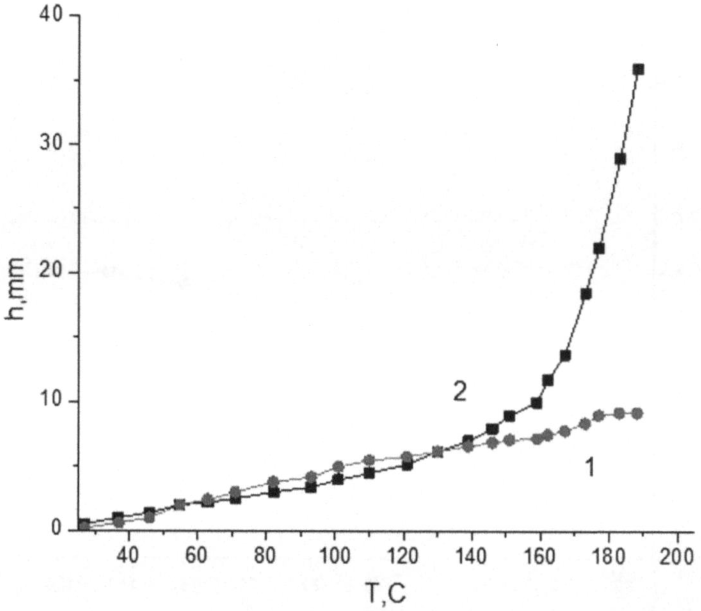

FIGURE 4.50 Dependence of softening on temperature for composites: 1 – Bamboo 90% + Wood glue (10 wt.%); and 2 – Bamboo 85% + Wood glue (15 wt.%) (composites 4.22 and 4.23).

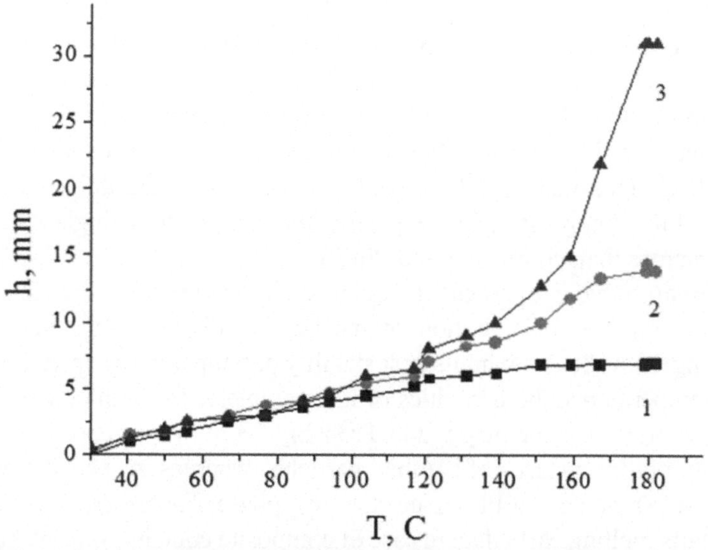

FIGURE 4.51 Dependence of softening on temperature for composites: 1 – Bamboo 95% + Liquid glass (5 wt.%); and 2 – Bamboo 95% + PhES-50 (5 wt.%), – Bamboo 95% + PhES-80 (5 wt.%) (composites 4.2, 4.6, and 4.10).

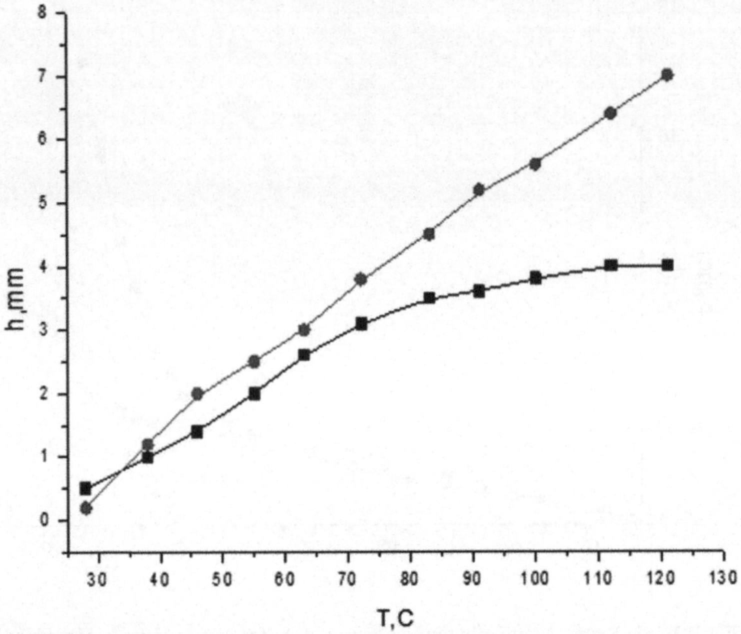

FIGURE 4.52 Dependence of softening on temperature for composites: 1 – Bamboo + Colophony (10 wt.%); and 2 – Bamboo + Colophony (15 wt.%) (composites 4.18 and 4.19).

The comparison of two curves of dependence of the softening for composites based on bamboo with liquid glass (10 wt.%) and PhES-80 (10 wt.%) shows that the TS of first composite is essentially higher than that for second one (Figure 4.48). Especially it is evident at temperatures higher 100–120°C. This fact may be explained in terms of the difference in the rigidity of the composites. Consequently, the composite with glass has more rigid structure than composite with PhES.

Thermal stable properties of composites contained PE are determined depending on the concentration of the binder (Figure 4.48). Namely, PE containing 10 wt.% PE exhibits best stability at temperatures up to 150°C. At higher temperatures, the quantities of this parameter for both composites are nearing to one and same magnitude. Probably PE in first composite is distributed fully on the surfaces of bamboo particles, whereas in second composite (PE 15 wt.%) the part of PE creates the own phase. Therefore, at heating this phase starts melting early than in case of composite containing 10 wt.% of PE.

Analogical situation we have in case of composites with wood glue (Figure 4.50). The composite containing 10 wt.% of binder is essentially stable than one containing 15 wt.% of binder. Here also the main reason of such behavior of the corresponding curves is explained in terms of present of the wood glue separate phase in composite, which melts more easily than the same binder adsorbed on the bamboo particles surfaces.

Comparison of the curves corresponding to composites, containing relatively low concentration (5 wt.%) of binders show that the TS of these composites increases in the raw PhES-80 – PhES-50 – Liquid glass (Figure 4.51). These results are in accordance with data of shock viscosity of these composites as it was noted above.

TS of the composite contained colophony as binder at 10 wt.% is higher than for composite with 15 wt.% colophony (Figure 4.52). The explanation of this fact is same as for composite contained 15 wt.% of a binder.

4.8 WATER ABSORPTION OF THE COMPOSITES ON THE BASIS OF BAMBOO

The numerical data of this parameter are presented in Table 4.8. First of all, it must be noted that the data on water absorption by obtained composites based on bamboo in general are essentially low in case of sawdust and straw (see accounts of previous quarters). This fact shows that the chemical structure (also microstructure) of bamboo favors to formation of chemical and physical bonds with binders (adhesives). Besides of the structure of composites based on bamboo may be contained the various micro- and macro-empties, don't connected each to other and therefore create the difficulties for diffusion of water in the composite body. Therefore, the composite matrix for the composites with bamboo is more solid than earlier analogs. The table data show that the coefficient of the water absorption depends on the concentration of binders. So, the water absorption in the composites contained liquid glass (4.1–4.4) this parameter decreases from 5.03 till 4.66%, when the concentration of the binder changes from 3 up to 15%. Although in case of composite with 5% of LG we have partially deviation from this rule. More high result was obtained for composites containing PhES-es (4.5–4.8 and 4.9–4.12). Namely the water absorption for composites with PhES-50 at 3–15% containing changes in the range 2.78–2.31 and for composites with PhES-80 the results are better: 2.28–2.70. With higher parameter are characterized the composites in case of composites with PE (composites 4.13–4.16). In this case the minimal water absorption reaches to 1.87. Although for composite with 5% of PE we have anomaly – the absorption increases. Here it may by supposed that at relatively low concentrations of this binder it is no conditions for creation of high rigid structure with low concentrations of empties. Good characteristic has the composites containing colophony (4.17–4.20). In this case the minimal parameter reaches to 1.79 (composite 4.17). On the basis of noted above it may be concluded that in the last case the degree of homogenization of the binder at low concentrations is higher in comparison with composites containing PE. It was unexpected relatively low magnitudes of the water absorption for composites based on wood glue (4.21–4.24). In this case probably the formation of the structural defects of type of penetrated empties are formatted, which will be the main reason of a relatively high absorption coefficients of these materials.

4.9 CONCLUSIONS TO COMPOSITES ON THE BASIS OF BAMBOO

There are obtained the composites based on bamboo and some organic-inorganic (PhES-50 and PhES-80, LG, PE, colophony, and wood glue) binders. The binders in composites were changed in the interval 3–20 wt.%. Powders of dry bamboo were blended with binders and the blends were pressed in the spatial press-forms at temperatures 110–150°C and pressures 100–150 MPa during 10 min.

Following investigations were provided: (i) Structural analysis with use of Fourier transformer infrared spectroscopy (FTIR), optical, and scanned electron microscopes (SEM), energy dispersive X-ray microanalysis (EDXA); (ii) physical-mechanical properties; (iii) TS by method Vica and thermogravimetry; and (iv) water absorption.

In result of Investigation with use of FTIR method it was proved the conception about formation of the chemical bounds between bamboo surface active groups and binders, thanks to which the strengthening of composites enhances.

Optical and electron microscopes data help us to establish the specific of formation of composite microstructure; EDXA we used for establishing of existence of the microelements in the bamboo.

Two types of the measure of physical-mechanical properties were used in the project–strengthening on bending and impact viscosity; it was shown that the mechanical strengthening of composites with one binder at both types of deformations are characterized with extreme dependence on the concentration of the binders. For example, the composites containing liquid glass PhES-50 and PhES-80 reveal the maximums of strengthening at 10 mas% of concentrations. With analogical behavior of the mechanical properties are characterized the composites with PE, colophony, and wood glue.

TABLE 4.8 Water Absorption of Composites Based on Bamboo and Different Binders

SL. No.	Composite (Weight %)	Mass (g)	Volume (cm^3)	Density (g/cm^3)	Weight After 3 h Exposition in Water (g)	Weight After 24 h Exposition in Water (g)	Water Absorption After 24 h Exposition in Water (%)
1.	Bamboo 97% + Liquid glass 3%	4.77	3.34	1.43	4.86	5.01	5.03
2.	Bamboo 95% + Liquid glass 5%	3.89	2.82	1.38	4.10	4.28	10.02
3.	Bamboo 90% + Liquid glass 10%	3.80	2.82	1.35	3.85	3.98	4.73
4.	Bamboo 85% + Liquid glass 15%	3.86	2.82	1.37	3.93	4.04	4.66
5.	Bamboo 97% + PhES-50 – 3%	3.95	2.82	1.40	3.98	4.06	2.78
6.	Bamboo 95% + PhES-50 – 5%	3.83	2.82	1.36	3.85	3.93	2.61
7.	Bamboo 90% + PhES-50 – 10%	3.78	2.82	1.34	3.80	3.86	2.11
8.	Bamboo 85% + PhES-50 – 15%	3.46	2.46	1.41	3.47	3.54	2.31
9.	Bamboo 97% + PhES-80 – 3%	3.94	2.82	1.40	3.97	4.03	2.28
10.	Bamboo 95% + PhES-80 – 5%	3.89	2.82	1.38	3.90	3.97	2.05
11.	Bamboo 90% + PhES-80 – 10%	3.66	2.64	1.39	3.67	3.74	2.18
12.	Bamboo 85% + PhES-80 – 15%	3.70	2.64	1.40	3.72	3.80	2.70
13.	Bamboo 95% + PE 5%	3.88	2.82	1.38	4.14	4.41	13.66
14.	Bamboo 90% + PE 10%	3.73	2.82	1.32	3.75	3.81	1.87
15.	Bamboo 85% + PE 15%	3.83	2.99	1.28	3.85	3.91	2.08
16.	Bamboo 80% + PE 20%	3.65	2.82	1.29	3.68	3.74	2.47
17.	Bamboo 95% + Colophony 5%	3.92	2.82	1.39	3.95	3.99	1.79
18.	Bamboo 90% + Colophony 10%	3.78	2.82	1.34	3.80	3.85	1.85
19.	Bamboo 85% + Colophony 15%	3.50	2.64	1.33	3.52	3.57	2.00
20.	Bamboo 80% + Colophony 20%	3.55	2.64	1.34	3.60	3.65	2.82

TABLE 4.8 *(Continued)*

SL. No.	Composite (Weight %)	Mass (g)	Volume (cm^3)	Density (g/cm^3)	Weight After 3 h Exposition in Water (g)	Weight After 24 h Exposition in Water (g)	Water Absorption After 24 h Exposition in Water (%)
21.	Bamboo 95% + Wood glue 5%	3.92	2.82	1.39	4.19	4.40	12.24
22.	Bamboo 90% + Wood glue 10%	3.95	2.82	1.40	4.00	4.09	3.54
23.	Bamboo 85% + Wood glue 15%	3.93	2.82	1.39	4.01	4.19	6.61
24.	Bamboo 80% + Wood glue 20%	3.85	2.64	1.46	4.05	4.54	17.92
25.	Bamboo 94% + PhES-50 – 3% + Liquid glass 3% + Hardener	3.96	2.82	1.40	4.02	4.08	3.03
26.	Bamboo 90% + PhES-50 – 5% + Liquid glass 5% + Hardener	3.96	2.82	1.40	4.00	4.07	2.77
27.	Bamboo 80% + PhES-50 – 10% + Liquid glass 10%	3.64	2.64	1.38	3.70	3.80	4.39
28.	Bamboo 80% + PhES-50 – 15% + Liquid glass 5%	3.60	2.64	1.36	3.64	3.73	3.61
29.	Bamboo 80% + PhES-50 – 5% + Liquid glass 15%	3.69	2.64	1.40	3.75	3.84	4.07
30.	Bamboo 94% + PhES-80 – 3% + Liquid glass 3% + Hardener	3.96	2.82	1.40	4.00	4.07	3.03
31.	Bamboo 90% + PhES-80 – 5% + Liquid glass 5%	3.78	2.82	1.34	3.81	3.89	2.91
32.	Bamboo 91% + PhES-50 – 3% + Liquid glass 3% + PE 3%	4.00	2.99	1.34	4.07	4.14	3.50
33.	Bamboo 85% + PhES-50 – 5% + Liquid glass 5% + PE 5%	3.81	2.82	1.35	3.86	3.93	3.14

TABLE 4.8 *(Continued)*

SL. No.	Composite (Weight %)	Mass (g)	Volume (cm^3)	Density (g/cm^3)	Weight After 3 h Exposition in Water (g)	Weight After 24 h Exposition in Water (g)	Water Absorption After 24 h Exposition in Water (%)
34.	Bamboo 91% + PhES-80 – 3% + Liquid glass 3% + PE 3%	3.83	2.82	1.36	3.87	3.99	4.17
35.	Bamboo 85% + PhES-80 – 5% + Liquid glass 5% + PE 5%	3.68	2.82	1.30	3.76	3.92	6.52
36.	Bamboo 94% + PhES-50 – 3% + PE 3%	3.85	2.82	1.37	3.89	3.99	3.63
37.	Bamboo 90% + PhES-50 – 5% + PE 5%	3.78	2.82	1.34	3.83	4.02	6.34
38.	Bamboo 85% + PhES-50 – 5% + PE 10%	3.75	2.82	1.33	3.80	3.91	4.26
39.	Bamboo 85% + PhES-50 – 10% + PE 5%	3.65	2.64	1.38	3.69	3.82	4.65
40.	Bamboo 85% + PhES-50 – 7.5% + PE 7.5%	3.69	2.82	1.31	3.72	3.83	3.79
41.	Bamboo 94% + PhES-80 – 3% + PE 3%	3.89	2.82	1.38	3.93	4.10	5.39
42.	Bamboo 90% + PhES-80 – 5% + PE 5%	3.80	2.82	1.35	3.82	3.90	2.63
43.	Bamboo 85% + PhES-80 – 5% + PE 10%	3.82	2.82	1.35	3.86	3.96	3.66
44.	Bamboo 85% + PhES-50 – 5% + Polyvinyl Acetate 10%	3.46	2.64	1.31	3.53	3.69	6.64
45.	Bamboo 88% + PhES-50 – 3% + Liquid glass 3% + PE 3% + $VinSi(OEt)_3$ – 3%	3.78	2.64	1.43	3.82	3.90	3.17
46.	Bamboo 80% + PhES-50 – 5% + Liquid glass 5% + PE 5% + $VinSi(OEt)_3$ – 5%	3.74	2.82	1.33	3.77	3.83	2.40
47.	Bamboo 88% + PhES-80 – 3% + Liquid glass 3% + PE 3% + $VinSi(OEt)_3$ – 3%	3.93	2.82	1.39	3.96	4.07	3.56
48.	Bamboo 80% + PhES-80 – 5% + Liquid glass 5% + PE 5% + $VinSi(OEt)_3$ – 5%	3.56	2.64	1.35	3.61	3.72	4.49

TABLE 4.8 *(Continued)*

SL. No.	Composite (Weight %)	Mass (g)	Volume (cm^3)	Density (g/cm^3)	Weight After 3 h Exposition in Water (g)	Weight After 24 h Exposition in Water (g)	Water Absorption After 24 h Exposition in Water (%)
49.	Bamboo 90% + Liquid glass 5% + PE 5%	3.76	2.82	1.33	3.82	3.93	4.52
50.	Bamboo 94% + Liquid glass 3% + PE 3%	3.89	2.82	1.38	3.93	4.03	3.59
51.	Bamboo 85% + Liquid glass 5% + PE 5% + Colophony 5%	3.65	2.2.64	1.38	3.69	3.77	3.28
52.	Bamboo 94% + Liquid glass 4% + PhES-80 – 2%	3.91	2.64	1.48	4.09	4.67	19.48
53.	Bamboo 94% + Liquid glass 2% + PhES-80 – 4%	3.88	2.82	1.37	3.95	4.12	6.18
54.	Bamboo 85% + Liquid glass 5% + PE 10%	3.54	2.82	1.25	3.59	3.71	4.85
55.	Bamboo 85% + Liquid glass 10% + PE 5%	3.71	2.82	1.31	3.77	3.80	5.12
56.	Bamboo 85% + Liquid glass 7.5% + PE 7.5%	3.59	2.64	1.38	3.63	3.76	4.73

There were provided the investigations on TS by method Vica and gravimetric analysis methods, with use of which it was shown, that thermostable are composites with high mechanical characteristics. Thermogravimetric analysis shows that intensive mass loss for all composites corresponds to temperature range near 300°C.

With relatively high hydrophobicity are characterized the composites based on bamboo with following binders: (a) PhES-50 (10%) – 2.1%; (b) PhES-80 (5%) – 2.1%; (c) PE-(10%) – 1.9%; (d) Colophony (5%) – 1.9; and (e) Colophony (10%) – 1.8%.

The data obtained for composites contained 3, 4, and 5 components (4.25–4.56) show that one of effective way for obtaining of the composites with low water absorption is the skill selection of the type and concentrations of binders at using of the modern methods the structural and morphological analysis of materials.

KEYWORDS

- **composite materials**
- **energy dispersion**
- **FTIR investigation**
- **microstructure**
- **micro-X-ray analysis**
- **physical-mechanical properties**
- **scanning electron microscope**

REFERENCES

1. *Kew World Checklist of Selected Plant Families*. August 25, 2017.
2. Dieter, O., (1999). *The Bamboos of the World* (pp. 279, 280). Elsevier, ISBN: 978-0-444-50020-5.
3. *Biology Pamphlets*, (1895). Vol. 741, p. 15. University of California.
4. Louppe, D., Oteng-Amoako, A. A., & Brink, M., (2008). *Timbers* (Vol. 1, pp. 100–103). PROTA, ISBN 978-90-5782-209-4.
5. Janssen, J. J. A., (1995). *Building with Bamboo* (2nd edn., p. 65). Intermediate Technology Publication Limited, London.
6. Amada, S., Ichikawa, Y., Munekata, T., Nagase, Y., & Shimizu, K., (1997). Fiber texture and mechanical graded structure of bamboo. *Composite Part B., 28*(B), 13–20.

7. Nishiyama, Y., Langan, P., & Chanzy, H., (2002). Crystal structure and hydrogen-bonding system in cellulose Iβ from synchrotron X-ray and neutron fiber diffraction. *J. Am. Chem. Soc., 124*(31), 9074–9082. doi:10.1021/ja0257319. PMID 12149011.
8. *NIOSH Pocket Guide to Chemical Hazards #0110*. National Institute for Occupational Safety and Health (NIOSH).
9. Crawford, R. L., (1981). *Lignin Biodegradation and Transformation*. New York: John Wiley and Sons. ISBN 0-471-05743-6.
10. Updegraff, D. M., (1969). Semimicro determination of cellulose in biological materials. *Analytical Biochemistry, 32*(3), 420–424. doi: 10.1016/S0003-2697(69)80009-6.
11. Romeo, T., (2008). *Bacterial Biofilms* (pp. 258–263). Berlin: Springer. ISBN 978-3-540-75418-3.
12. Klemm, D., Heublein, B., Hans-Peter, F., & Bohn, A., (2005). Cellulose: Fascinating biopolymer and sustainable raw material. *Angew. Chem. Int. Ed., 44*(22). doi: 10.1002/anie.200460587.
13. Cellulose, (2008). In: *Encyclopædia Britannica*. Retrieved from: Encyclopedia Britannica Online.
14. *Chemical Composition of Wood*. ipst.gatech.edu.
15. Piotrowski, S., & Carus, M., (2011). *Multi-criteria Evaluation of Lignocellulosic Niche Crops for Use in Biorefinery Processes*. Nova-Institut GmbH, Hürth, Germany.
16. Scheller, H. V., & Ulvskov, P., (2010). Hemicelluloses. *Annu Rev Plant Biol., 61*, 263–289. doi: 10.1146/annurev-arplant-042809-112315.
17. Martone, P. T., Estevez, J. M., Lu, F., Ruel, K., Denny, M. W., Somerville, C., & Ralph, J., (2009). Discovery of lignin in seaweed reveals convergent evolution of cell-wall architecture. *Current Biology: CB, 19*(2), 169–175. doi: 10.1016/j.cub.2008.12.031. ISSN 0960-9822. PMID 19167225.
18. Lebo, S. E. Jr., Gargulak, J. D., & McNally, T. J., (2001). Lignin. *Kirk-Othmer Encyclopedia of Chemical Technology*. John Wiley & Sons, Inc.
19. Haynes, W. M., (2011). *CRC Handbook of Chemistry and Physics* (92nd edn.). Boca Raton, FL: CRC Press. ISBN 1439855110.
20. Iler, R. K., (1979). *The Chemistry of Silica*. Plenum Press. ISBN 0-471-02404-X.
21. Fernández, L. D., Lara, E., & Mitchell, E., (2015). Checklist, diversity and distribution of testate amoebae in Chile. *European Journal of Protistology, 51*, 409–424.
22. Fedko, I. V., Kitapova, R. R., Kambalina, M. G., & Khvashchevskaia, A. A., (2013). Data element of the bamboo (*Bambusa arundinaceae*). *Bashkirian Chemical J., 20*(2), 96–99.
23. Guoqi, X., Lihai, W., Junliang, L., & Jinzhuo, W., (2013). FTIR and XPS analysis of the changes in bamboo chemical structure decayed by white-rot and brown-rot fungi. *J. Appl. Surface Sci., 280*, 799–805.
24. 24. Hui-Ting, C., Ting-Feng, Y., Fu-Lan, H., Ling-Long, Kuo-Huang., Chin-Mai, L., Yan-San, H., & Shang-Tzen, C., (2015). Profiling the chemical composition and growth strain of Giang bamboo, (*Dendrocalamus giganteus* Munro). *BioResource, 10*(1), 1260–1270.
25. Muraganatham, S., Anbalagan, G., & Ramamurthy, N., (2009). FT-IR and SEM-EDS comparative analysis of medicinal plants. *Eclipta alba* Hassk and *Eclipta prostrate* Linn. *Rom. J. Biophysics, 19*, 285–294.
26. Shirwaikar, A., Malini, S., & Kumari, S. C., (2003). Protective effect of *Pongamia pinnata* flowers against cisplatin and gentamicin induced nephrotoxicity in rats. *Indian J Exp Biol., 1*, 58–62.

27. Aneli, J., Buzaladze, G., Markarashvili, E., Tatrishvili, T., Esartia, I., & Mukbaniani, O., (2016). Composites based on mineral raw materials. VIII International Scientific-Technical Conference Advance in Petroleum and Gas Industry and Petrochemistry (APGIP-8). Lviv, Ukraine (in Press).
28. Omari Mukbaniani, Witold Brostow, Haley E. Hagg Lobland, Jimsher Aneli, Tamara Tatrishvili, et al. (2018). "Composites containing bamboo with different Binders." *Pure Appl. Chem.* https://doi.org/10.1515/pac-2017-0804.

PART II
Biomedical Applications

CHAPTER 5

Ph-Responsive Nanocomposite Hydrogels and Their Potential Scope in the Biomedical Field

RABIA KOUSER,[1] ASIF HUSAIN,[1,2] MOHD IRFAN,[3]
SHAHIDUL ISLAM BHAT,[4] and ABDUL WAHIED KHAN[5]

[1]*Baba Ghulam Shah Badshah University, Rajouri, Jammu and Kashmir, India*

[2]*Jamia Millia Islamia University, New Delhi, India*

[3]*Archaeological Survey Chemist, Jammu and Kashmir, India*

[4]*Corrosion Research Lab, Department of Chemistry, AMU, Aligarh, Uttar Pradesh, India*

[5]*Department of Electrical Engineering, Mewar University, Chittorgarh, Rajasthan, India*

ABSTRACT

Hydrogels imitate native tissue microenvironments because of the spongy as well as hydrated molecular ingredients. Traditional hydrogels usually possess inferior physical properties as well as a lack of multi-functionalities. Promising progress toward reinforcing traditional polymeric hydrogels, and including unique chemical, physical, and biological functions considerations on assimilating nanoparticles within the hydrogel materials. An ample of nanoparticles such as carbon nanotubes, graphene, and inorganic nanomaterials such as hydroxyapatite (HA), clay, gold, etc., have unique chemical, physical, and biological functions, and have been extensively studied as biomaterials or bio-functional materials. These nanoparticles, when incorporated within

Biocomposites: Environmental and Biomedical Applications. Omar Mukbaniani, Tamara Tatrishvili, Neha Kanwar Rawat, & A. K. Haghi (Eds.)

the hydrogel networks, attain nanocomposite materials with advanced properties and customized functionality. The biopolymer-based hydrogels are biodegradable, biocompatible, economically sound, easily available, and nontoxic without creating any toxic effect on the surrounding environment. Nanocomposite hydrogels, which combine the advantages of both nano-fillers and hydrogel matrices, may result in improved mechanical and biological properties and find their potential applications in the biomedical field. In this chapter, we focus on the development of sustainable polymer-based nanocomposite hydrogels with prominence in their biomedical applications. We also discuss various nanofiller (carbon nanotubes, graphene, gold, hydroxyl apatite, clay) based nanocomposite hydrogels, their mechanical properties, and applications, especially biomedical such as drug delivery, tissue engineering, and wound healing.

5.1 INTRODUCTION

Hydrogels are cross-linked networks of hydrophilic polymers capable of retaining large amounts of water yet remaining insoluble and maintaining their three-dimensional structure. Since their discovery by Wichterle et al. in the early 1950s and applications in the field of biomedical sciences among others [1]. An abundant variety of sustainable polymer-based nanocomposite hydrogels (Figure 5.1) have been reported for a wide range of pharmaceutical applications [2]. However, upon dehydration, they become mechanically fragile, weak, and brittle in nature, which limits their potential application in the field of the biomedical field. In regard to this shortcoming, Considerable progress in the synthesis and technology of nanocomposite hydrogels has gained great attention in recent years [3]. Nanocomposite/hybrid hydrogels may be defined as cross-linked polymer networks swollen with water in the presence of nanoparticles or nanostructures (10–100 nm), exhibiting superior properties as compared to traditionally made hydrogels [4]. The hard inorganic nanoparticles (carbon nanotubes, graphene, gold, hydroxyl apatite, clay) strength the soft organic polymer matrix, so the final nanocomposite hydrogel can display novel or enhanced mechanical, optical, electrically conductive, magnetic, or biological properties with diverse functionality [5]. The features of nanocomposite hydrogels can be tuned by choosing a specific combination of polymer and nanoparticles. Inspired by the synergistic effect of nanofillers within the hydrogel composition, researchers incorporate carbon-based, polymeric, ceramic, or metallic nanofillers to give nanocomposite hydrogel superior characteristics [6], which can find a pivotal role in drug delivery, tissue engineering, wound healing, etc. (Figure 5.2). Because the nanoparticles combine the various synergistic properties (i.e., size, shape, electrical, optical, mechanical, etc.) with those of hydrogel networks [7]. For example, the dispersion of nanoparticles within the hydrophilic polymer chains led to the generation of mechanically enhanced or stable nanocomposite hydrogels, with reference to their application in sustained drug release, better cell adhesion, cell proliferation, and cell migration than conventional hydrogels [7, 8]. For the first time, nanocomposite hydrogels can be synthesized by Harachugi using poly-(PNIPAAm) and montmorillonite nanoclay with improvement in their mechanical strength, swelling ability, ocular precision, and stimulus responsiveness. Here, montmorillonite can act as a cross-linking agent without using any other chemical cross-linker. Gaharwar et al. developed tough and elastic nanocomposite hydrogels using poly(ethylene glycol) PEG and hydroxyapatite (nHA) nanoparticles. The accumulation of nHA in polymeric networks enhanced their characteristic properties such as high flexibility, mechanical strength, resistance power, and cell holding capacity [9]. Gaharwar et al. also developed nanocomposite hydrogels using silica nanofiller and PEG-based nanocomposite hydrogels with improved mechanical and cell binding properties [10].

FIGURE 5.1 Various sources of Sustainable polymer-based nanocomposite hydrogels.

Recent trends have focused on nanocomposite hydrogels with modified functionality has opened up new opportunities for developing advanced biomaterials for various biomedical and biotechnological applications. In this chapter, we will focus on the recent advances in the designing and developing of pH-responsive nanocomposite hydrogels with tailored properties. In addition, we also focus on various nanofillers (Carbon nanotubes, Graphene, clays, metal, hydroxyapatite) based nanocomposite hydrogels with reference to their biomedical applications.

5.2 NANOCOMPOSITE HYDROGELS WITH SPECIFIC PROPERTIES

Nanocomposite hydrogels can be obtained by either using chemical cross-linking polymers (due to covalent bonds) or by physically cross-linking polymers (non-covalent bonds) in the presence of nanoparticles [11]. The physical cross-linking interactions including hydrogen bonding, Vander Waals forces, and electrostatic interactions are alterable, and thus the nanocomposite hydrogels formed can be disrupted with external stimuli such as pH, temperature, magnetic [12], etc. However, chemically crosslinked materials can exhibit stronger mechanical properties as compared to physically crosslinked counterparts. However, chemical cross-linking is their poor mechanical strength, unreacted organic cross-linkers and monomers can leach out and cause toxicity [13]. The dispersion of nanoparticles into physically and chemically crosslinked polymer hydrogels is another important approach. In this approach, instead of using synthetic cross-linkers, nanoparticles are used as multifunctional cross-linkers in hydrogel composition and the polymer chains strongly interact with nanoparticles. These types of nanocomposite hydrogels display synergistic properties like toughness, resistance against compression, bending, tearing, twisting, knotting, and elongation [14–16].

5.2.1 CHARACTERISTICS OF PH-RESPONSIVE NANOCOMPOSITE HYDROGELS

Characteristics of nanocomposite hydrogels such as soft or elastic nature, water uptake ability, low roughness, and high mechanical, optical, and electrical properties gained much attention recently. The precursor (Sustainable polymers) used for the fabrication of nanocomposite hydrogels are usually easily available in the market, non-toxic, easily modified further, biodegradable, and biocompatible in nature due to their origin from flora and fauna. Nanocomposite hydrogels once prepared can be evaluated for their characteristic properties such as mechanical, swelling, biocompatible, and toxicity for their successful biomedical application [17].

5.2.1.1 SWELLING PROPERTIES

The swelling behavior of nanocomposite hydrogels can be enhanced in the presence of various hydrophilic nano-filler in the polymer matrices and maintain the three-dimensional structure [18]. All the polymeric chains in the nanocomposite hydrogels are cross-linked with each other by physical and chemical interactions and therefore it is called super macromolecules. The enhancement in swelling ratio can be increased by increasing the concentration of nanofiller in the polymeric matrices but up to some prominent level [19].

5.2.1.2 BIODEGRADABILITY

Biodegradability is the process of the chemical dissolution of material by microorganisms or other biological means. Biodegradation takes place aerobically with oxygen or aerobically without oxygen [20]. Biodegradability is the main characteristic property of nanocomposite hydrogels, which exhibit significant contributions in the biomedical field. Inorganic nanoparticle-based biodegradable materials have good ability in the biomedical field due to their variety of applications. Nanocomposite hydrogels have labile linkages in their structure that allow the degradation of materials in various pH solutions as well as enzymes present in the body [21]. Biodegradable nanocomposite hydrogels also termed bionanocomposite when implanted or injected in the body do not need to be removed after their useful applications. Various natural biopolymers that are protein based (such as elastin, collagen, and gelatin) are used for the fabrication of biodegradable, biocompatible, and non-toxic materials. Based on these properties' nanocomposite hydrogels are used for encapsulation and controlled drug delivery applications [22].

5.2.1.3 BIOCOMPATIBILITY

Biocompatibility is the ability of materials to perform effectively at their desired sites without creating a toxic environment in the surrounding. Natural polymer-based nanocomposite hydrogels are biocompatible and less toxic as compared to that conventional hydrogels [23]. Nanocomposite hydrogels used for biomedical applications must pass cytotoxicity and *in vitro* toxicity tests. Due to good osmotic properties, excellent mechanical properties and good biocompatibility have potential applications for nanocomposite hydrogels in biomedical and tissue engineering [24].

5.2.1.4 MECHANICAL PROPERTIES

Fabrication of mechanically reinforced hydrogels is important for their specific use as biomaterials. The uniform dispersion of nano-fillers and their strong interactions with polymer matrix in the nanocomposite hydrogels can exhibit excellent mechanical properties [25]. The presence of nanofiller in the polymeric matrices enhanced the mechanical strength of nanocomposite hydrogels, which are very significant from a pharmaceutical and biomedical point of view. Various biomedical applications would benefit from mechanically strong materials such as wound healing, drug delivery, tissue engineering, and sensor applications [26]. Recently an enormous number of biocompatible nanocomposite hydrogels have been prepared with exceptional properties such as extraordinary mechanical strength, controllable TS, self-healing ability, and ease to synthesize to overcome the limitation of conventional hydrogels [27–29].

5.2.1.5 STIMULI-RESPONSIVE NANOCOMPOSITE HYDROGELS

The nanocomposite hydrogels undergo reversible physical and chemical changes after being exposed to external stimuli, e.g., temperature, pH, light, electric, and magnetic fields. The fabrication of stimuli-responsive nanocomposite hydrogels can be depending on the nanomaterials, polymers, or nanomaterials/polymers mixtures. These stimuli-responsive hydrogels have been used in the field of biomedical sciences like tissue engineering, drug delivery, and wound healing.

5.3 SUSTAINABLE POLYMER-BASED NANOCOMPOSITE HYDROGELS

5.3.1 GUAR GUM (GG)-BASED NANOCOMPOSITE HYDROGELS

Guar gum (GG) is a galactomannan extracted from the seed of the leguminous shrub *Cyamopsis tragonoloba*. Among all of the naturally occurring biopolymers, GG is one of the most important water-soluble biopolymers with the highest molecular weight (Mw). India accounts for 80% world's production of GG and it is used in the paper, textile, cosmetic, and food industries. GG has prime importance for the controlled delivery of drugs in the gastrointestinal tract, such as a carrier for colon-targeted drugs, for the treatment of colorectal cancer, and cholera in adults due to its prominent properties like biocompatibility, biodegradability, high viscosity, inoffensive or nontoxic [27, 30]. However, the GG-based plain hydrogel materials show some drawbacks in packing materials due to their poor mechanical strength. The incorporation of various nanofillers in the polymer matrices can improve their properties. For instant, Giri et al. (2011) reported nanocomposite hydrogels from carboxymethyl GG and carbon nanotubes, with improved mechanical properties and also useful for sustained trans-dermal release of diclofenac sodium [31]. Similarly, Giri et al. (2012) synthesized guar gum-based nanocomposite hydrogel membranes by using acrylic acid and nano-silica for controlled release of diclofenac sodium [32]. Yang et al. (2011) synthesized nanocomposite beads using pH-sensitive Guar gum-g-polyacrylic acid-attapulgite (GG-g-PAA/APT) through a facile ionic gelation method for drug delivery system. The release rate of diclofenac sodium drug from the beads can depend upon the concentration of APT. The incorporation of APT into the polymeric matrix can control the burst release of drugs [33].

Kono et al. (2014) synthesized GG-based hydrogels using 1,2,3,4-butane tetracarboxylic di anhydride (BTCA) through esterification. The addition of BTCA in the polymer matrix increases the degree of cross-linking, which further affect enzymatic degradation, swelling behavior, and rheological properties. These materials exhibit good adsorption of bovine serum albumin and lysosome as well as protein, which shows the potential ability to be used in drug delivery applications [34].

Abdel-Halim et al. (2014) synthesize electrically conductive silver-guar gum-poly(acrylic acid) based composite hydrogels. There is a grafting between acrylic acid monomers and natural GG by the use of ammonium persulfate as an initiator. GG grafted poly-acrylic acid can be further subjected to add epichlorohydrin as a cross-linking agent followed by the addition of the solution of silver nitrate [35].

Menon et al. [37] synthesized superabsorbent hydrogels by using guar gum-grafted-poly-acrylic acid/cloisite [36]. Cloisite is an organically modified nanoclay that can be used to enhance the properties of nanocomposite hydrogels. The grafting of acrylic acid in the guar gum can be enhanced by using the microwave method. FTIR and XRD showed excellent grafting can occur between polyacrylic acid into the backbone of guar gum and exfoliation of cloisite in guar gum. The dispersion of cloisite (2 wt.%.) in polymer matrices enhanced the swelling rates and equilibrium swelling. The water-withholding properties and biodegradable nature make hydrogels a good candidature that can be used as eco-friendly superabsorbent.

5.4 HYDROGEL CROSS-LINKING IN THE PRESENCE OF NANOPARTICLES

One of the typical approaches for manufacturing nanocomposite hydrogels consists in mixing hydrogel precursors with the colloidal nanofiller suspensions earlier to hydrogel formation [37].

Generally, this technique offers relatively homogeneous dispersion than other assembly techniques, like nanoparticle diffusion in the polymeric matrix. For instance, through the application of magnetic fields, the position of nanoparticles can be controlled during hydrogel cross-linking, thus allowing the construction 3D network pattern [38]. The basic benefit of this method is that it permits a wide range of nanoparticles incorporated into the hydrogel matrices, which can facilitate the translational potential of these hybrid hydrogels. This versatile property enables researchers to design tough and complicated nanocarriers [39]. The crosslinking pattern and other conditions used for the fabrication of nanocomposite hydrogels do not significantly affect nanoparticle stability and physicochemical characteristic. However, this method requires extra optimization of nanomaterials, if the nanoparticles within the matrix are prone to aggregate, i.e., with native liposomes. To overcome these issues, liposomes-carboxy-modified gold nanoparticles have been engineered with improved properties and stability for antimicrobial delivery [40].

Similarly, nanoparticles with hydrophobic characteristics to that of hydrogel constituents, i.e., polymers, proteins, etc.) lead to cluster formation and restrict uniform distribution of nanoparticle distribution within the polymer matrices produced by this strategy. Another important issue related to this approach is that increases in the concentration of nanoparticles increase the possibility of particle-particle aggregation and may also affect hydrogel network formation. So, manufacturing homogeneous nanocomposite hydrogels with synergistic properties is essential for standardizing production and stimuli-responsive [41, 42].

5.5 BIOPOLYMER-BASED NANOCOMPOSITE HYDROGEL

With the concept of eco-friendly protection and the rising demands of sustainable materials, biodegradability, and biocompatibility has been extensively focused on. In recent years, researchers focus much on biopolymer-based materials due to their tremendous advantages in the technological and biomedical fields especially in drug delivery, tissue engineering, food packaging, etc. In general, biopolymer-based hydrogels are biodegradable, biocompatible, economically sound, easily available, and nontoxic without creating any toxic effect on the surrounding environment. Mostly natural polymers can be divided into major classes according to their chemical structure: (i) Polysaccharides (chitosan, guar gum, agar, starch, gum tragacanth, aloe vera gel, hyaluronic acid, alginate, and agarose) (ii) Polyamides (or Polypeptides-collagen, gelatin, fibroin, wheat, soya, and fibrin) and (iii) Polyesters (poly(3-hydroxy alkanoates) [43]. For example, Tanpichai et al. prepared nanocomposite hydrogels using cellulose nanocrystal and poly(vinyl alcohol) and showed potential for use in biomedical and tissue engineering applications) [44]. Peng et al. nanocomposite hydrogel using cellulose-clay with superabsorbent properties and high mechanical strength has been used as adsorbent for methylene blue [45].

5.6 STIMULI-RESPONSIVE NANOCOMPOSITE HYDROGELS AND BIOMEDICAL APPLICATIONS

Moreover, the traditional drug delivery system with sustained release has provided a significant advantage over conventional therapeutics administration by extending the release period of various drugs and decreasing administration frequency. The invention of stimuli-responsive biomaterials is visualized to progress the biomedical field to a large extent [46, 47]. This can be attributed to the remarkable progress in the designing to installation multi-stimuli responsive nanomaterials, with the establishment of electrostatic interactions that respond to a wide array of external or internal clues [48, 49]. For instance, the internal stimuli present within the human body have been widely explored for stimuli-responsive delivery like pH, redox, enzyme, glucose, and temperature (Figure 5.2).

Nanocomposite hydrogels can also be fabricated to respond to external stimuli that are not present in our body except for some external applications, namely high temperature, magnetic fields, ultrasound, and light [50–54].

This allows researchers to propose platforms that are stable in biological conditions but react vigorously on demand upon an externally applied sign.

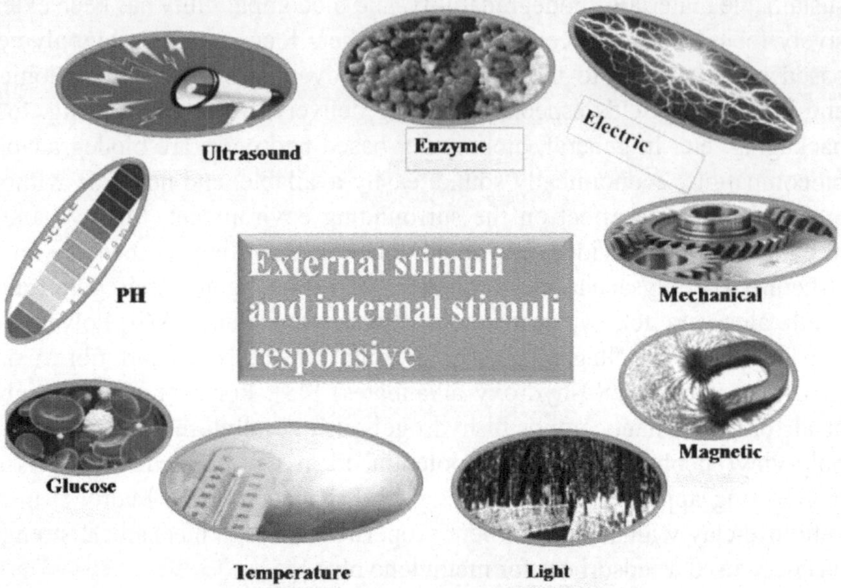

FIGURE 5.2 Internal and external stimuli are responsive.

5.6.1 THERMORESPONSIVE NANOCOMPOSITE HYDROGELS

Thermoresponsive nanocomposite hydrogels are manufactured above a lower critical solution temperature (LCST) in which polymer matrix solutions undergo phase separation [55, 56]. They have a wide range of hydrophobic functional groups, and the ratios of LCST/UCST can be modified by changing the concentration of hydrophilic and hydrophobic functional groups of the polymers [57–59]. Thermosensitive nanocomposite hydrogels have a wide range of potential scope in the biomedical field like tissue engineering, drug delivery, and wound dressing. Many nanocomposites' hydrogels form gel formation at physiological conditions (at 37°C) especially for drug delivery applications as many hydrogels show gel formation at a universally accepted physiological temperature of 37°C, and also, several easy modifications are available to control gel formation at physiological temperature [58–63].

5.6.2 PH-RESPONSIVE NANOCOMPOSITE HYDROGELS

pH-responsive nanocomposite hydrogels show characteristic features like swelling/shrinking performance in response to changes in the environmental pH [43, 64], and this class of stimuli-responsive nanocomposite hydrogels finds a potential scope in biomedical applications as significant pH changes occur in different organs or locations in the body, which are mandatory for normal body function such as the gastrointestinal tract [43], blood vessels, intracellular vesicles [44], and female genital tract, etc. [45]. The various changes or modifications in pH within the body can also take place due to abnormal body functions or the creation of certain cancerous diseases within the body [46] tumors and inflammation [47]. pH-responsive platforms can drastically change their structural properties to activate specific bioactive molecule delivery with respect to pathologically promoted at sharp ph. Instead of its pathological disorder, pH usually changes in certain anatomical sites like along the gastrointestinal tract, i.e., mouth saliva, stomach HCl secretions, intestines, etc. In addition, intracellular endosomal sections are highly acidic that also facilitate specific cargo release when nanoparticles are being released from the nanocomposite hydrogel materials.

The pH-responsive nanocomposite hydrogel systems have been widely applicable for developing a wide variety of drug delivery systems [41, 48–50, 65, 66].

KEYWORDS

- **biocompatibility**
- **biodegradability**
- **guar gum**
- **hydrogels**
- **nanocomposite**
- **pH-responsive**
- **sustainable polymer**
- **thermosensitive**

REFERENCES

1. Hlaing, N. N., & Mya, M., (2008). Manufacture of alkyd resin from castor oil. In: *Proceedings of World Academy of Science, Engineering and Technology*.
2. Davis, J. P., Sweigart, D. S., Price, K. M., Dean, L. L., & Sanders, T. H., (2013). Refractive index and density measurements of peanut oil for determining oleic and linoleic acid contents. *Journal of the American Oil Chemists' Society, 90*, 199–206.
3. Pathan, S., & Ahmad, S., (2013). Synthesis, characterization and effect of s-triazine ring on physico-mechanical and electrochemical corrosion resistance properties of waterborne castor oil alkyd. *J. Mater. Chem. A*.
4. Ghosal, A., Shah, J., Kotnala, R. K., & Ahmad, S., (2013). Facile green synthesis of nickel nanostructures using natural polyol and morphology dependent dye adsorption properties. *J. Mater. Chem. A*.
5. Mukherjee, A. K., Mondal, K., Akhan, M. A. I., & Biswas, S., (2013). *Effects of Phospholipase A2 Degumming on Palm Oil Components*.
6. Baravkar, A., KALE, R., Sawant, S. *FT-IR Spectroscopy: Principle, Technique and Mathematics*.
7. Hauser, M., & Oelichmann, J., (1988). A critical comparison of solid sample preparation techniques in infrared spectroscopy. *Microchimica Acta, 94*, 39–43.
8. Xiaomin, X., Yang, L., Wenbo, F., Mingyu, Y., Zhen, D., Jiaming, X., et al. *Poly(N-isopropylacrylamide)-Based Thermoresponsive Composite Hydrogels for Biomedical Applications*.
9. Akhilesh, G. K., Sandhya, D. A., Jamie, C. M., Chia-Jung, W., & Gudrun, S. *Highly Extensible, Tough, and Elastomeric Nanocomposite Hydrogels from Poly(ethylene glycol) and Hydroxyapatite Nanoparticles*.
10. Akhilesh, G. K., Christian, R., Chia-Jung, W., Burke, C. K., Gudrun, S. *Photocrosslinked Nanocomposite Hydrogels from PEG and Silica Nanospheres: Structural, Mechanical and Cell Adhesion Characteristics*.
11. Gill, P., Moghadam, T. T., & Ranjbar, B., (2010). Differential scanning calorimetry techniques: Applications in biology and nanoscience. *Journal of Biomolecular Techniques: JBT, 21*, 167.
12. Hancock, B. C., & Zografi, G., (1997). Characteristics and significance of the amorphous state in pharmaceutical systems. *Journal of Pharmaceutical Sciences, 86*, 1–12.
13. Ravichandran, P., Shantha, K., & Rao, K. P., (1997). Preparation, swelling characteristics and evaluation of hydrogels for stomach specific drug delivery. *International Journal of Pharmaceutics, 154*, 89–94.
14. Hoare, T. R., & Kohane, D. S., (2008). Hydrogels in drug delivery: Progress and challenges. *Polymer, 49*, 1993–2007.
15. Kim, S. J., Park, S. J., & Kim, S. I., (2003). Swelling behavior of interpenetrating polymer network hydrogels composed of poly (vinyl alcohol) and chitosan. *Reactive and Functional Polymers, 55*, 53–59.
16. Vashist, A., Gupta, Y. K., & Ahmad, S., (2012). Interpenetrating biopolymer network based hydrogels for an effective drug delivery system. *Carbohydrate Polymers, 87*, 1433–1439.
17. Quercioli, F., (2011). Fundamentals of optical microscopy. In: *Optical Fluorescence Microscopy* (pp. 1–36). Springer.
18. Kerns, C., Zarakov, E., & Gilley, T. S., (1994). *Digital Camera with Time Bracketing Feature*. In Google Patents.

19. Goldstein, J., Newbury, D. E., Joy, D. C., Lyman, C. E., Echlin, P., Lifshin, E., Sawyer, L., & Michael, J. R., (2003). *Scanning Electron Microscopy and X-ray Microanalysis*. Springer.
20. Vashist, A., Shahabuddin, S., Gupta, Y. K., & Ahmad, S., (2013). Polyol induced interpenetrating networks: Chitosan-methylmethacrylate based biocompatible and pH responsive hydrogels for drug delivery system. *Journal of Materials Chemistry B, 1*, 168–178.
21. Williams, D. B., & Carter, C. B., (1996). *The Transmission Electron Microscope*. Springer.
22. Ding, F., Nie, Z., Deng, H., Xiao, L., Du, Y., & Shi, X., (2013). Antibacterial hydrogel coating by electrophoretic co-deposition of chitosan/alkynyl chitosan. *Carbohydrate Polymers, 98*, 1547–1552.
23. Förster, H., (2004). UV/vis spectroscopy. In: *Characterization I* (pp. 337–426). Springer.
24. Lin, C. C., & Metters, A. T., (2006). Hydrogels in controlled release formulations: Network design and mathematical modeling. *Advanced Drug Delivery Reviews, 58*, 1379–1408.
25. Sannino, A., Demitri, C., & Madaghiele, M., (2009). Biodegradable cellulose-based hydrogels: Design and applications. *Materials, 2*, 353–373.
26. Dong, F., Zhao, C., Hou, Z., Deng, F., Li, R., Li, Y., Bai, Y., et al., (2013). *Biodegradable Medical Hydrogel and its Manufacture and Application* (p. 32). In Shandong Success Medicine Industry Science and Technology Co., Ltd., Peop. Rep. China.
27. Goldberg, M., Langer, R., & Jia, X., (2007). Nanostructured materials for applications in drug delivery and tissue engineering. *Journal of Biomaterials Science, Polymer Edition, 18*, 241–268.
28. Riaz, U., Vashist, A., Ahmad, S. A., Ahmad, S., & Ashraf, S., (2010). Compatibility and biodegradability studies of linseed oil epoxy and PVC blends. *Biomass and Bioenergy, 34*, 396–401.
29. Silva-Correia, J., Zavan, B., Vindigni, V., Silva, T. H., Oliveira, J. M., Abatangelo, G., & Reis, R. L., (2012). Biocompatibility evaluation of ionic- and photo-crosslinked methacrylated gellan gum hydrogels: *In vitro* and *in vivo* study. *Adv. Healthc. Mater., 2*, 201200256.
30. Aurand, E. R., Wagner, J., Lanning, C., & Bjugstad, K. B., (2012). Building biocompatible hydrogels for tissue engineering of the brain and spinal cord. *J. Funct. Biomater., 3*, 839–863.
31. Karadağ, E., Saraydin, D., Çetinkaya, S., & Güven, O., (1996). *In vitro* swelling studies and preliminary biocompatibility evaluation of acrylamide-based hydrogels. *Biomaterials, 17*, 67–70.
32. Arindam, G., Manas, B., Sagar, P., & Abhijit, B. *Polymer Hydrogel from Carboxymethyl Guar Gum and Carbon Nanotube for Sustained Trans-dermal Release of Diclofenac Sodium*.
33. Arindam, G., Totan, G., Asit, B. P., Sagar, P., & Abhijit, B. *Tailoring Carboxymethyl Guargum Hydrogel with Nanosilica for Sustained Transdermal Release of Diclofenac Sodium*.
34. Peng, L., Liping, J., Longxiang, Z., Jinshan, G., & Aiqin, W. *Synthesis of Covalently Crosslinked Attapulgite/Poly (Acrylic Acid-co-Acrylamide) Nanocomposite Hydrogels and their Evaluation as Adsorbent for Heavy Metal Ions*.
35. Hiroyuki, K., Fumihiro, O., & Masato, O. *Preparation and Characterization of Guar Gum Hydrogels as Carrier Materials for Controlled Protein Drug Delivery*.

36. Hossein, H., & Barghi, A. *Synthesis of Poly(AN)/Poly(AA-co-AM) Hydrogel Nanocomposite with Electrical Conductivity and Antibacterial Properties.*
37. Menon, S., Deepthi, M. V., Sailaja, R. R. N., & Ananthapadmanabha, G. S. *Study on Microwave Assisted Synthesis of Biodegradable Guar Gum Grafted Acrylic Acid Superabsorbent Nanocomposites.*
38. Lee, S. C., Kwon, I. K., & Park, K., (2013). Hydrogels for delivery of bioactive agents: A historical perspective. *Advanced Drug Delivery Reviews, 65*, 17–20.
39. Kopeček, J., (2010). Biomaterials and drug delivery: Past, present, and future. *Molecular Pharmaceutics, 7*, 922.
40. Kamath, K. R., & Park, K., (1993). Biodegradable hydrogels in drug delivery. *Advanced Drug Delivery Reviews, 11*, 59–84.
41. Mastropietro, D. J., Omidian, H., & Park, K., (2012). Drug delivery applications for superporous hydrogels. *Expert Opin. Drug Delivery, 9*, 71–89.
42. Ganji, F., & Vasheghani-Farahani, E., (2009). Hydrogels in controlled drug delivery systems. *Iran Polym. J, 18*, 63.
43. Henriksen, I., Green, K. L., Smart, J. D., Smistad, G., & Karlsen, J., (1996). Bioadhesion of hydrated chitosans: An *in vitro* and *in vivo* study. *International Journal of Pharmaceutics, 145*, 231–240.
44. Supachok, T., & Kristiina, O. *Cross-Linked Nanocomposite Hydrogels Based on Cellulose Nanocrystals and PVA: Mechanical Properties and Creep Recovery.*
45. Na, P., Danning, H., Jian, Z., Yu, L., Lei, L., & Chunyu, C. *Superabsorbent Cellulose–Clay Nanocomposite Hydrogels for Highly Efficient Removal of Dye in Water.*
46. Cheung, H. Y., Lau, K. T., Lu,T. P., & Hui, D., (2007). A critical review on polymer-based bio-engineered materials for scaffold development. *Composites Part B: Engineering, 38*, 291–300.
47. Srinivasan, A., Karchmer, T., Richards, A., Song, X., & Perl, T. M., (2006). A prospective trial of a novel, silicone-based, silver-coated foley catheter for the prevention of nosocomial urinary tract infections. *Infection Control and Hospital Epidemiology, 27*, 38–43.
48. Balakrishnan, B., Mohanty, M., Umashankar, P., & Jayakrishnan, A., (2005). Evaluation of an *in situ* forming hydrogel wound dressing based on oxidized alginate and gelatin. *Biomaterials, 26*, 6335–6342.
49. De Groot, J. H., Van, B. F. J., Haitjema, H. J., Dillingham, K. A., Hodd, K. A., Koopmans, S. A., & Norrby, S., (2001). Injectable intraocular lens materials based upon hydrogels. *Biomacromolecules, 2*, 628–634.
50. Koh, W. G., & Pishko, M. V., (2006). Fabrication of cell-containing hydrogel microstructures inside microfluidic devices that can be used as cell-based biosensors. *Analytical and Bioanalytical Chemistry, 385*, 1389–1397.
51. Lin, C. C., & Metters, A. T., (2006). Hydrogels in controlled release formulations: Network design and mathematical modeling. *Advanced Drug Delivery Reviews, 58*, 1379–1408.
52. Kim, S. J., Shin, S. R., Kim, N. G., & Kim, S. I., (2005). Swelling behavior of semi-interpenetrating polymer network hydrogels based on chitosan and poly (acryl amide). *Journal of Macromolecular Science, Part A: Pure and Applied Chemistry, 42*, 1073–1083.
53. Kalyani, S., Smitha, B., Sridhar, S., & Krishnaiah, A., (2006). Blend membranes of sodium alginate and hydroxyethylcellulose for pervaporation-based enrichment of *t*-butyl alcohol. *Carbohydrate Polymers, 64*, 425–432.

54. Tozaki, H., Komoike, J., Tada, C., Maruyama, T., Terabe, A., Suzuki, T., Yamamoto, A., & Muranishi, S., (1997). Chitosan capsules for colon-specific drug delivery: Improvement of insulin absorption from the rat colon. *Journal of Pharmaceutical Sciences, 86*, 1016–1021.
55. Gong, C., Qi, T., Wei, X., Qu, Y., Wu, Q., Luo, F., & Qian, Z., (2013). Thermosensitive polymeric hydrogels as drug delivery systems. *Current Medicinal Chemistry, 20*, 79–94.
56. Tiwari, S., Singh, S., Rawat, M., Tilak, R., & Mishra, B., (2009). L9 orthogonal design assisted formulation and evaluation of chitosan based buccoadhesive films of miconazole nitrate. *Current Drug Delivery, 6*, 305–316.
57. Bhattarai, N., Gunn, J., & Zhang, M., (2010). Chitosan-based hydrogels for controlled, localized drug delivery. *Advanced Drug Delivery Reviews, 62*, 83–99.
58. Qu, X., Wirsen, A., & Albertsson, A. C., (2000). Novel pH-sensitive chitosan hydrogels: Swelling behavior and states of water. *Polymer, 41*, 4589–4598.
59. Neto, C. D. T., Giacometti, J., Job, A., Ferreira, F., Fonseca, J., & Pereira, M., (2005). Thermal analysis of chitosan based networks. *Carbohydrate Polymers, 62*, 97–103.
60. Naik, A., Kalia, Y. N., & Guy, R. H., (2000). Transdermal drug delivery: Overcoming the skin's barrier function. *Pharmaceutical Science & Technology Today, 3*, 318–326.
61. Sannino, A., Demitri, C., & Madaghiele, M., (2009). Biodegradable cellulose-based hydrogels: Design and applications. *Materials, 2*, 353–373.
62. Singh, A., Narvi, S., Dutta, P., & Pandey, N., (2006). External stimuli response on a novel chitosan hydrogel crosslinked with formaldehyde. *Bulletin of Materials Science, 29*, 233–238.
63. Göpferich, A., (1996). Mechanisms of polymer degradation and erosion. *Biomaterials, 17*, 103–114.
64. Peanasky, J. S., Long, J., & Wool, R., (1991). Percolation effects in degradable polyethylene-starch blends. *Journal of Polymer Science Part B: Polymer Physics, 29*, 565–579.
65. Silva-Correia, J., Zavan, B., Vindigni, V., Silva, T. H., Oliveira, J. M., Abatangelo, G., & Reis, R. L., (2012). Biocompatibility evaluation of ionic- and photo-crosslinked methacrylated gellan gum hydrogels: *In vitro* and *in vivo* study. *Adv. Healthc. Mater, 2*, 201200256.
66. Vashist, A., Shahabuddin, S., Gupta, Y. K., & Ahmad, S., (2013). Polyol induced interpenetrating networks: Chitosan-methylmethacrylate based biocompatible and pH responsive hydrogels for drug delivery system. *Journal of Materials Chemistry B, 1*, 168–178.

CHAPTER 6

Polymer-Based Hybrid Composites for Tissue Engineering Applications

G. SANTHOSH[1] and G. P. NAYAKA[2]

[1]*Department of Mechanical Engineering, NMAM Institute of Technology, Nitte, Karnataka, India*

[2]*Physical and Materials Chemistry Division, CSIR–National Chemistry Laboratory, Pune, Maharashtra, India*

ABSTRACT

Materials used as implants with single composition suffer from properties limiting their life and restricting their applications. Therefore, composite materials have been developed by combining various materials of different properties. The hybrid materials prepared are tailored to provide the properties of the tissues. Various hybrid materials have been developed and tested for tissue repair and/or replacement. The tissue replacement may include joint replacements, materials to heal fractured bones, restorative materials, and tissue scaffolds and implants. This chapter provides an overview of hybrid material-based implants/scaffolds and their essential requirements for tissue repair or replacements.

6.1 INTRODUCTION

Polymers have great capability and feasibility as implants in biomedical applications. The application and use of such polymers have seen steady growth in recent decades [1–4]. The use of natural and synthetic polymers is not limited to fewer areas of tissue engineering, but they have been used as

Biocomposites: Environmental and Biomedical Applications. Omar Mukbaniani, Tamara Tatrishvili, Neha Kanwar Rawat, & A. K. Haghi (Eds.)

hemostats, absorbable sutures, bone fixation aids, and also for nerve-guided tubes [1, 2]. Tissue engineering treatments are essential due to the shortage of clinical therapies which can restore the defects. An important policy to promote the revival of healthy cells involves the use of hybrid materials capable of self-repair. Hybrid composites are considered as "materials of the decade" and one of the major ways of fabricating biomaterials. The use of traditional materials such as metals, ceramics, and synthetic polymers is restricted due to their compatibility and immunological rejection by the body [5–8]. Many researchers have adopted composite materials/hybrid materials to resolve compatibility challenges. The development of composite materials with an anticipated behavior to help the recovery of damaged tissues. Composite materials can establish ultimate contact with living cells. The main intention of the use of composite materials as implantable devices is to replace the damaged tissues or structures, most used implants consist of heart, bone, knees, hips, and cardiovascular system implants. In biomedical and dental applications, calcium phosphates (CaP) are widely employed. Materials and bio-glasses based on CaP bond firmly to both hard and soft tissues [9].

For many patients, replacing damaged joints with prostheses has brought about a pain-free life. Permanent implants are total joint replacements (TJRs), and their fitting involves considerable bone and cartilage removal. The success of joint replacement prostheses depends on many factors, most of which are related to the materials used to make them. The failure of TJRs usually ends up with revision surgery, which has less satisfactory clinical results. Aseptic loosening has been found to be the most prevalent cause. Because of the younger patients involved in these operations, there is still a need for longer-lasting *in vitro* performance. Based on the specific application, the material requirements must be determined. Several characteristics must be met for all materials for joint replacements [10–12].

This chapter provides an overview of permanent implants and/or prostheses with necessary materials details and applications. The chapter also addresses the use of composite biomaterials in hard tissue replacements.

6.2 REQUIREMENTS OF POLYMERS HYBRIDS AS IMPLANTS [13]

The hybrid polymer system can be any natural or synthetic materials used as implants, which are modified and designed to interact with the biological system:

- Biostability: The hybrid system should not enable the soluble compounds to enter the living system.
- Implants must be sterilizable.
- It should remain non-toxic and non-allergic in response to the living tissues.
- Hybrid implants must have enough mechanical and physical properties to serve the intended purpose and or until surrounding tissues heal.
- Hybrid systems should be biocompatible viz.; implants should not adversely affect the living tissues and should not interact with the biological fluids.
- It should have an appropriate shelf life.

6.3 THE PROPERTIES OF AN IDEAL IMPLANT

Tissue loss and tissue defects, caused by destruction and failure to heal immediately, have been major health concern that affects the quality and length of human life. The engineering scaffolds should have the ability to replicate the biological functions and structural support of the native tissues. The scaffolds use are largely dependent on clinical success and material compositions. The major concern in adopting the scaffolds is the use appropriate selection and use of raw materials. Organic, inorganic, and metals were used to prepare tissue scaffolds for many years. However, the monotonic materials used as scaffolds are not providing satisfactory results. In many instances, the scaffolds prepared by single material can hardly meet the requirements of repairing or replacing the tissues. The composites, especially bio-composites have been more effective with minimum adverse effects are being used. In addition to the composition, the major deciding factor in tissue repair or replacement is its structure. An ideal scaffold must possess structural porosity with good mechanical properties. However, regardless of the tissue type, the scaffolds must be biocompatible, biodegradable, and provide sufficient strength and stiffness while handling, later with the host tissue after implantation. The tissue engineering materials with distinct conditions for repairing and replacing the tissues are shown in Figure 6.1. The material's strength is the most important parameter of the implants used. Usually, the scaffolds comprise many components. Materials with adequate strength are necessary for prostheses to avoid failure/fracture. The fracture can occur in the implant itself, or at the interfaces of the cemented prostheses. However, if the hard tissue interface begins to fail, the growth of soft fibrous tissue allows "micro-motion" of the prostheses relative to the bone. This can cause aseptic loosening of the implant and need subsequent revision procedures [14]. Hence, it is very much essential to consider specific mechanical strength, which would be as same as the original hard tissue [15].

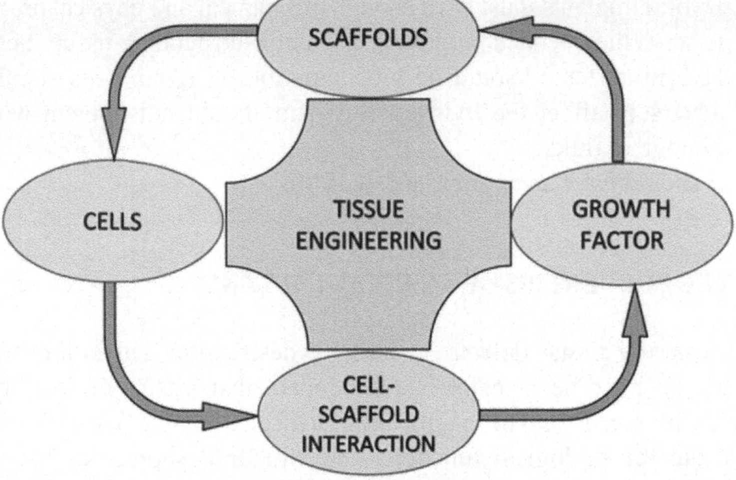

FIGURE 6.1 The scaffolds for tissue engineering applications.

Modulus of elasticity, fatigue, creep resistance, wear resistance, corrosion resistance, biocompatibility, and bioactivity are the most important parameters in tissue engineering applications. Elasticity or stiffness of the prosthesis materials are major parameters involved in load distribution, prostheses shield the tissues and help distribute the load that leads to bone resorption [16–18]. Fatigue is a phenomenon where the materials become weak due to continuous loading. The prostheses are used at knee joints; hip joints are exposed to cyclic loading due to routine activities. The cyclic loading creates issues like implant loosening and debris formation, these adverse effects can generate fatigue [19]. Further, the material's exposure to high-stress results in permanent deformation. The material's resistance to permanent deformation is coined as creep resistance. With aging, osteoporosis becomes a major health issue in elderly people. This condition reduces bone strength and density. Thus, increasing the demand for the use of bone grafts.

The polymeric hybrids as bone grafts/scaffolds selected as the tissue replacements need to have good strength and creep behavior. The implant materials used should be corrosion-resistant and should operate with minimum friction under loading [20–24]. Biocompatibility is one of the important parameters considered while using implants, it is the ability of the scaffolds to interact with living tissues [25]. Polymer hybrids exhibit better compatibility by eliminating adverse effects [26]. However, the use of polymers also reveals various deficiencies when compared to host tissues and these difficulties can be controlled by creating polymer hybrids/scaffolds resembling the original tissues' behavior.

6.4 POLYMER IN TISSUES ENGINEERING

The need to develop effective regeneration and/or replacement materials for damaged tissues has become a major concern these days. Polymers are such materials has outstanding flexibility and functionalities, making them unique compared to other materials. In this regard, both natural and synthetic polymers have the ability to tailor the needs driven by applications. With all the benefits and advantages of design, the soft nature of the polymers makes them weak in many applications. However, with tissue engineering, this translates to the fact that some polymers can mimic the behavior of soft tissues. Further, it is difficult to tune the polymer's properties to match the hardness of hard and soft tissues. The mechanical strength of the materials affects and drives tissue growth, enhancement of such properties is essential. To overcome such challenges hybrid materials are used [27]. The hybrid material used must meet the biological requirements in designing the extra-cellular matrix for various tissue engineering applications. Further, the hybrid systems must be biocompatible and/or biodegradable which is obtained by polymeric or inorganic content. The stability and mechanical properties are also necessary to match the desired tissues, thus can be achieved by tuning the type, concentration, and distribution of inorganic content [28].

6.4.1 NATURAL AND SYNTHETIC POLYMERS IN TISSUE ENGINEERING

Naturally, derived polymers have seen huge success in the field of tissue engineering due to their specific structure and compatibility. The natural biomaterials used as tissue scaffolds have a greater ability to encapsulate animal cells [29].

Synthetic polymers have gained more attention compared to other polymers. Unique properties such as biocompatibility, cell adherence, controlled drug release, and stimuli-response are sometimes necessary for biodegradable polymers. Among various natural and synthetic polymers hydroxyapatite (HA) and tricalcium phosphates (TCP) ceramic materials are the most promising candidates for scaffolds [30, 31]. HA and TCP possess osteoconductive properties resembling natural inorganic components of bone, however, due to their brittle nature, the mechanical strength cannot be the same as the hard tissues [32]. Collagen, the most abundant protein in mammals is used to create hybrid biomaterial to improve cell adhesion [33]. Chitosan is derived from shrimps, mollusks, and fungi as its applications in medicine and as a biomaterial due to its degradable nature [34]. The use of chitosan is limited as it dissolves in a higher acidic medium, intern restricts the use of chiton when working with cell and PH-dependent molecules [35]. Regarding synthetic polymers, the most widely used polymers for tissue engineering scaffold applications are aliphatic polyesters such as polyglycolic acid (PGA), polylactic acid (PLA), copolymers such as lactic-co-glycolic acid (PLGA) (also known as polylactide-co-glycolide) and polycaprolactone (PCL) [36–40]. As mentioned, the scaffolds were fabricated using either ceramics or polymers. The ceramic scaffolds are proffered over polymeric scaffolds as they are too flexible and appeared to be fragile [41, 42]. However, polymeric composite materials reinforced with inorganic ceramic fillers have been widely adopted to reconstruct the hard and soft tissues [41, 42].

One can easily understand, that polymer-based composites have been instrumental in designing scaffolds for tissue engineering with enhanced functionality and regeneration. The details of various synthetic and natural polymers as implants and their advantages are listed in Table 6.1.

6.5 NANOMATERIALS IN TISSUE ENGINEERING

Nanotechnology is the most promising area which enhances the bioactivity of the implants due to the high surface area of the nanomaterials [43]. The use of nanoparticles improves the overall functioning of the scaffolds by functionalizing them with proteins and initiators which are capable of stimulating bone cells.

Nanoparticles of size below 100 nm with uniform distribution in the matrix yield the best results as scaffolds [44]. Nanoparticles used in the scaffolds were selected either based on their bactericidal activity and encapsulation efficiencies [45]. Nanocomposites used in tissue engineering applications exhibit an intricate architecture constructed with an organic structure [46]. Nanomaterials such as "calcium" and "phosphate (HA ($Ca_{10}(PO_4)_6(OH)_2$),"

TABLE 6.1 Polymers and Their Use as Implants

Polymers	Devices	Advantages
Polysaccharides, e.g., Alginates	• Wound healing. • Tissue engineering.	• Non-inflammatory. • Non-toxic. • Biocompatible.
Starch	• Tissue engineering scaffolds.	• Non-toxic. • Excellent biocompatibility.
Cellulose	• Skin burns, wound healing, implants, and tissue engineering.	Excellent biocompatibility and, mechanical properties.
Chitin and Chitosan	• Tissue engineering. • Absorbable sutures.	• Biocompatibility.
Collagen	• Cartilage, skin, cornea, bone, and blood vessels.	• Biocompatibility.
Gelatin	• Wound dressing and tissue engineering.	• Biodegradability, availability, and low cost.
Fibrin	• Numerous tissue engineering scaffolds.	• Excellent biocompatibility.
Silk fiber	• Sutures	• Good flexibility and elasticity. • Excellent tensile strength.
Hyaluronic acid (HA)	• Scaffolds and wound dressing. • Temporary implants.	• Good structural and biological properties.
Aliphatic polyester, e.g., PLA and PGA	• Sutures. • Tissue engineering scaffolds. • Wound closure staples. • Orthopedic fixation devices.	• Ease of processing. • Good mechanical, chemical, and physical properties. • Biocompatible.

TABLE 6.1 *(Continued)*

Polymers	Devices	Advantages
Polycaprolactone	• Long-term implants. • Tissue engineering scaffolds.	• Non-toxic. • Extremely good biocompatibility.
Poly(β-hydroxybutyrate)	Long-term tissue engineering scaffolds.	• Good processability and biocompatibility.
Poly(phosphoesters)	• Tissue engineering scaffolds.	• Excellent biocompatibility. • Good cytocompatibility.
Polyether urethanes (PEU)	• Catheters, artificial heart, ligaments, pacemaker leads, etc.	• Biocompatible. • Biostable. • Good flexibility and durability.

"biphasic calcium phosphate" and "b-TCP ($Ca_3(PO4)_2$)" can exhibit the properties of bone minerals [47].

Metal nanoparticles such as "zirconia," "titanium," "silver," and "zinc" are widely used as tissue engineering scaffolds due to their antimicrobial, antifungal, and enhanced mechanical behaviors [48, 49]. Titanium oxide is a well-known organic ceramic material with a tri-crystalline architecture used as tissue scaffolds due to its non-toxic, photocatalytic, and bactericidal attributes [50].

6.6 SYNTHESIS AND PROPERTIES OF FEW HYBRIDS' POLYMERS

Osteoconductive hybrid polymers can be fabricated by using osteoconductive materials into biopolymers [51]. Bioactive glass (BG) and calcium phosphate (CP) are widely used as bone-bonding materials to enhance bone regeneration [52–56]. BG and CP-based materials have been used in osteoconductive hybrid polymers for tissue regeneration [57–61]. The hybrid polymer systems have been fabricated by melting, solvent-casting, and *in situ* precipitation [62]. The polymer hybrids with low CP content can significantly improve the mechanical strength of the polymer systems with better bioactivity [62]. The amorphous BG particles enable controlled biodegradation and high bone bonding [63]. The BG is fabricated by solvent-casting technique with various polymers such as PLA, PCL gelatin, and PLGA [64–66]. The polymer hybrid prepared with BG improves compressive strength, tensile strength, and osteoblast biocompatibility [61].

BG particles agglomerate in the polymer matrix exhibiting unfavorable properties. This problem can be resolved by using silica-based bioactive glass sol [61]. The use of carbon-based polymer hybrids has been reported as osteoconductive scaffolds. Carbon nanotubes with PLLA and graphene with PLLA enhanced *in vivo* bone regeneration [67].

Many tissues of the human body possess elastomeric behavior, and the development of polymeric scaffolds with good elastomeric behavior has gained huge attention. The polymer crosslinked elastomers are of particular interest as they exhibit biomimetic properties [68]. Polyurethanes and polyesters are widely crosslinked polymers and have shown high biodegradability and moderate biocompatibility [26]. Scaffolds of this kind have demonstrated good regeneration behavior despite having low mechanical strength and bioactivity [26]. PGS-PCL-based elastomers fabricated by solvent electrospinning can mimic the mechanical properties of the human aortic valve [69]. Conducting polymers are a class of electroactive hybrid polymers that possess good optical, electrical, magnetic, and mechanical properties [70, 71]. Nowadays, conducting polymers has drawn the attention of researchers and scientists as they could tune the properties of the cells under electrical stimulation [72–74]. 3D conductive implants were prepared by poly(3,4-ethylenedioxythiophene) poly(4-styrene sulfonate) (PEDOT:PSS), gelatin, and bioactive glass. The hybrid scaffold enhances the physical stability with the improved mechanical behavior of the composites [75].

6.7 HYBRID MATERIALS FOR INTERSTITIAL AND CONNECTIVE TISSUES

Polymers have been used in various tissue engineering applications as they provide a variety of advantages for scaffold development. The polymer hybrids used as tissue scaffolds should be biocompatible, biodegradable, and have superior mechanical properties. The polymer matrix is the most important part of the hybrid system. The matrix may have synthetic or natural polymers such as PGA, PLA, polycaprolactone (PCL), etc. [76–79] and chitosan, alginate, collagen, hyaluronic acid, gellan gum, and gelatin [80–83].

6.7.1 HYBRID MATERIALS AS CONNECTIVE TISSUES

Tendons and ligaments are musculoskeletal connective tissues. Tendons serve to move the bone or structure, while ligaments connect/hold bones together. The collagen fibrils of tendons give them enough strength and elasticity. To mimic the nature and behavior of tendons and ligaments hybrid materials with engineered matrix inhomogeneities are used. Synthetic and natural polymers such as PCL, PGA, polyurethane, collagen, and silk fibroin are used [84]. Various scaffold techniques have been used to fabricate functional matrices for connective tissues [85, 86]. The hybrid polymeric materials for tendons are obtained by electrospinning techniques [87, 88]. $CaCO_3$-coated PCL nanofibers particles exhibited improved adhesion and integration of the bone implants [89].

6.7.2 HYBRID MATERIALS FOR VASCULAR TISSUE ENGINEERING

Vascular tissue engineering is one of the interesting fields of tissue engineering, where diseased cardiovascular vessels are replaced by artificial ones. Polytetrafluoroethylene (PTFE) is one of the useful polymers used in vascular vessels. PTFE can be used as a medium-large blood vessel due to its crystallinity and hydrophobicity, However, for small-diameter vessels, PTFE can be dangerous because of calcification and thrombosis [90].

6.7.3 HYBRID MATERIALS FOR NERVE TISSUE ENGINEERING

The tissue engineering of nerves remains crucial in repairing the central nervous system (CNS) and peripheral nervous system (PNS), the reason being both nervous systems have limited self-regeneration capacity after an injury/damage [91]. Impairments to CNS can be caused by various reasons such as car accidents, falls, and trauma which leads to long-term disability [92]. In addition, impairments in PNS are also caused by penetrating, crush, and traction injuries [93, 94]. In neural tissue engineering the use of natural polymers such as collagen, gelatin, elastin, etc., has become more advantageous due to their compatibility and degradation kinetics, natural polymers can mimic the behavior of host tissues, reducing the risk of cytotoxicity and immunogenic reactions [95].

6.7.4 HYBRID MATERIALS FOR SKELETAL MUSCLES

Skeletal muscles recover fast from minor injuries, but they fail to recover if the damage to the tissues is serious. Hybrid polymer systems used can provide both physical and chemical support in the reconstruction of damaged tissue. Hybrid aerogels are one of the potential candidates for skeletal tissue repair, due to their porous nature with extremely low density, low thermal conductivity, and high specific surface area (SSA). It is very much important to engineering implants to mimic the natural extracellular matrix (ECM) as it can provide structural support and regulate cell functioning. Various types of polymer hybrid systems are used in skeletal tissue engineering such as electro-spun nanofibers, hydrogels, aerogels, pattern scaffolds, and decellularized tissues.

Hydrogels are a class of materials prepared from both natural and synthetic polymers with high water content. Hydrogels have been widely used in biomedical applications. Hybrid aerogels have also been in the medical field for a quite long time. The alginate-chitosan hybrid aerogel fibers prepared for wound healing applications exhibited better biocompatibility and non-cytotoxic nature [96].

6.7.5 *HYBRID MATERIALS FOR CARTILAGE*

Cartilage tissue engineering provides many surgical strategies to ameliorate cartilage dysfunction or devasting deficits. Cartilage scaffolds provide an appropriate environment to facilitate cell migration and development. Polymer hybrids with adequate mechanical properties and biocompatibility are used as cartilage scaffolds. Further, porosity, permeability, and regeneration properties of hybrid systems play an important role in cartilage development [97, 98]. Chondrotissue with a resorbable membrane composed of PGA and HA to relieve the pain and to support cartilage regeneration [99, 100]. PLGA and collagen-based hybrid scaffolds with high porosity and good mechanical strength have been used, the hybrid scaffolds help in the regeneration of cartilage tissues in rats. Hybrid scaffolds such as PLCL-COLI, C2C1H, and ECM-coated polylactic-co-glycolic acid (ECM-PLGA) were also used. The scaffolds had high porosity, good biocompatibility, and mechanical strength mimicking an appropriate environment for cartilage regeneration [101–103].

From the above sections, it is evident that polymers and hybrid materials are used in many applications of tissue engineering. The applications are often unique, and they are stand-alone areas. A few selected applications of the polymers and hybrids used in tissue engineering are listed in Table 6.2.

6.8 HYBRID MATERIALS FOR HARD TISSUE REPLACEMENTS

Bone deterioration causes significant pain, loss of motion, and sometimes angular deformity, especially at joints. In severe cases, total joint arthroplasty (TJR) is the only treatment [10, 118]. The success of joint replacement prostheses depends on many factors, most of which are related to the materials used for manufacturing them. Traditionally, implants with insufficient knowledge of biomaterials have been designed. The expansion of research activities into biomaterials has resulted in improved designs. There is still a need for longer-lasting *in vitro* performance. Involved in these surgeries because of higher life expectancy [136]. Based on the specific application, the material requirements must be determined. The clinical outcomes were not, therefore, highly promising. The biomechanics of joint replacements have resulted in better designs, but they still need to be improved.

TABLE 6.2 Polymers Hybrid Materials in Various Tissue Engineering Applications

SL. No.	Polymer as Base	Polymer Coating for Specific Tissue	Tissue	Advantages	References
1.	PCL	Collagen, gelatin	Tendons	Improve cell adhesion	[84, 85, 104]
2.	ePTFE	Fibrin glue	Blood vessels	Increasing capillarization, increased collagen content	[105]
3.	PEA, PMA	VEGF based		Improves vascularization	[106]
4.	Chitosan	Laminin peptides	Nerve	Improves adhesion of neural cells	[107]
5.	Poly(sialic acid)	Poly-L-lysine and poly-L-ornithine		Improves cell adhesivity	[108]
6.	PGA/PLA	bFGF and AA2P based	Lungs	Improves cell adhesivity and stimulates cells proliferation	[109]
7.	PEG	Laminin and RGD based	Skeletal muscle	Improves cells adhesion	[110, 111]
8.	Alginate	Agrin or laminin based		Exhibited acetylcholine receptor (AChR) clustering	[112, 113]
9.	PLGA	Hyaluronic acid based	Cartilage	Improves cells adhesion	[77]
10.	PGA	Fibrin		Improves tissue regeneration	[114, 115]
11.	PHB and PHBV	$CaCO_3$/PHB and PHBV based	Bone	Improvs cell adhesion and ossification	[116]
12.	Alginate	$CaCO_3$ with loaded ALP			[117]

6.8.1 HYBRIDS MATERIALS FOR TOTAL TISSUES REPLACEMENTS

Most metallic and ceramic materials, however, are not biodegradable or biocompatible [14]. This shortcoming has made it possible for polymeric materials with biodegradability and biocompatibility to receive greater attention in tissue engineering. There are many biodegradable and biocompatible polymers, including polymers that are hydrolytically and enzymatically degradable [4]. There are still some drawbacks, despite the progress achieved since the use of biodegradable polymers. For instance, acidic products released from the biodegradation process of all polyesters may affect biocompatibility unfavorably [4]. The materials used for bone repair and regeneration enthusiastically attract considerable interest because their biodegradable property avoids the need for their removal surgery and additional costs and pain for patients. The metal exemptions are magnesium (Mg), iron (Fe), and zinc (Zn) used in cardiovascular and hard tissue regeneration applications (in combination with other materials) [119, 120]. Tricalcium phosphate and HA are from ceramic materials that are biodegradable [121].

Polymers, proteins, and polysaccharides have been used in orthopedic applications. However, for load-bearing applications Poly(alpha-esters) (e.g., polyglycolide, polylactides, poly(lactide-co-glycolide), polydioxanone), polyurethanes, poly(ester amide), poly(ortho esters), polyanhydrides, poly(anhydride-co-imide), cross-linked polyanhydrides are used. These polymers are much more suitable for hard tissue replacements due to their high stiffness and biological mismatch with the soft tissues.

In all groups of polymer materials, deficiencies exist and make them not suitable for tissue replacement or tissue engineering applications, hence a new set of hybrid materials is developed. One of the best methods adopted to develop hybrid composites is by incorporating tricalcium phosphate ceramic [122]. Further, by using stiff ceramic materials in ductile polymers the limitations of polymers/single-material implants can be reduced [23, 123]. To make composites that control the biodegradability rate and with superior mechanical strength and structural integrity, degradable metals are used as reinforcing materials in polymers [124]. The hybrid composites developed using Mg metal particles and poly-L lactic acid (PLLA) have shown promising results as implants. The Mg metal particle used increases the degradation of PLLA scaffolds by increasing the pH of the degradation medium [124].

The hybrid composite materials fabricated for hard tissue engineering require more effort and approvals due to their biocompatibility and cytotoxicity.

Hybrid composites with carbon fibers in PEEK matrix have also been developed for hip prostheses shown better wear properties [125–128]. The hybrid composites with carbon and Kevlar also being fabricated with HDP, composites exhibit strong anti-wear behaviors. Carbon fiber-reinforced epoxy composites were used for acetabular cups against alumina femoral heads. The wear rate of the hybrid composites decreased about 30 times compared to UHMWPE articulating surface coupled with alumina and metal [129].

For metallic implants, composites are used as coatings to reduce wear and friction by enhancing cohesion between the implant and bone tissue [130, 131]. However, stress shielding is the major issue associated with metal and ceramic implants as their elastic modulus is higher compared to human bone. The composites with elastic modulus close to human bone are considered potential candidates for tissue replacement. Several hybrid systems have been fabricated with carbon fiber-reinforced polyamide 12 and carbon fiber-reinforced PEEK, the results obtained prove their load-bearing ability as scaffolds [132–134]. However, as these composites are more flexible, it should be noted that higher strains are produced inside the implants. The stiffness of the composites can be controlled by having uniform stress distribution at the interface by reducing the stress shielding effect. These types of "advanced composites" are designed to provide selectively a gradient in composition in the component. Some examples of advanced composite materials for the various components of a hip prosthesis are HA-collagen, stainless steel-bioglass, and cellular-graded Co-Cr alloy [118].

The development of implants for total knee replacement is also a challenging task because of the complex geometry, loading conditions, and kinematics. For knee joint replacement, countable composite hybrids have been developed. Carbon fibers reinforced with UHMWPE show improved tensile strength, creep, and fatigue behaviors. However, the composite has not shown promising results in wear [135]. Only a few of the different composite materials developed for joint replacements have been commercially used and distributed as approved medical implants. Further research and development are still required to ensure successful commercial application of these materials.

6.9 CONCLUSIONS

The chapter highlights the use of different scaffold materials for tissue engineering applications. The selection of functional implants is made based on the degree of impairments caused by injuries to the tissues. Several natural and synthetic polymers have been used as implants. However, the polymer's compatibility with tissues is an issue that must be overcome. Thus, different polymeric systems as tissues replacement with modifications are used. The hybrid composites fabricated with different reinforcing elements have been used. The hybrid materials must fulfill the mechanical and physical properties of the tissues providing better interaction and cell adhesion. Hybrid systems provide a tailor-based methodology with more flexibility to change their properties as desired. The hybrid systems with different constituents and reinforcing phases may improve the interaction with the tissues and cells. The use of bioactive fillers in the polymer matrix creates a structure that can mimic the actual tissue behaviors. It is found that the composites prepared with UHMWPE have shown attractive results as bone implants. Nonetheless, UHMWPE composites still have problems associated with wear. For this reason, hybrid materials have been widely researched for bone tissue replacement.

KEYWORDS

- **composites**
- **hybrid materials**
- **implants**
- **materials requirements**
- **nerve tissue engineering**
- **polymer**
- **scaffolds**
- **tissue**

REFERENCES

1. Ulery, B. D., Nair, L. S., & Laurencin, C. T., (2011). Biomedical applications of biodegradable polymers. In: *Journal of Polymer Science Part B: Polymer Physics* (Vol. 49, No. 12, pp. 832–864). https://doi.org/10.1002/polb.22259.
2. Gross, R. A., (2002). Biodegradable polymers for the environment. In: *Science* (Vol. 297, No. 5582, pp. 803–807). https://doi.org/10.1126/science.297.5582.803.
3. Chandra, R., (1998). Biodegradable polymers. In: *Progress in Polymer Science* (Vol. 23, No. 7, pp. 1273–1335). https://doi.org/10.1016/s0079-6700(97)00039-7.

4. Nair, L. S., & Laurencin, C. T., (2007). Biodegradable polymers as biomaterials. In: *Progress in Polymer Science* (Vol. 32, No. 8, 9, pp. 762–798). https://doi.org/10.1016/j.progpolymsci.2007.05.017.
5. Sheikh, Z., Najeeb, S., Khurshid, Z., Verma, V., Rashid, H., & Glogauer, M., (2015). Biodegradable materials for bone repair and tissue engineering applications. In: *Materials* (Vol. 8, No. 9, pp. 5744–5794). https://doi.org/10.3390/ma8095273.
6. Lee, S. Y., & Jiang, C. P., (2015). Development of a three-dimensional slurry printing system using dynamic mask projection for fabricating zirconia dental implants. In: *Materials and Manufacturing Processes* (Vol. 30, No. 12, pp. 1498–1504). https://doi.org/10.1080/10426914.2014.984208.
7. Kuroda, K., Nakamoto, S., Ichino, R., Okido, M., & Pilliar, R. M., (2005). Hydroxyapatite coatings on a 3D porous surface using thermal substrate method. In: *Materials Transactions* (Vol. 46, No. 7, pp. 1633–1635). https://doi.org/10.2320/matertrans.46.1633.
8. Dorozhkin, S. V., (2015). Calcium orthophosphate bioceramics. In: *Ceramics International* (Vol. 41, No. 10, pp. 13913–13966). https://doi.org/10.1016/j.ceramint.2015.08.004.
9. Vaquette, C., Ivanovski, S., Hamlet, S. M., & Hutmacher, D. W., (2013). Effect of culture conditions and calcium phosphate coating on ectopic bone formation. *Biomaterials, 34*(22), 5538–5551.
10. Sadoghi, P., Liebensteiner, M., Agreiter, M., Leithner, A., Böhler, N., & Labek, G., (2013). Revision surgery after total joint arthroplasty: A complication-based analysis using worldwide arthroplasty registers. *The Journal of Arthroplasty, 28*(8), 1329–1332.
11. Bahraminasab, M., Sahari, B. B., Edwards, K. L., Farahmand, F., Arumugam, M., & Hong, T. S., (2012). Aseptic loosening of femoral components – A review of current and future trends in materials used. In: *Materials & Design* (Vol. 42, pp. 459–470). https://doi.org/10.1016/j.matdes.2012.05.046.
12. Luther, C., Germann, G., & Sauerbier, M., (2010). Proximal interphalangeal joint replacement with surface replacement arthroplasty (SR–PIP): Functional results and complications. In: *Hand* (Vol. 5, No. 3, pp. 233–240). https://doi.org/10.1007/s11552-009-9246-z.
13. Ratner, B. D., (1989). Biomedical applications of synthetic polymers. In: *Comprehensive Polymer Science and Supplements* (pp. 201–247). https://doi.org/10.1016/b978-0-08-096701-1.00210-x.
14. Bahraminasab, M., Sahari, B. B., Edwards, K. L., Farahmand, F., & Arumugam, M., (2013). Aseptic loosening of femoral components – materials engineering and design considerations. In: *Materials & Design* (Vol. 44, pp. 155–163). https://doi.org/10.1016/j.matdes.2012.07.066.
15. Farag, M. M., (2007). *Materials and Process Selection for Engineering Design*. CRC Press.
16. Christen, P., Ito, K., Ellouz, R., Boutroy, S., Sornay-Rendu, E., Chapurlat, R. D., & Van, R. B., (2014). Bone remodeling in humans is load-driven but not lazy. In: *Nature Communications* (Vol. 5, No. 1). https://doi.org/10.1038/ncomms5855.
17. Geetha, M., Singh, A. K., Asokamani, R., & Gogia, A. K., (2009). Ti based biomaterials, the ultimate choice for orthopaedic implants: A review. In: *Progress in Materials Science* (Vol. 54, No. 3, pp. 397–425). https://doi.org/10.1016/j.pmatsci.2008.06.004.
18. Denard, P. J., Raiss, P., Gobezie, R., Edwards, T. B., & Lederman, E., (2018). Stress shielding of the humerus in press-fit anatomic shoulder arthroplasty: Review and recommendations for evaluation. *Journal of Shoulder and Elbow Surgery / American Shoulder and Elbow Surgeons, 27*(6), 1139–1147..

19. Sozen, T., Ozisik, L., & Basaran, N. C., (2017). An overview and management of osteoporosis. In: *European Journal of Rheumatology* (Vol. 4, No. 1, pp. 46–56). https://doi.org/10.5152/eurjrheum.2016.048.
20. Broomfield, J. A. J., Malak, T. T., Thomas, G. E. R., Palmer, A. J. R., Taylor, A., & Glyn-Jones, S., (2017). The relationship between polyethylene wear and periprosthetic osteolysis in total hip arthroplasty at 12 years in a randomized controlled trial cohort. In: *The Journal of Arthroplasty* (Vol. 32, No. 4, pp. 1186–1191). https://doi.org/10.1016/j.arth.2016.10.037.
21. Ingham, E., & Fisher, J., (2005). The role of macrophages in osteolysis of total joint replacement. In: *Biomaterials* (Vol. 26, No. 11, pp. 1271–1286). https://doi.org/10.1016/j.biomaterials.2004.04.035.
22. Özcan, M., & Hämmerle, C., (2012). Titanium as a reconstruction and implant material in dentistry: Advantages and pitfalls. In: *Materials* (Vol. 5, No. 9, pp. 1528–1545). https://doi.org/10.3390/ma5091528.
23. Bahraminasab, M., & Edwards, K. L., (2018). Biocomposites for hard tissue replacement and repair. In: *Futuristic Composites* (pp. 281–296). https://doi.org/10.1007/978-981-13-2417-8_14.
24. Okazaki, Y., & Gotoh, E., (2005). Comparison of metal release from various metallic biomaterials *in vitro*. *Biomaterials, 26*(1), 11–21.
25. Williams, D. F., (2009). On the nature of biomaterials. In: *Biomaterials* (Vol. 30, No. 30, pp. 5897–5909). https://doi.org/10.1016/j.biomaterials.2009.07.027.
26. Chen, Q., Liang, S., & Thouas, G. A., (2013). Elastomeric biomaterials for tissue engineering. In: *Progress in Polymer Science* (Vol. 38, No. 3, 4, pp. 584–671). https://doi.org/10.1016/j.progpolymsci.2012.05.003.
27. Vining, K. H., & Mooney, D. J., (2017). Mechanical forces direct stem cell behavior in development and regeneration. *Nature Reviews Molecular Cell Biology, 18*(12), 728–742.
28. Mitrousis, N., Fokina, A., & Shoichet, M. S., (2018). Biomaterials for cell transplantation. In: *Nature Reviews Materials* (Vol. 3, No. 11, pp. 441–456). https://doi.org/10.1038/s41578-018-0057-0.
29. Andersen, T., Auk-Emblem, P., & Dornish, M., (2015). 3D cell culture in alginate hydrogels. In: Microarrays (Vol. 4, No. 2, pp. 133–161). https://doi.org/10.3390/microarrays4020133.
30. Gloria, A., Russo, T., De Santis, R., & Ambrosio, L., (2009). 3D fiber deposition technique to make multifunctional and tailor-made scaffolds for tissue engineering applications. *Journal of Applied Biomaterials & Biomechanics: JABB, 7*(3), 141–152.
31. Burg, K. J. L., Porter, S., & Kellam, J. F., (2000). Biomaterial developments for bone tissue engineering. In: *Biomaterials* (Vol. 21, No. 23, pp. 2347–2359). https://doi.org/10.1016/s0142-9612(00)00102-2.
32. LeGeros, R. Z., (2002). Properties of osteoconductive biomaterials: Calcium phosphates. In: *Clinical Orthopaedics and Related Research* (Vol. 395, pp. 81–98). https://doi.org/10.1097/00003086-200202000-00009.
33. Trappmann, B., Gautrot, J. E., Connelly, J. T., Strange, D. G. T., Li, Y., Oyen, M. L., Cohen, S. M. A., et al., (2012). Extracellular-matrix tethering regulates stem-cell fate. *Nature Materials, 11*(7), 642–649..
34. Lin, C. C., Fu, S. J., Lin, Y. C., Yang, I. K., & Gu, Y., (2014). Chitosan-coated electrospun PLA fibers for rapid mineralization of calcium phosphate. I*nternational Journal of Biological Macromolecules, 68*, 39–47.

35. Chander, V., Singh, A. K., & Gangenahalli, G., (2018). Cell encapsulation potential of chitosan-alginate electrostatic complex in preventing natural killer and CD8 cell-mediated cytotoxicity: An *in vitro* experimental study. In: *Journal of Microencapsulation* (Vol. 35, No. 6, pp. 522–532). https://doi.org/10.1080/02652048.2018.1516827.
36. Cheung, H. Y., Lau, K. T., Lu, T. P., & Hui, D., (2007). A critical review on polymer-based bio-engineered materials for scaffold development. In: *Composites Part B: Engineering* (Vol. 38, No. 3, pp. 291–300). https://doi.org/10.1016/j.compositesb.2006.06.014.
37. Causa, F., Netti, P. A., & Ambrosio, L., (2007). A multi-functional scaffold for tissue regeneration: The need to engineer a tissue analogue. In: *Biomaterials* (Vol. 28, No. 34, pp. 5093–5099). https://doi.org/10.1016/j.biomaterials.2007.07.030.
38. Hutmacher, D. W., (2001). Scaffold design and fabrication technologies for engineering tissues-state of the art and future perspectives. In: *Journal of Biomaterials Science, Polymer Edition* (Vol. 12, No. 1, pp. 107–124). https://doi.org/10.1163/156856201744489.
39. Kyriakidou, K., Lucarini, G., Zizzi, A., Salvolini, E., Mattioli, B. M., Mollica, F., Gloria, A., & Ambrosio, L., (2008). Dynamic co-seeding of osteoblast and endothelial cells on 3D polycaprolactone scaffolds for enhanced bone tissue engineering. In: *Journal of Bioactive and Compatible Polymers* (Vol. 23, No. 3, pp. 227–243). https://doi.org/10.1177/0883911508091905.
40. Freed, L. E., Vunjak-Novakovic, G., Biron, R. J., Eagles, D. B., Lesnoy, D. C., Barlow, S. K., & Langer, R., (1994). Biodegradable polymer scaffolds for tissue engineering. *Bio/technology, 12*(7), 689–693.
41. Devin, J. E., Attawia, M. A., & Laurencin, C. T., (1996). Three-dimensional degradable porous polymer-ceramic matrices for use in bone repair. *Journal of Biomaterials Science Polymer Edition, 7*(8), 661–669.
42. Mathieu, L., Bourban, P., & Manson, J., (2006). Processing of homogeneous ceramic/polymer blends for bioresorbable composites. In: *Composites Science and Technology* (Vol. 66, No. 11, 12, pp. 1606–1614). https://doi.org/10.1016/j.compscitech.2005.11.012.
43. Marks, L. J., & Michael, J. W., (2001). Science, medicine, and the future: Artificial limbs. *BMJ, 323*(7315), 732–735.
44. Mathieu, L. M., Montjovent, M. O., Bourban, P. E., Pioletti, D. P., & Månson, J. A. E., (2005). Bioresorbable composites prepared by supercritical fluid foaming. *Journal of Biomedical Materials Research Part A, 75*(1), 89–97.
45. Mathieu, L. M., Mueller, T. L., Bourban, P. E., Pioletti, D. P., Müller, R., & Månson, J. A. E., (2006). Architecture and properties of anisotropic polymer composite scaffolds for bone tissue engineering. *Biomaterials, 27*(6), 905–916.
46. Hing, K. A., Best, S. M., Tanner, K. E., Bonfield, W., & Revell, P. A., (2004). Mediation of bone ingrowth in porous hydroxyapatite bone graft substitutes. *Journal of Biomedical Materials Research. Part A, 68*(1), 187–200.
47. Cesur, S., Küçükgöksel, Y., Taşdemir, Ş., & Ürkmez, A. Ş., (2018). Polycaprolactone-hydroxy apatite composites for tissue engineering applications. In: *Journal of Vinyl and Additive Technology* (Vol. 24, No. 3, pp. 248–261). https://doi.org/10.1002/vnl.21569.
48. Liu, Y. T., Lee, T. M., & Lui, T. S., (2013). Enhanced osteoblastic cell response on zirconia by bio-inspired surface modification. *Colloids and Surfaces. B, Biointerfaces, 106*, 37–45.
49. Okamoto, M., & John, B., (2013). Synthetic biopolymer nanocomposites for tissue engineering scaffolds. In: *Progress in Polymer Science* (Vol. 38, No. 10, 11, pp. 1487–1503). https://doi.org/10.1016/j.progpolymsci.2013.06.001.

50. Matsuno, H., Yokoyama, A., Watari, F., Uo, M., & Kawasaki, T., (2001). Biocompatibility and osteogenesis of refractory metal implants, titanium, hafnium, niobium, tantalum and rhenium. *Biomaterials, 22*(11), 1253–1262.
51. Liu, X., Holzwarth, J. M., & Ma, P. X., (2012). Functionalized synthetic biodegradable polymer scaffolds for tissue engineering. *Macromolecular Bioscience, 12*(7), 911–919.
52. Jones, J. R., (2013). Review of bioactive glass: From Hench to hybrids. *Acta Biomaterialia, 9*(1), 4457–4486.
53. Lei, B., Chen, X., Wang, Y., & Zhao, N., (2009). Synthesis and *in vitro* bioactivity of novel mesoporous hollow bioactive glass microspheres. In: *Materials Letters* (Vol. 63, No. 20, pp. 1719–1721). https://doi.org/10.1016/j.matlet.2009.04.041.
54. Lei, B., Chen, X., Wang, Y., Zhao, N., Du, C., & Fang, L., (2010). Surface nanoscale patterning of bioactive glass to support cellular growth and differentiation. *Journal of Biomedical Materials Research Part A, 94*(4), 1091–1099.
55. Chen, X., Lei, B., Wang, Y., & Zhao, N., (2009). Morphological control and *in vitro* bioactivity of nanoscale bioactive glasses. In: *Journal of Non-Crystalline Solids* (Vol. 355, No. 13, pp. 791–796). https://doi.org/10.1016/j.jnoncrysol.2009.02.005.
56. Zakaria, S. M., Sharif, Z. S. H., Othman, M. R., Yang, F., & Jansen, J. A., (2013). Nanophase hydroxyapatite as a biomaterial in advanced hard tissue engineering: A review. *Tissue Engineering Part B, Reviews, 19*(5), 431–441.
57. Boccaccini, A. R., Erol, M., Stark, W. J., Mohn, D., Hong, Z., & Mano, J. F., (2010). Polymer/bioactive glass nanocomposites for biomedical applications: A review. In: *Composites Science and Technology* (Vol. 70, No. 13, pp. 1764–1776). https://doi.org/10.1016/j.compscitech.2010.06.002.
58. Lei, B., Chen, X., Han, X., & Zhou, J., (2012). Versatile fabrication of nanoscale sol–gel bioactive glass particles for efficient bone tissue regeneration. In: *Journal of Materials Chemistry* (Vol. 22, No. 33, p. 16906). https://doi.org/10.1039/c2jm31384g.
59. Lei, B., Shin, K. H., Noh, D. Y., Jo, I. H., Koh, Y. H., Kim, H. E., & Kim, S. E., (2013). Sol–gel derived nanoscale bioactive glass (NBG) particles reinforced poly(ε-caprolactone) composites for bone tissue engineering. In: *Materials Science and Engineering: C* (Vol. 33, No. 3, pp. 1102–1108). https://doi.org/10.1016/j.msec.2012.11.039.
60. Lei, B., Shin, K. H., Noh, D. Y., Koh, Y. H., Choi, W. Y., & Kim, H. E., (2012). Bioactive glass microspheres as reinforcement for improving the mechanical properties and biological performance of poly(ε-caprolactone) polymer for bone tissue regeneration. In: *Journal of Biomedical Materials Research Part B: Applied Biomaterials* (Vol. 100B, No. 4, pp. 967–975). https://doi.org/10.1002/jbm.b.32659.
61. Lei, B., Wang, L., Chen, X., & Chae, S. K., (2013). Biomimetic and molecular level-based silicate bioactive glass-gelatin hybrid implants for loading-bearing bone fixation and repair. *Journal of Materials Chemistry B: Materials for Biology and Medicine, 1*(38), 5153–5162.
62. Roohani-Esfahani, S. I., Nouri-Khorasani, S., Lu, Z., Appleyard, R., & Zreiqat, H., (2010). The influence hydroxyapatite nanoparticle shape and size on the properties of biphasic calcium phosphate scaffolds coated with hydroxyapatite-PCL composites. *Biomaterials*, *31*(21), 5498–5509.
63. Rahaman, M. N., Day, D. E., Bal, B. S., Fu, Q., Jung, S. B., Bonewald, L. F., & Tomsia, A. P., (2011). Bioactive glass in tissue engineering. *Acta Biomaterialia, 7*(6), 2355–2373.
64. Peter, M., Binulal, N. S., Nair, S. V., Selvamurugan, N., Tamura, H., & Jayakumar, R., (2010). Novel biodegradable chitosan–gelatin/nano-bioactive glass ceramic composite

scaffolds for alveolar bone tissue engineering. In: *Chemical Engineering Journal* (Vol. 158, No. 2, pp. 353–361). https://doi.org/10.1016/j.cej.2010.02.003.

65. Mozafari, M., Moztarzadeh, F., Rabiee, M., Azami, M., Maleknia, S., Tahriri, M., Moztarzadeh, Z., & Nezafati, N., (2010). Development of macroporous nanocomposite scaffolds of gelatin/bioactive glass prepared through layer solvent casting combined with lamination technique for bone tissue engineering. In: *Ceramics International* (Vol. 36, No. 8, pp. 2431–2439). https://doi.org/10.1016/j.ceramint.2010.07.010.
66. Hong, Z., Reis, R. L., & Mano, J. F., (2008). Preparation and *in vitro* characterization of scaffolds of poly(L-lactic acid) containing bioactive glass ceramic nanoparticles. *Acta Biomaterialia, 4*(5), 1297–1306.
67. Duan, S., Yang, X., Mei, F., Tang, Y., Li, X., Shi, Y., Mao, J., et al., (2015). Enhanced osteogenic differentiation of mesenchymal stem cells on poly(L-lactide) nanofibrous scaffolds containing carbon nanomaterials. *Journal of Biomedical Materials Research: Part A, 103*(4), 1424–1435.
68. Nettles, D. L., Chilkoti, A., & Setton, L. A., (2010). Applications of elastin-like polypeptides in tissue engineering. In: *Advanced Drug Delivery Reviews* (Vol. 62, No. 15, pp. 1479–1485). https://doi.org/10.1016/j.addr.2010.04.002.
69. Sant, S., Hwang, C. M., Lee, S. H., & Khademhosseini, A., (2011). Hybrid PGS-PCL microfibrous scaffolds with improved mechanical and biological properties. *Journal of Tissue Engineering and Regenerative Medicine, 5*(4), 283–291.
70. Guimard, N. K., Gomez, N., & Schmidt, C. E., (2007). Conducting polymers in biomedical engineering. In: *Progress in Polymer Science* (Vol. 32, No. 8, 9, pp. 876–921). https://doi.org/10.1016/j.progpolymsci.2007.05.012.
71. Patil, A. O., Heeger, A. J., & Wudl, F., (1988). Optical properties of conducting polymers. In: *Chemical Reviews* (Vol. 88, No. 1, pp. 183–200). https://doi.org/10.1021/cr00083a009.
72. Guo, B., Glavas, L., & Albertsson, A. C., (2013). Biodegradable and electrically conducting polymers for biomedical applications. In: *Progress in Polymer Science* (Vol. 38, No. 9, pp. 1263–1286). https://doi.org/10.1016/j.progpolymsci.2013.06.003.
73. Xie, M., Wang, L., Ge, J., Guo, B., & Ma, P. X., (2015). Strong electroactive biodegradable shape memory polymer networks based on star-shaped polylactide and aniline trimer for bone tissue engineering. In: *ACS Applied Materials & Interfaces* (Vol. 7, No. 12, pp. 6772–6781). https://doi.org/10.1021/acsami.5b00191.
74. Xie, M., Wang, L., Guo, B., Wang, Z., Chen, Y. E., & Ma, P. X., (2015). Ductile electroactive biodegradable hyperbranched polylactide copolymers enhancing myoblast differentiation. *Biomaterials, 71*, 158–167.
75. Yazdimamaghani, M., Razavi, M., Mozafari, M., Vashaee, D., Kotturi, H., & Tayebi, L., (2015). Biomineralization and biocompatibility studies of bone conductive scaffolds containing poly(3,4-ethylenedioxythiophene): Poly(4-styrene sulfonate) (PEDOT:PSS). *Journal of Materials Science: Materials in Medicine, 26*(12), 274.
76. Wei, J., Cai, J., Li, Y., Wu, B., Gong, X., & Ngai, T., (2015). Investigation of cell behaviors on thermo-responsive PNIPAM microgel films. *Colloids and Surfaces: B, Biointerfaces, 132*, 202–207.
77. Siclari, A., Mascaro, G., Kaps, C., & Boux, E., (2014). A 5-year follow-up after cartilage repair in the knee using a platelet-rich plasma-immersed polymer-based implant. *The Open Orthopaedics Journal, 8*, 346–354.
78. Lopes, M. S., Savioli, L. M., Jardini, A. L., & Maciel, F. R., (2012). Poly (lactic acid) production for tissue engineering applications. In: *Procedia Engineering* (Vol. 42, pp. 1402–1413). https://doi.org/10.1016/j.proeng.2012.07.534.

79. Gentile, P., McColgan-Bannon, K., Gianone, N. C., Sefat, F., Dalgarno, K., & Ferreira, A. M., (2017). Biosynthetic PCL-graft-collagen bulk material for tissue engineering applications. *Materials, 10*(7). https://doi.org/10.3390/ma10070693.
80. Sehgal, R. R., Roohani-Esfahani, S. I., Zreiqat, H., & Banerjee, R., (2017). Nanostructured gellan and xanthan hydrogel depot integrated within a baghdadite scaffold augments bone regeneration. *Journal of Tissue Engineering and Regenerative Medicine, 11*(4), 1195–1211.
81. Duan, X., McLaughlin, C., Griffith, M., & Sheardown, H., (2007). Biofunctionalization of collagen for improved biological response: Scaffolds for corneal tissue engineering. *Biomaterials, 28*(1), 78–88.
82. Douglas, T. E. L., Łapa, A., Samal, S. K., Declercq, H. A., Schaubroeck, D., Mendes, A. C., Van, D. V. P., et al., (2017). Enzymatic, urease-mediated mineralization of gellan gum hydrogel with calcium carbonate, magnesium-enriched calcium carbonate and magnesium carbonate for bone regeneration applications. *Journal of Tissue Engineering and Regenerative Medicine, 11*(12), 3556–3566.
83. He, Y., Liu, W., Guan, L., Chen, J., Duan, L., Jia, Z., Huang, J., et al., (2018). A 3D-printed PLCL scaffold coated with collagen type I and its biocompatibility. *BioMed Research International, 2018*, 5147156.
84. Fuller, K. P., Gaspar, D., Delgado, L. M., Shoseyov, O., & Zeugolis, D. I., (2019). *In vitro* and preclinical characterization of compressed, macro-porous and collagen coated poly-ε-caprolactone electro-spun scaffolds. *Biomedical Materials, 14*(5), 055007.
85. Bölgen, N., (2017). Electrospun materials for bone and tendon/ligament tissue engineering. In: *Electrospun Materials for Tissue Engineering and Biomedical Applications* (pp. 233–260). https://doi.org/10.1016/b978-0-08-101022-8.00004-1.
86. Webb, W. R., Dale, T. P., Lomas, A. J., Zeng, G., Wimpenny, I., El Haj, A. J., Forsyth, N. R., & Chen, G. Q., (2013). The application of poly(3-hydroxybutyrate-co-3-hydroxyhexanoate) scaffolds for tendon repair in the rat model. *Biomaterials, 34*(28), 6683–6694.
87. Li, X., Xie, J., Lipner, J., Yuan, X., Thomopoulos, S., & Xia, Y., (2009). Nanofiber scaffolds with gradations in mineral content for mimicking the tendon-to-bone insertion site. *Nano Letters, 9*(7), 2763–2768.
88. Prabhakaran, M. P., Venugopal, J., & Ramakrishna, S., (2009). Electrospun nanostructured scaffolds for bone tissue engineering. *Acta Biomaterialia, 5*(8), 2884–2893.
89. Saveleva, M. S., Ivanov, A. N., Kurtukova, M. O., Atkin, V. S., Ivanova, A. G., Lyubun, G. P., Martyukova, A. V., et al., (2018). Hybrid PCL/CaCO3 scaffolds with capabilities of carrying biologically active molecules: Synthesis, loading and *in vivo* applications. In: *Materials Science and Engineering: C* (Vol. 85, pp. 57–67). https://doi.org/10.1016/j.msec.2017.12.019.
90. Famaey, N., Verhoeven, J., Jacobs, S., Pettinari, M., & Meyns, B., (2014). *In situ* evolution of the mechanical properties of stretchable and non-stretchable ePTFE vascular grafts and adjacent native vessels. *The International Journal of Artificial Organs, 37*(12), 900–910.
91. Lombardi, V. R. M., (2012). New challenges in CNS repair: The immune and nervous connection. In: *Current Immunology Reviews* (Vol. 8, No. 1, pp. 87–93). https://doi.org/10.2174/157339512798991272.
92. Li, M., Zhao, Z., Yu, G., & Zhang, J., (2016). Epidemiology of traumatic brain injury over the world: A systematic review. In: *General Medicine: Open Access* (Vol. 4, No. 5). https://doi.org/10.4172/2327-5146.1000275.

93. Reichert, P., Wnukiewicz, W., Witkowski, J., Bocheńska, A., Mizia, S., Gosk, J., & Zimmer, K., (2016). Causes of secondary radial nerve palsy and results of treatment. *Medical Science Monitor: International Medical Journal of Experimental and Clinical Research, 22*, 554–562.
94. Adiguzel, E., Yaşar, E., Tecer, D., Güzelküçük, Ü., Taşkaynatan, M. A., Kesikburun, S., & Özgül, A., (2016). Peripheral nerve injuries: Long term follow-up results of rehabilitation. *Journal of Back and Musculoskeletal Rehabilitation, 29*(2), 367–371.
95. Ai, J., Kiasat-Dolatabadi, A., Ebrahimi-Barough, S., Ai, A., Lotfibakhshaiesh, N., Norouzi-Javidan, A., Saberi, H., et al., (2013). Polymeric scaffolds in neural tissue engineering: A review. In: *Archives of Neuroscience* (Vol. 1, No. 1, pp. 15–20). https://doi.org/10.5812/archneurosci.9144.
96. Batista, M. P., Gonçalves, V. S. S., Gaspar, F. B., Nogueira, I. D., Matias, A. A., & Gurikov, P., (2020). Novel alginate-chitosan aerogel fibres for potential wound healing applications. In: *International Journal of Biological Macromolecules* (Vol. 156, pp. 773–782). https://doi.org/10.1016/j.ijbiomac.2020.04.089.
97. Hollister, S. J., (2005). Porous scaffold design for tissue engineering. *Nature Materials, 4*(7), 518–524.
98. Kalkan, R., Nwekwo, C. W., & Adali, T., (2018). The use of scaffolds in cartilage regeneration. *Critical Reviews in Eukaryotic Gene Expression, 28*(4), 343–348.
99. Tsai, M. C., Hung, K. C., Hung, S. C., & Hsu, S. H., (2015). Evaluation of biodegradable elastic scaffolds made of anionic polyurethane for cartilage tissue engineering. Colloids and *Surfaces: B, Biointerfaces, 125*, 34–44.
100. He, Y., Jin, Y., Wang, X., Yao, S., Li, Y., Wu, Q., Ma, G., et al., (2018). An antimicrobial peptide-loaded gelatin/chitosan nanofibrous membrane fabricated by sequential layer-by-layer electrospinning and electrospraying techniques. *Nanomaterials (Basel, Switzerland), 8*(5). https://doi.org/10.3390/nano8050327.
101. Haaparanta, A. M., Järvinen, E., Cengiz, I. F., Ellä, V., Kokkonen, H. T., Kiviranta, I., & Kellomäki, M., (2014). Preparation and characterization of collagen/PLA, chitosan/PLA, and collagen/chitosan/PLA hybrid scaffolds for cartilage tissue engineering. *Journal of Materials Science: Materials in Medicine, 25*(4), 1129–1136..
102. Nogami, M., Kimura, T., Seki, S., Matsui, Y., Yoshida, T., Koike-Soko, C., Okabe, M., et al., (2016). A human amnion-derived extracellular matrix-coated cell-free scaffold for cartilage repair: *In vitro* and *in vivo* studies. *Tissue Engineering: Part A, 22*(7, 8), 680–688.
103. Yamanaka, K., Yamamoto, K., Sakai, Y., Suda, Y., Shigemitsu, Y., Kaneko, T., Kato, K., et al., (2015). Seeding of mesenchymal stem cells into inner part of interconnected porous biodegradable scaffold by a new method with a filter paper. *Dental Materials Journal, 34*(1), 78–85.
104. Sheng, D., Li, J., Ai, C., Feng, S., Ying, T., Liu, X., Cai, J., et al., (2019). Electrospun PCL/Gel-aligned scaffolds enhance the biomechanical strength in tendon repair. *Journal of Materials Chemistry: B, Materials for Biology and Medicine, 7*(31), 4801–4810.
105. Gray, J. L., Kang, S. S., Zenni, G. C., Kim, D. U., Kim, P. I., Burgess, W. H., Drohan, W., et al., (1994). FGF-1 affixation stimulates ePTFE endothelialization without intimal hyperplasia. *The Journal of Surgical Research, 57*(5), 596–612.
106. Moulisová, V., Gonzalez-García, C., Cantini, M., Rodrigo-Navarro, A., Weaver, J., Costell, M., Sabater, I. S. R., et al., (2017). Engineered microenvironments for synergistic VEGF - Integrin signaling during vascularization. *Biomaterials, 126*, 61–74.

107. Suzuki, M., Itoh, S., Yamaguchi, I., Takakuda, K., Kobayashi, H., Shinomiya, K., & Tanaka, J., (2003). Tendon chitosan tubes covalently coupled with synthesized laminin peptides facilitate nerve regeneration *in vivo*. *Journal of Neuroscience Research, 72*(5), 646–659.
108. Haile, Y., Berski, S., Dräger, G., Nobre, A., Stummeyer, K., Gerardy-Schahn, R., & Grothe, C., (2008). The effect of modified polysialic acid based hydrogels on the adhesion and viability of primary neurons and glial cells. In: *Biomaterials* (Vol. 29, No. 12, pp. 1880–1891). https://doi.org/10.1016/j.biomaterials.2007.12.030.
109. Ramaswamy, S., Gottlieb, D., Engelmayr, G. C. Jr., Aikawa, E., Schmidt, D. E., Gaitan-Leon, D. M., Sales, V. L., et al., (2010). The role of organ level conditioning on the promotion of engineered heart valve tissue development *in-vitro* using mesenchymal stem cells. *Biomaterials, 31*(6), 1114–1125.
110. Smythe, G. M., Hodgetts, S. I., & Grounds, M. D., (2001). Problems and solutions in myoblast transfer therapy. *Journal of Cellular and Molecular Medicine, 5*(1), 33–47.
111. Hall, J. K., Banks, G. B., Chamberlain, J. S., & Olwin, B. B., (2010). Prevention of muscle aging by myofiber-associated satellite cell transplantation. *Science Translational Medicine, 2*(57), 57ra83.
112. Wang, L., Shansky, J., & Vandenburgh, H., (2013). Induced formation and maturation of acetylcholine receptor clusters in a defined 3D bio-artificial muscle. In: *Molecular Neurobiology* (Vol. 48, No. 3, pp. 397–403). https://doi.org/10.1007/s12035-013-8412-z.
113. Ko, I. K., Lee, B. K., Lee, S. J., Andersson, K. E., Atala, A., & Yoo, J. J., (2013). The effect of *in vitro* formation of acetylcholine receptor (AChR) clusters in engineered muscle fibers on subsequent innervation of constructs *in vivo*. In: *Biomaterials* (Vol. 34, No. 13, pp. 3246–3255). https://doi.org/10.1016/j.biomaterials.2013.01.029.
114. Moutos, F. T., Freed, L. E., & Guilak, F., (2007). A biomimetic three-dimensional woven composite scaffold for functional tissue engineering of cartilage. *Nature Materials, 6*(2), 162–167.
115. Duarte, C. D. F., Drescher, W., Rath, B., Tingart, M., & Fischer, H., (2012). Supporting biomaterials for articular cartilage repair. *Cartilage, 3*(3), 205–221.
116. Chernozem, R. V., Surmeneva, M. A., Shkarina, S. N., Loza, K., Epple, M., Ulbricht, M., Cecilia, A., et al., (2019). Piezoelectric 3-D fibrous poly(3-hydroxybutyrate)-based scaffolds ultrasound-mineralized with calcium carbonate for bone tissue engineering: Inorganic phase formation, osteoblast cell adhesion, and proliferation. *ACS Applied Materials & Interfaces, 11*(21), 19522–19533.
117. Muderrisoglu, C., Saveleva, M., Abalymov, A., Van Der, M. L., Ivanova, A., Atkin, V., Parakhonskiy, B., & Skirtach, A. G., (2018). Nanostructured biointerfaces based on bioceramic calcium carbonate/hydrogel coatings on titanium with an active enzyme for stimulating osteoblasts growth. In: *Advanced Materials Interfaces* (Vol. 5, No. 19, p. 1800452). https://doi.org/10.1002/admi.201800452.
118. Bahraminasab, M., & Farahmand, F., (2017). State of the art review on design and manufacture of hybrid biomedical materials: Hip and knee prostheses. *Proceedings of the Institution of Mechanical Engineers: Part H, Journal of Engineering in Medicine, 231*(9), 785–813.
119. Liu, Y. J., Su, W. T., & Chen, P. H., (2018). Magnesium and zinc borate enhance osteoblastic differentiation of stem cells from human exfoliated deciduous teeth *in vitro*. *Journal of Biomaterials Applications, 32*(6), 765–774.

120. Qi, Y., Qi, H., He, Y., Lin, W., Li, P., Qin, L., Hu, Y., et al., (2018). Strategy of metal-polymer composite stent to accelerate biodegradation of iron-based biomaterials. *ACS Applied Materials & Interfaces, 10*(1), 182–192.
121. Sheikh, Z., Sima, C., & Glogauer, M., (2015). Bone replacement materials and techniques used for achieving vertical alveolar bone augmentation. In: *Materials* (Vol. 8, No. 6, pp. 2953–2993). https://doi.org/10.3390/ma8062953.
122. Chu, T. M. G., Warden, S. J., Turner, C. H., & Stewart, R. L., (2007). Segmental bone regeneration using a load-bearing biodegradable carrier of bone morphogenetic protein-2. *Biomaterials, 28*(3), 459–467.
123. Kim, M. H., Yun, C., Chalisserry, E. P., Lee, Y. W., Kang, H. W., Park, S. H., Jung, W. K., et al., (2018). Quantitative analysis of the role of nanohydroxyapatite (nHA) on 3D-printed PCL/nHA composite scaffolds. In: *Materials Letters* (Vol. 220, pp. 112–115). https://doi.org/10.1016/j.matlet.2018.03.025.
124. Shuai, C., Li, Y., Feng, P., Guo, W., Yang, W., & Peng, S., (2018). Positive feedback effects of Mg on the hydrolysis of poly-l-lactic acid (PLLA): Promoted degradation of PLLA scaffolds. In: *Polymer Testing* (Vol. 68, pp. 27–33). https://doi.org/10.1016/j.polymertesting.2018.03.042.
125. Marques, N., & Paulo, D. J., (2002). Tribological comparative study of conventional and composite materials in biomedical applications. In: *Key Engineering Materials* (Vols. 230–232, pp. 487–490). https://doi.org/10.4028/www.scientific.net/kem.230-232.487.
126. Saha, D., Dhabal, S., Bose, P. K., & Banthia, A. K., (2007). Production and biocompatibility evaluation of carbon fiber reinforced polyethylene composite for acetabular cup. In: *Science and Engineering of Composite Materials* (Vol. 14, No. 1, pp. 47–56). https://doi.org/10.1515/secm.2007.14.1.47.
127. Scholes, S. C., & Unsworth, A., (2009). Wear studies on the likely performance of CFR-PEEK/CoCrMo for use as artificial joint bearing materials. In: *Journal of Materials Science: Materials in Medicine* (Vol. 20, No. 1, pp. 163–170). https://doi.org/10.1007/s10856-008-3558-3.
128. Geringer, J., Tatkiewicz, W., & Rouchouse, G., (2011). Wear behavior of PAEK, poly(aryl-ether-ketone), under physiological conditions, outlooks for performing these materials in the field of hip prosthesis. In: *Wear* (Vol. 271, No 11, 12, pp. 2793–2803). https://doi.org/10.1016/j.wear.2011.05.034.
129. Liu, J. L., Jin-Long, L. I. U., Yuan-Yuan, Z. H. U., Wang, Q. L., & Shi-Rong, G. E., (2008). Biotribological behavior of ultra-high molecular weight polyethylene composites containing bovine bone hydroxyapatite. In: *Journal of China University of Mining and Technology* (Vol. 18, No. 4, pp. 606–612). https://doi.org/10.1016/s1006-1266(08)60303-x.
130. Ghorbel, H. F., Guidara, A., Danlos, Y., Bouaziz, J., & Coddet, C., (2017). Alumina-fluorapatite composite coating deposited by atmospheric plasma spraying: An agent of cohesion between bone and prostheses. *Materials Science & Engineering: C, Materials for Biological Applications, 71*, 1090–1098.
131. Qadir, M., Li, Y., Munir, K., & Wen, C., (2018). Calcium phosphate-based composite coating by micro-arc oxidation (MAO) for biomedical application: A review. In: *Critical Reviews in Solid State and Materials Sciences* (Vol. 43, No. 5, pp. 392–416). https://doi.org/10.1080/10408436.2017.1358148.
132. Campbell, M., Denault, J., Yahia, L., & Martin, N. B., (2008). CF/PA12 composite femoral stems: Manufacturing and properties. In: *Composites Part A: Applied Science*

and Manufacturing (Vol. 39, No. 5, pp. 796–804). https://doi.org/10.1016/j.compositesa.2008.01.016.
133. Sridhar, I., Adie, P. P., & Ghista, D. N., (2010). Optimal design of customised hip prosthesis using fiber reinforced polymer composites. In: *Materials & Design* (Vol. 31, No. 6, pp. 2767–2775). https://doi.org/10.1016/j.matdes.2010.01.016.
134. Bougherara, H., Bureau, M., Campbell, M., Vadean, A., & Yahia, L., (2007). Design of a biomimetic polymer-composite hip prosthesis. *Journal of Biomedical Materials Research: Part A, 82*(1), 27–40.
135. Salernitano, E., & Migliaresi, C., (2003). Composite materials for biomedical applications: A review. *Journal of Applied Biomaterials & Biomechanics: JABB, 1*(1), 3–18.
136. Hernigou, P., Nogier, A., Manicom, O., Poignard, A., De Abreu, L., & Filippini, P. (2004). Alternative femoral bearing surface options for knee replacement in young patients. *The Knee, 11*(3), 169–172. DOI: 10.1016/j.knee.2004.04.001.

CHAPTER 7

Polymer-Based Composite Hybrids for Drug/Gene Delivery Applications

G. SANTHOSH

Department of Mechanical Engineering, NMAM Institute of Technology, Nitte, Karnataka, India

ABSTRACT

Drug delivery is an attractive process among various medical applications. The amalgamation of new technology and new materials into drug delivery systems resulting a unique feature of medicine. The existing polymeric research with the integration of nanotechnology is offering a controlled release of therapeutic agents. Various bio-polymeric systems are also in place, and these biomaterials offer large diversity for modification with nanofillers. The conventional methods used for drug delivery have a lot of drawbacks and side effects, causing severe damage to healthy tissues. Novel and innovative delivery methods based on nanotechnology have shown matured results and grown as promising techniques in achieving the desired therapeutic efficiencies. In this chapter, we are providing a brief overview of polymer-based drug delivery systems and their applications.

7.1 INTRODUCTION

Presently many researchers are working on nanomaterials integrated with polymeric systems in medicine for drug and gene delivery applications. The search for new drug and gene delivery systems has led to the invention of new modes of research areas. The research with a Multidisciplinary

Biocomposites: Environmental and Biomedical Applications. Omar Mukbaniani, Tamara Tatrishvili, Neha Kanwar Rawat, & A. K. Haghi (Eds.)

approach provides a breakthrough in various therapeutic applications [1, 2]. Bio-based polymers have proved to be a promising choice for drug delivery. The biopolymers like poly-lactic acid, polyglycolic acid, and hydrogels are widely used in creating a delivery component [3, 4]. Polymer-based drug delivery systems are used as devices that assist in delivering the drug into a biological system. A wide range of polymer-based materials in biomedical sectors has been fabricated and investigated for drug delivery with nano-fillers. Nanoparticles integrated with various polymeric systems provide significant results in the diagnosis and treatment of diseases [5–9]. These polymeric systems are safe and efficient as they can control the rate, time, and place of release of drugs in the biological system.

Gene therapy is also a rapidly growing research area in medicine and has been considered a promising strategy to cure various infectious diseases [10–14]. However, researchers face many difficulties during gene delivery, the problems include gene packaging, cellular uptake, and translocation in the nucleus [15]. The gene transfection into the cells demands the use of appropriate vectors like viral and non-viral [16]. Compared to viral vectors non-viral vectors have shown huge potential in gene delivery due to their low immunogenicity, large production, tunable surface, and structure [17–19]. Several non-viral vectors such as lipids, polymers, and dendrimers have been widely used in gene delivery applications [18, 20, 22–24]. Polymer-based polycations are non-viral carriers that have gained huge attention due to their chemical versatility [25, 26]. The polycations used as gene carriers should meet several clinical requirements such as protecting the genetic material, degrading, and eliminating it from the biological system, and entering the targeted cells/systems without any side effects.

7.2 DRUG DELIVERY SYSTEMS

Drug delivery is an approach for the delivery of drugs to a specific biological system [27]. Many drug delivery carriers like nanoparticles, biopolymers, hydrogels, scaffolds, and implants are used for drug delivery [27, 33]. Biopolymers are naturally derived polymeric biomolecules made up of several monomeric units [28–32]. Biopolymers have been used for fabricating various drug delivery systems, but it is very difficult to choose a biopolymer as an agent in controlled drug delivery applications due to its structural complexity.

7.3 BIOPOLYMERS FOR DRUG DELIVERY

Biopolymers have been in use as drug carriers for many decades [33]. At present, more than 6 crore patients around the globe have been treated with innovative drug delivery methods to encounter various diseases [34].

Polysaccharides are widely used and investigated biopolymers in drug delivery applications. Polysaccharides are composed of long-chain carbohydrates with the general formula $(CH_2O)_n$. Cellulose is one of the commonly used polysaccharides derived from natural resources [35]. Cellulose is a linear polysaccharide with crystalline microfibril structures [36]. The microfibril structure of the cellulose provides strong resistance to enzymatic actions with an aligned cell wall structure [37]. Cellulose extracted from plant fibers can be reinforced with powder cellulose which is being used as pharmaceutical tablets [38].

Chitosan is also a naturally available polymer and is the second most abundant biopolymer (after cellulose) which is found in the exoskeleton of crustaceans. It can be obtained from partial/complete deacetylation of chitin with a preferable deacetylation degree (fraction of D-glucosamine units) of more than 60%. The Mw and deacetylation degree has a high influence on the biocompatibility, hydrophilicity, and solubility of chitosan in an aqueous environment [39]. The modified chitosan is widely used in drug delivery applications especially for the oral delivery of insulin [40]. Many other biopolymers like alginate, gellan gum, pectin, gum arabica, gaur gum, locust bean gum, and tamarind gum are also in drug delivery applications.

7.4 SYNTHETIC POLYMER FOR DRUG DELIVERY

Over the last few years, we have seen an increasing demand for the use of synthetic polymers for drug delivery applications. Synthetic polymers have shown better compatibility for the use of pharmaceutical applications with better therapeutic efficiencies. Controlled drug delivery is growing as one of the challenging research areas in modern-day pharmaceutical and medical applications [41–43]. The controlled drug delivery technology is more effective as the drug can be delivered to the infected site eliminating under or overdosing of the drug. Further, controlled drug delivery helps to maintain the drug level within the desired range and reduces the side effects with optimum use and delivery of the drug. However, controlled drug delivery has some specific problems that cannot be ignored: toxicity, non-biocompatibility, production of undesired by-products, and high cost incurred for the drug delivery systems.

The ideal polymeric drug delivery system should be inert, non-leachable, biocompatible, should be strong, should be capable of loading higher amounts of the drug, and easy to fabricate. The polymeric system with the mentioned properties has been used in the past are polyester, poly(ortho esters), polyanhydrides, poly(amides), polyurethanes, and many more.

Polyesters are the most widely used aliphatic degradable polymers in drug delivery applications [43, 44]. The degradation of aliphatic polyesters takes place due to the hydrolysis of ester links present in the polymer chain. Lactic acid and glycolic acid-based aliphatic polyesters such as polylactides, polyglycolide, and poly(lactide-co-glycolide) are obtained by ring-opening polymerization and are best suited for controlled drug delivery applications [45–47]. Poly(orthoesters) are another class of synthetic hydrophobic polymer synthesized by the addition of polyols to diketene acetals for drug delivery applications. Few researchers have synthesized hydrolytically degradable poly(orthoesters) for controlled drug delivery [48, 49]. Polyanhydrides are also essential bioabsorbable polymers that are preferred in controlled drug delivery applications due to their hydrophobicity [50–52]. Polyanhydrides contain water-sensitive linkages which undergo hydrolytic bond cleavage to generate water-soluble products [53]. Polyamides are yet another class of polymeric materials with a structural resemblance to polypeptides used widely for the transport of drugs. The easy metabolism, nonantigenic, and non-toxic nature of these polymers make them suitable for drug delivery applications [54–57]. Polyurethanes having structural characterizations like polyesters and polyamides have gained a lot of attention and scope in drug delivery applications. Polyurethanes' biodegradability and biocompatibility motivated Iskakov and co-workers to fabricate a novel polyurethane drug delivery system loaded with various antitumor drugs [58, 59].

7.5 GENE DELIVERY SYSTEMS

7.5.1 *POLYMERS IN GENE DELIVERY*

Gene therapy is the Nobel and stable modality for many inherited and acquired diseases [60]. Yet, its clinical is delayed due to challenges faced in selecting a suitable delivery medium [61]. These days the advances in understanding the molecular and genetic basis of diseases have led to the discovery of polymeric gene delivery systems [62]. Polymer-based gene systems have gained huge attention due to their structural and chemical stability with the ability to control physicochemical properties and essentially, with larger gene-carrying capacity [63–65].

7.5.2 SYNTHETIC POLYMERS FOR GENE DELIVERY

Synthetic polymers as gene carriers are of prime importance due to the presence of cationic molecules. The polymers used as carriers are cationic lipids, polyamides, polyethyleneimine, and dendrimers [66]. The cationic polymers can interact with DNA molecules and enflame the formation of polyplexes. The polyplexes formed can protect the nucleic acid by maintaining stability and integrity till the cellular uptake process happens [67].

Among various cationic polymers, polyethyleneimine is the most widely studied polymer, and polyethyleneimine is the most effective gene-carrying agent. Polyethyleneimine can be synthesized in two forms: (i) linear form and (ii) branched form, and both forms have a high rate of transfection capabilities [68]. The transfection ability of polyethyleneimine is decided on factors like target cell type, structure, and Mw of the polymer [69]. The poly-l-lysine is another important cationic polymer with low Mw that has been used to collapse DNA into nanoparticles and transport the DNA to the cells.

7.5.3 NATURAL POLYMER FOR GENE DELIVERY

Natural polymers play a very significant role as carrier materials in gene delivery applications. Biopolymers such as chitosan, collagen, pullulan, dextran, and Hyaluronic Acid are widely used as gene-delivery vehicles.

Chitosan is the most appreciated and widely used biopolymer as an oral and nasal delivery vehicle due to its mucoadhesive nature. The unique ability of chitosan to condense DNA molecules to nano-sized particles in acidic and neutral pH makes it unique among other natural polymers. Apart from this behavior, the low immunogenicity, low cytotoxicity, and biocompatibility of chitosan make it a suitable candidate for gene delivery application [70]. Pullulan is yet another water-soluble biopolymer with α-1,4-glucopyranose and α-1,6-glucopyranose units. Pullulan's non-toxic, non-immunogenic, non-carcinogenic, and non-mutagenic nature makes it a suitable candidate for pharmaceutical industries. Pullulan polymer as a gene delivery vehicle can protect DNA molecules from degradation and can show excellent cellular viability.

Dextran is another carbohydrate polymer with applications in biomedicine and gene delivery. It is composed predominantly of α-1,6-linked glucopyranose units with a low degree of 1,3-branching. Cationic, biodegradable dextran hydrogel nanoparticles could be prepared by derivatization with cationic methacrylate monomers for siRNA delivery. Along with this, negatively charged dextran sulfate may perhaps form polyelectrolyte complexes with positively charged polymers, such as poly-L-arginine for siRNA delivery. Dextran-grafted branched polyethyleneimine was found to be very effective to rally the steadiness of the polyethyleneimine complexes with DNA. Dextran-polyethyleneimine conjugates were found to be less toxic than unmodified polyethyleneimine. Nanoparticles formed by the complexation of polyallylamine-dextran conjugate with DNA were more capably transfected than that polyallylamine-DNA nanoparticles [21].

7.6 CONCLUSION

The recent developments in the polymers towards the development of novel drug and gene delivery systems have seen a study growth in the development of novel and smart delivery systems, which can produce advanced treatment with better patient compliance. A suitable deliberation of various parameters like extraction technique, surface characteristics, biocompatibility, biodegradability, and the chemistry of the polymers can help in choosing and designing a better carrier vehicle for drug and gene delivery applications. It is expected that various improved and versatile carrier systems are formulated and used in the future as a result of continuous exploration for the development of carrier systems.

KEYWORDS

- **biopolymers**
- **drug delivery**
- **drug delivery systems**
- **gene delivery systems**
- **nanofillers**
- **nanomedicine**
- **nanotechnology**

REFERENCES

1. Din, F., Aman, W., Ullah, I., Quereshi, O. S., Mustapha, O., Shafique, S., & Zeb, A., (2017). Effective use of nano-carriers as drug delivery systems for the treatment of selected tumors. *Int. J. Nanomedicine, 12*, 7291–7309.
2. Tiwari, G., Tiwari, R., Sriwastawa, B., Bhati, L., Pandey, S., Pandey, P., & Bannerjee, S. K., (2012). Drug delivery systems: An updated review. *Int. J. Pharm. Investig., 2*, 2–11.
3. Sinha, V. R., & Khosla, L. (1998). Bio-absorbable polymers for implantable therapeutic systems. *Drug Dev. Ind Pharm., 24*, 1129–1138. DOI: 10.3109/03639049809108572.
4. Basua, A., Kundurua, K. R., Doppalapudib, S., Domba, A. J., & Khanb, W., (2016). Poly (lactic acid) based hydrogels. *Adv. Drug Deliv. Rev., 107*, 192–205.
5. Duncan, R., (2003). The dawning era of polymer therapeutics. *Nat. Rev. Drug Discov., 2*, 347–360.
6. De Jong, W. H., Geertsma, R. E., & Roszek, B., (2005). *Possible Risks for Human Health*. Report 265001002/2005. Bilthoven, The Netherlands: National Institute for Public Health and the Environment (RIVM). Nanotechnology in medical applications.
7. *European Science Foundation*, (2005). Policy Briefing (ESF), ESF Scientific Forward Look on Nanomedicine IREG Strasbourg, France. ISBN: 2-912049-520.
8. *European Technology Platform on Nanomedicine*, (2005). Vision paper and basis for a strategic research agenda for nanomedicine. European Commission Luxembourg, Office for Official Publications of the European Commission. ISBN: 92-894-9599-5.
9. Ferrari, M., (2005). Cancer nanotechnology: Opportunities and challenges. *Nat. Rev. Cancer, 5*, 161–171.
10. Hayakawa, K., Uchida, S., Ogata, T., Tanaka, S., Kataoka, K., & Itaka, K., (2015). Intrathecal injection of a therapeutic gene-containing polyplex to treat spinal cord injury. *J. Control. Release, 197*, 1–9.
11. Karimi, M., Ghasemi, A., Zangabad, P. S., Rahighi, R., Basri, S. M. M., Mirshekari, H., Amiri, M., Pishabad, Z. S., Aslani, A., Bozorgomid, M., et al., (2016). Smart micro/nanoparticles in stimulus-responsive drug/gene delivery systems. *Chem. Soc. Rev., 45*, 1457–1501.
12. Kumar, M. D., Dravid, A., Kumar, A., & Sen, D., (2016). Gene therapy as a potential tool for treating neuroblastoma-a focused review. *Cancer Gene Ther., 23*, 115–124.
13. Naldini, L., (2015). Gene therapy returns to center stage. *Nature, 526*, 351–360.
14. Tan, X. Y., Li, B. B., Lu, X. G., Jia, F., Santori, C., Menon, P., Li, H., et al., (2015). Light-triggered, self-immolative nucleic acid-drug nanostructures. *J. Am. Chem. Soc., 137*, 6112–6115.
15. Ullah, I., Muhammad, K., Akpanyung, M., Nejjari, A., Neve, A. L., Guo, J. T., Feng, Y. K., & Shi, C. C., (2017). Bioreducible, hydrolytically degradable and targeting polymers for gene delivery. *J. Mater. Chem. B, 5*, 3253–3276.
16. Wicki, A., Witzigmann, D., Balasubramanian, V., & Huwyler, J., (2015). Nanomedicine in cancer therapy: Challenges, opportunities, and clinical applications. *J. Controlled Release, 200*, 138–157.
17. Eltoukhy, A. A., Chen, D. L., Alabi, C. A., Langer, R., & Anderson, D. G., (2013). Degradable terpolymers with alkyl side chains demonstrate enhanced gene delivery potency and nanoparticle stability. *Adv. Mater., 25*, 1487–1493.
18. Pandey, A. P., & Sawant, K. K., (2016). Polyethylenimine: A versatile, multifunctional non-viral vector for nucleic acid delivery. *Mat. Sci. Eng. C-Mater., 68*, 904–918.

19. Machitani, M., Yamaguchi, T., Shimizu, K., Sakurai, F., Katayama, K., Kawabata, K., & Mizuguchi, H., (2011). Adenovirus vector-derived VA-RNA-mediated innate immune responses. *Pharmaceutics, 3*, 338.
20. Pack, D. W., Hoffman, A. S., Pun, S., & Stayton, P. S., (2005). Design and development of polymers for gene delivery. *Nat. Rev. Drug Discovery, 4*, 581–593.
21. Ribeiro, L. N. M., Franz-Montan, M., Breitkreitz, M. C., Alcântara, A. C. S., Castro, S. R., Guilherme, V. A., Barbosa, R. M., & De Paula, E., (2016). Nanostructured lipid carriers as robust systems for topical lidocaine-prilocaine release in dentistry. *European Journal of Pharmaceutical Sciences: Official Journal of the European Federation for Pharmaceutical Sciences, 93*, 192–202.
22. Zhou, X. Y., Zheng, Q. Q., Wang, C. Y., Xu, J. K., Wu, J. P., Kirk, T. B., Ma, D., & Xue, W., (2016). Star-shaped amphiphilic hyperbranched polyglycerol conjugated with dendritic poly(L-lysine) for the codelivery of docetaxel and MMP-9 siRNA in cancer therapy. *ACS Appl. Mater. Interfaces, 8*, 12609–12619.
23. Pan, J. J., Yuan, Y. Q., Wang, H. W., Liu, F., Xiong, X. H., Chen, H., & Yuan, L., (2016). Efficient transfection by using PDMAEMA-modified SINWAs as a platform for Ca^{2+}-dependent gene delivery. *ACS Appl. Mater. Interfaces, 8*, 15138–15144.
24. Samanta, K., Jana, P., Backer, S., Knauer, S., & Schmuck, C., (2016). Guanidiniocarbonyl pyrrole (GCP) conjugated PAMAM-G2, a highly efficient vector for gene delivery: The importance of DNA condensation. *Chem. Commun., 52*, 12446–12449.
25. Palivan, C. G., Fischer-Onaca, O., Delcea, M., Itel, F., & Meier, W., (2012). Protein-polymer nanoreactors for medical applications. *Chem. Soc. Rev., 41*, 2800–2823.
26. Kataoka, K., Harada, A., & Nagasaki, Y., (2012). Block copolymer micelles for drug delivery: Design, characterization and biological significance. *Adv. Drug Delivery Rev., 64*, 37–48.
27. Nayak, A. K., Ahmad, S. A., Beg, S., Ara, T. J., & Hasnain, M. S., (2018). Drug delivery: Present, past and future of medicine. In: Inamuddin, Asiri, A. M., & Mohammad, A., (eds.), *Applications of Nanocomposite Materials in Drug Delivery* (pp. 255–282). Elsevier Inc., Amsterdam.
28. Hasnain, M. S., & Nayak, A. K., (2019). Natural polysaccharides: Sources and extraction methodologies. In: Hasnain, M. S., & Nayak, A. K., (eds.), *Natural Polysaccharides in Drug Delivery and Biomedical Applications* (pp. 1–14). Academic Press, Elsevier Inc., Cambridge, Amsterdam.
29. Nayak, A. K., Pal, D., & Santra, K., (2015). Screening of polysaccharides from tamarind, fenugreek and jackfruit seeds as pharmaceutical excipients. *Int. J. Biol. Macromol., 79*, 756–760.
30. Nayak, A. K., Pal, D., Pradhan, J., & Ghorai, T., (2012). The potential of *Trigonella foenum-graecum* L. seed mucilage as suspending agent. *Indian J. Pharm. Educ., 46*, 312–317.
31. Nayak, A. K., Pal, D., Pany, D. R., & Mohanty, B., (2010). Evaluation of *Spinacia oleracea* L. leaves mucilage as innovative suspending agent. *J. Adv. Pharm. Technol. Res., 1*(3), 338–341.
32. Pal, D., Saha, S., Nayak, A. K., & Hasnain, M. S., (2019). Marine-derived polysaccharides: Pharmaceutical applications. In: Nayak, A. K., Hasnain, M. S., & Pal, D., (eds.), *Natural Polymers for Pharmaceutical Applications: Marine and Microbiologically Derived Polymers,* (Vol. II, pp. 1–36). Apple Academic Press, USA.
33. Gemma, V., Judit, T., & Fernando, A., (2012). Polymers and drug delivery systems. *Curr. Drug. Deliv., 9*, 1–28.

34. Mahammad, R. S., Madhuri, K., & Dinakar, P., (2012). Polymers in controlled drug delivery systems. *Int. J. Pharm. Sci., 2*, 112–116.
35. Pal, D., Nayak, A. K., & Saha, S., (2019). Cellulose-based hydrogels: Present and future. In: Akhtar, M. S., Swamy, M. K., & Sinnaih, U. R., (eds.), *Natural Bio-active Compounds: Production and Applications* (Vol. 1, pp. 285–332). Springer Nature Pvt. Ltd., Singapore.
36. Keshipour, S., & Maleki, A., (2019). Modification of cellulose. In: Akhtar, M. S., Swamy, M. K., & Sinnaih, U. R., (eds.), *Natural Bio-active Compounds: Production and Applications* (Vol. 1, pp. 435–486). Springer Nature Pvt. Ltd., Singapore.
37. Cosgrove, D. J., (2005). Growth of the plant cell wall. *Nat. Rev. Mol. Cell Biol., 6*, 850–861.
38. Ogaji, I., Nep, E., & Audu-Peter, J. D., (2011). Advances in natural polymers as pharmaceutical excipients. *Pharm. Anal. Acta, 3*, 1–16.
39. Huang, M., Khor, E., & Lim, L. Y., (2004). Uptake and cytotoxicity of chitosan molecules and nanoparticles: Effects of molecular weight and degree of deacetylation. *Pharm. Res., 21*, 344–353.
40. Ubaidulla, U., Khar, R. K., Ahmad, F. J., & Tripathi, P., (2009). Optimization of chitosan succinate and chitosan phthalate microspheres for oral delivery of insulin using response surface methodology. *Pharm. Dev. Technol., 14*(1), 96–105.
41. Stevenson, T. K. W., & Sefton, M. V., (1994). *Trends in Polym. Sci., 2*, 98.
42. McCullosch, I., & Shalaby, S. W., (1998). Tailored polymeric materials for controlled delivery systems. *Am. Chem. Soc.* Washington, DC.
43. Uhrich, K. E., Cannizzaro, S. M., Langer, R. S., & Shakesheff, K. M., (1999). *Chem. Rev., 99*, 3181.
44. Chandra, R., & Rustgi, R., (1998). *Prog. Polym. Sci., 23*, 1273.
45. Lanza, R. P., Langer, R., & Chick, W. L., (1997). *Principles of Tissue Eng*. R.G. Landes Co. and Academic Press: Austin, TX.
46. Wong, W. H., & Mooney, D. J., (1997). In: Atata, & Mooney, (eds.), *Synthetic Biodegradable Polymer Scaffolds*. Birkhauser: Boston, MA.
47. Griffith, L. G., (2000). *Acta Mater., 48*, 263.
48. Ng, S. Y., Vandamme, T., Taylor, M. S., & Heller, J., (1997). *Macromolecules, 30*, 770.
49. Bernatchez, S. F., Merkli, A., Tabatabay, C., Gurny, R., Zhao, Q. H., Anderson, J. M., & Heller, J., (1993). *J. Biomed. Mater. Res., 27*, 677.
50. Domb, A. J., Amselem, S., Shah, J., & Maniar, M., (1992). In: Peppas, N. A., & Langer, R., (eds.), *Adv. Polym. Sci.* (Vol. 107, p. 93). Springer, Berlin.
51. Kumar, N., Langer, R. S., & Domb, A. J., (2002). *Adv. Drug. Delv. Rev., 54*, 889.
52. Katti, D. S., Lakshmi, S., Langer, R., & Laurencin, C. T., (2002). *Adv. Drug. Delv. Rev., 54*, 933.
53. Gopferich, A., & Tessmar, J., (2002). *Adv. Drug Delivery Rev., 54*, 911.
54. Dimitriu, S., (1994). *Polymeric Biomaterials*. Marcel Dekker New York.
55. Nathan, A., & Kohn, J., (1994). In: Shalaby, S., (ed.), *Biomedical Polymers: Designed-to-Degrade Systems*. Hanser/Gardner: Cincinati, OH.
56. General, S., & Thunemann, A. F., (2001). *Int. J. Pharm., 230*, 11.
57. Brocchini, S., Schachter, D. M., & Kohn, J., (1997). *Am. Chem. Soc. Symp. Ser., 675*, 154.
58. Iskakov, R., Batyrbekov, E., Zhubanov, B., Teleuova, T., & Volkova, M., (1998). *Polym. Adv. Tech., 9*, 266.
59. Iskakov, R., Batyrbekov, E., Leonova, M. B., & Zhubanov, B. A., (2000). *J. Appl. Polym. Sci., 75,* 35.

60. Gori, J. L., et al., (2015). Delivery and specificity of CRISPR/Cas9 genome editing technologies for human gene therapy. *Hum. Gene Ther., 26*, 443–451.
61. Jacobson, S. G., et al., (2015). Improvement and decline in vision with gene therapy in childhood blindness. *N. Engl. J. Med., 372*, 1920–1926.
62. Peng, L. H., et al., (2016). Integration of antimicrobial peptides with gold nanoparticles as unique non-viral vectors for gene delivery to mesenchymal stem cells with antibacterial activity. *Biomaterials, 103*, 137–149.
63. Pack, D. W., et al., (2005). Design and development of polymers for gene delivery. *Nat. Rev. Drug Discov., 4*, 581–593.
64. Bishop, C. J., et al., (2016). Quantification of cellular and nuclear uptake rates of polymeric gene delivery nanoparticles and DNA plasmids via flow cytometry. *Acta Biomater., 37*, 120–130.
65. Lin, G., et al., (2015). Smart polymeric nanoparticles for cancer gene delivery. *Mol. Pharm. 12*, 314–321.
66. Pattni, B. S., Chupin, V. V., & Torchilin, V. P., (2015). New developments in liposomal drug delivery. *Chemical Reviews, 115*(19), 10938–10966.
67. Gubernator, J., (2011). Active methods of drug loading into liposomes: Recent strategies for stable drug entrapment and increased *in vivo* activity. *Expert Opinion on Drug Delivery, 8*(5), 565–580.
68. Pozzi, D., Colapicchioni, V., Caracciolo, G., Piovesana, S., Capriotti, A. L., Palchetti, S., De Grossi, S., et al., (2014). Effect of polyethyleneglycol (PEG) chain length on the bio-nano-interactions between PEGylated lipid nanoparticles and biological fluids: From nanostructure to uptake in cancer cells. *Nanoscale, 6*(5), 2782–2792.
69. Varga, Z., Wacha, A., Vainio, U., Gummel, J., & Bóta, A., (2012). Characterization of the PEG layer of sterically stabilized liposomes: A SAXS study. *Chemistry and Physics of Lipids, 165*(4), 387–392.
70. Caddeo, C., Díez-Sales, O., Pons, R., Carbone, C., Ennas, G., Puglisi, G., Fadda, A. M., & Manconi, M., (2016). Cross-linked chitosan/liposome hybrid system for the intestinal delivery of quercetin. *Journal of Colloid and Interface Science, 461*, 69–78.

Index

D

E

F

G

N

O